Green Energy and Technology

For further volumes:
http://www.springer.com/series/8059

Luis Puigjaner
Editor

Syngas from Waste

Emerging Technologies

Springer

Luis Puigjaner
ETSEIB
Universitat Politècnica de Catalunya
Diagonal 647
08028 Barcelona
Spain
e-mail: luis.puigjaner@upc.edu

ISSN 1865-3529 e-ISSN 1865-3537
ISBN 978-0-85729-539-2 e-ISBN 978-0-85729-540-8
DOI 10.1007/978-0-85729-540-8
Springer London Dordrecht Heidelberg New York

British Library Cataloguing in Publication Data
A catalogue record for this book is available from the British Library

Cover design: eStudio Calamar, Berlin/Figueres

Printed on acid-free paper

Springer is part of Springer Science+Business Media (www.springer.com)

Foreword

"Syngas from Waste: Emerging Technologies"—An Example of What is Required for Today's New Green and Sustainable Chemical and Process Engineering

Oil, chemical and related industries are today evolving considerably at the beginning of this new century, because of unprecedented globalization market demands and constraints stemming from public and media concerns over environmental and safety issues, and sustainable considerations, in combination with tools like stakeholders analysis, indicators, and Life Cycle Assessment covering a cradle-to-grave approach of the supply chain.

A great number of requirements are clearly focused on societal exigencies such as carbon dioxide sequestration, chemical looping combustion, methane CPO and reforming, biodiesel synthesis and hydrogen production, as mentioned in the USA and European roadmaps related to the mise en oeuvre of the required green chemistry and the green chemical and process engineering. And it is obvious that the existing processes and new processes, both involving the need of emerging technologies and computer aided process engineering, must be progressively adapted to the principles of the green chemistry [1].

The present book is a typically good illustration of the modern chemical engineering in the framework of globalization, sustainability, and acceleration of innovation that includes a computer aided multi-scale product design and engineering approach [2], and process intensification reducing material usage, energy consumption, and waste generation, i.e., producing much more and better with using much less, thus also enhancing the corporate image of chemistry [3].

Indeed this book, based on the competences of scientists confronted with industry practice is a reference tool. It concerns the topic of production of syngas from low-cost solid biomass and organic wastes from any origin and their blends with coal, coke, and petcoke for electricity production or pure hydrogen production, incorporating advanced technologies for carbon dioxide concentration. The clear objective of the book is to emphasize how production and cleaning technologies can be improved by developing perfected models and simulations that are validated, interlinked in optimized physical and chemical processes and economically designed and assessed in the framework of today's energetic context.

The content presents successively (a) the raw material handling involving the raw materials characterization and the green supply chain management modeling, (b) the process modeling and simulation with the modeling of the reactors encountered in syngas generation and purification, with the emerging technologies on syngas purification by adsorption processes, with the modeling of the hydrogen production and the carbon dioxide separation with the Water Gas Shift Membrane Reactor, and with the appropriate representation by means of a process superstructure mathematic tool of all possible flowsheets for the production of a given product from different raw materials using different process units, and (c) the conceptual design and decision-making support for the global clean gas process synthesis and optimization and for the selection of the best designs for specific applications, and finally industrial data collection and several examples of industrial applications are presented with successful case studies and also limitations, ranging from clean power generation to complex combined heat and power systems and high purity hydrogen for use in fuel cells.

Summarizing the book, contains two main areas. The first area focuses on the improvement of the conceptual design of syngas production and purification both with the scientific advancement of improved and emerging innovative technologies and with the development of better performing models. The other area focuses on the modeling, validation, and technico-economical evaluation of improved and innovative systems for clean syngas production, targeted to enhance the industrial design of gasification processes at a preliminary stage. Moreover it has to be underlined and emphasized that a state of the art systems approach is used throughout the book: a process superstructure is introduced to describe the flowsheet configurations for intermediate syngas production aiming at power and hydrogen production. This process superstructure is based on commercial and custom-made simulation tools, which are standards-compliant, interoperable and open to future developments. And as explained by the author of the book, Professor Puigjaner, the reader is guided in the use of the superstructure, which acts as a versatile and flexible tool for the preliminary design of novel process alternatives and redesign of existing ones, and serves to improve the operating conditions of the real process plant. And in all cases, several process-wide performance indicators related to the simultaneous consideration of economic and environmental aspects are considered for optimization of the best design.

As mentioned previously, this book illustrates what I call the modern green and sustainable chemical and process engineering in presenting for the first time the most recent concepts, methods, and techniques for the preliminary conceptual design and optimization of most promising emerging technologies for the production of clean quality syngas upon demand from low-cost solid biomass and organic wastes under simultaneous economical and environmental constraints. Moreover the design strategy presented in this guide allows a tailor-made design for several processes other than syngas production, where an appraisal of technico-economical feasibility and environmental improvement in the system approach of green and sustainable process is required. Therefore the book may be considered as a vade mecum which may serve also as a decision-making support for the chemical engineer at the plant to assess or improve its performance.

It is clear that this book entitled "Syngas from Waste: Emerging Technologies" which contain high quality works from leading experts in the field, is intended as a textbook for postgraduate students, scientists, academy and industry researchers, and industry practitioners in the up-to-date topic of syngas production and applications. It will especially interest those who are concerned and involved in the design, retrofit design and evaluation activities of alternative solutions. Such a vade mecum should also be strongly recommended to lecturers and teachers delivering advanced courses in process design and engineering focusing on engineers and scientists working in industry and being involved with the crucial problems and challenges encountered in the evaluation and improvement of existing installations or in the design of a new one that require emerging technologies and the use of better performing models.

Nancy, 31 October 2010

Prof. Dr. Ing. Jean-Claude Charpentier
Past-President of the European Federation
of Chemical Engineers
CNRS/ENSIC/INPL Nancy-Université, France

References

1. Anastas T, Zimmerman JB (2003) Design of green engineering, through the 12 principles. Environ Sci Technol 37(23):423A
2. Puigjaner L, Heyen G (2006) Computer aided process and product engineering. Wiley
3. Charpentier JC (2007) Modern chemical engineering in the framework of globalization, sustainability, and technical innovation, Ind Eng Chem Res 57:3465–3485

Contents

Contributors

R. Álvarez-Rodríguez ETSI Minas, Alenza 4, 28003 Madrid, Spain

M. Bogataj Faculty of Chemistry and Chemical Engineering, University of Maribor, Smetanova ulica 17, 2000 Maribor, Slovenia

A. D. Bojarski Universitat Politècnica de Catalunya, ETSEIB, Diagonal 647, 08028 Barcelona, Spain

C. Clemente-Jul Alenza 4, 28003 Madrid, Spain

P. Coca ELCOGAS, S. A.Ctra. Calzada de Calatrava a Puertollano, Km. 27, Apdo. Correos 200, Puertollano—(Ciudad Real) 13500, Spain

A. Di Donato Centro Sviluppo Materiali, S.p.A., Via di Castel Romano, 100, 00128 Rome, Italy

Z. Kravanja Faculty of Chemistry and Chemical Engineering, University of Maribor, Smetanova ulica 17, 2000 Maribor, Slovenia

J. M. Laínez Universitat Politècnica de Catalunya, ETSEIB, Diagonal 647, 08028 Barcelona, Spain

C. R. A. Medina Universitat Politècnica de Catalunya, ETSEIB, Diagonal 647, 08028 Barcelona, Spain

J. M. Nougués Universitat Politècnica de Catalunya, ETSEIB, Diagonal 647, 08028 Barcelona, Spain

M. Pérez-Fortes Universitat Politècnica de Catalunya, ETSEIB, Diagonal 647, 08028 Barcelona, Spain

C. Pevida Instituto Nacional del Carbón (CSIC), Apartado 73, 33080 Oviedo, Spain

J. J. Pis Instituto Nacional del Carbón (CSIC), Apartado 73, 33080 Oviedo, Spain

L. Puigjaner Universitat Politècnica de Catalunya, ETSEIB, Diagonal 647, 08028 Barcelona, Spain

F. Rubiera Instituto Nacional del Carbón (CSIC), Apartado 73, 33080 Oviedo, Spain

A. Soršak Javni zdravstveni zavod Mariborske lekarne Maribor, Minarikova ulica 6, 2000 Maribor, Slovenia

Introduction

Luis Puigjaner

Abstract This chapter briefly introduces the main components that play a role in syngas production technologies. First, the thermochemical principles underlying the syngas production process are enunciated. Second, the suitable raw materials for gasification, the products obtained, and their end uses are identified. Third, the advantages, opportunities, present commercial status, and future perspectives of gasification are discussed. Finally, the need for enhanced technologies supported by the development of an appropriate framework for a robust design and optimization is justified. These issues are covered in detail in this book and should prove useful for assessing current technologies, conveying novel concepts and providing the basis for the future technological advancement of sustainable clean syngas production at a competitive advantage.

Notation

BGC	British Gas Corporation
CCS	Carbon Capture and Storage
ERG	Eastern Research Group
IGCC	Integrated Gasification Combined Cycle
MSW	Municipal Solid Waste

1 Thermochemical Principles

Gasification is a thermal process that converts a combustible material or residual into gas through partial oxidation at elevated temperatures. A solid is normally converted into a moderate heating value gas consisting of hydrogen and carbon

L. Puigjaner (✉)
ETSEIB, Universitat Politècnica de Catalunya, Diagonal 647, 08028 Barcelona, Spain
e-mail: Luis.puigjaner@upc.edu

L. Puigjaner (ed.), *Syngas from Waste*, Green Energy and Technology,
DOI: 10.1007/978-0-85729-540-8_1,

monoxide in varying ratios. The gasification agent can be air, oxygen, and/or steam.

Air gasification produces a gas with a poor higher-heating value in the range of 4–7 MJ/Nm^3 (950–1,600 kcal/Nm^3), which can be burned in a steam generator, an internal combustion engine, or a steam turbine to create plants with an integrated gasification combined cycle (IGCC). It is not usually economical when used at a distance from the gasifier because of the higher cost of transportation. Gasification with oxygen produces a higher-quality gas with a higher-heating value in the range of 10–18 MJ/Nm^3 (2,400–4,300 kcal/Nm^3). In this case, limited transportation is a possibility and it can even be used as a raw material for the synthesis of organic compounds, such as methanol or gasoline.

The process of biomass or organic waste gasification has three stages: drying (evaporation of moisture contained in the solid), pyrolysis (thermal decomposition in the absence of oxygen), and gasification (partial oxidation of pyrolysis products).

Pyrolysis occurs when a solid fuel is heated to temperatures between 300 and 500°C without an oxidizing agent. The products of pyrolysis are coal, condensable hydrocarbons or tar, and gases. The relative proportions of different products depend largely on the heating rate and final temperature. In general, pyrolysis is much faster than gasification, so the latter stage controls the speed of the process. Gas, liquid, and coal produced by pyrolysis react with the oxidant (usually air) and generate permanent gases (CO, CO_2, and H_2) and smaller quantities of hydrocarbons and olefins. The gasification of coal pyrolysis is the interactive combination of various gas–solid and gas–gas reactions, which oxidizes and becomes carbon monoxide and carbon dioxide, and hydrogen is produced by the shift reaction between the gas and water. Gas–solid reactions are slower and limit the speed of the process. The final composition of the gas depends on many factors, such as the composition of feed, water content, reaction temperature (between 600 and 1,100°C), and the degree of oxidation of pyrolysis products [1, 2].

When the products of pyrolysis are not completely converted into gas, tar contaminants appear. These contaminants, which are more likely to occur in biomass gasification than in coal, are difficult to remove by means of thermal, catalytic, or physical processes. Several research teams are therefore working on ways to overcome this problem by cracking or removing the tars.

2 Raw Materials and Products

Raw materials with high carbon content (coal, petcoke, biomass, and organic waste) are considered suitable for gasification. The last two are especially relevant to make the syngas production process an economically competitive alternative source of energy. Biomass is widely considered the greatest potential fuel for energy in the future. As an energy source, could represent 50% of total energy demand in Europe, using biomass crop specific field unnecessary for food, and also considering other wastes and residues from agriculture, trade, and consumers. Biomass can be classified by origin as follows [3, 4]:

(a) *Agriculture* includes a wide range of ligno-cellulosic materials generated by the food industry.
(b) *Forestry* includes wood from forest clearing, logging, and other wood-processing industries.
(c) *Industry* typically originates from packaging and pallets, which are generated in large quantities and often include paper, cardboard, and wood (all with a high heating value).
(d) *Municipal solid waste* can be gasified after glass and metal are removed.
(e) *Mix* includes sewage sludge and wood from the construction and demolition of buildings. Although it has a high carbon content, sewage sludge poses serious difficulties related to high moisture content, high ash content, and the presence of heavy metals. The wood from demolished buildings can also contain significant amounts of heavy metals.

A classic way of characterizing these types of solids is by immediate analysis (moisture, fixed carbon, volatiles, and ash) and elemental analysis (C, O, N, H, and S) from which the material balance and scope of certain environmental effects can be set.

The most common processes of transformation of biomass for further use of fuel in electricity generation or cogeneration are thermal combustion, gasification, and pyrolysis (Fig. 1). Of the three processes described for the assessment of biomass energy, gasification is one that involves more advantages [5, 6]:

(i) First, the volume of gases produced in gasification is much lower and with lower concentrations of pollutants; thus, cleaning systems are smaller and operate more efficiently.
(ii) Second, with the gasification yields a fuel that can be used in a wide variety of applications with conventional equipment designed for fuel gases appropriately tailored so that it can be transported to a distance from the site of generation.
(iii) Third, if the goal is to produce electricity and steam, the overall thermodynamic efficiency by using a synthetic gas that expands in heat engines after combustion and uses the excess energy in the hot gases to produce steam is much higher for gasification than in combustion.
(iv) Finally, gasification is much more developed than pyrolysis-level industrial operation, and both equipment design and the operation are simpler for the case of the gasification.

From an energy point of view, the main feature that differentiates gasification from combustion is that although all the energy of the gas is in the form of sensible heat during combustion, part of the chemical energy contained in the feedstock is transferred to the gas during gasification. However, not all the energy is used as

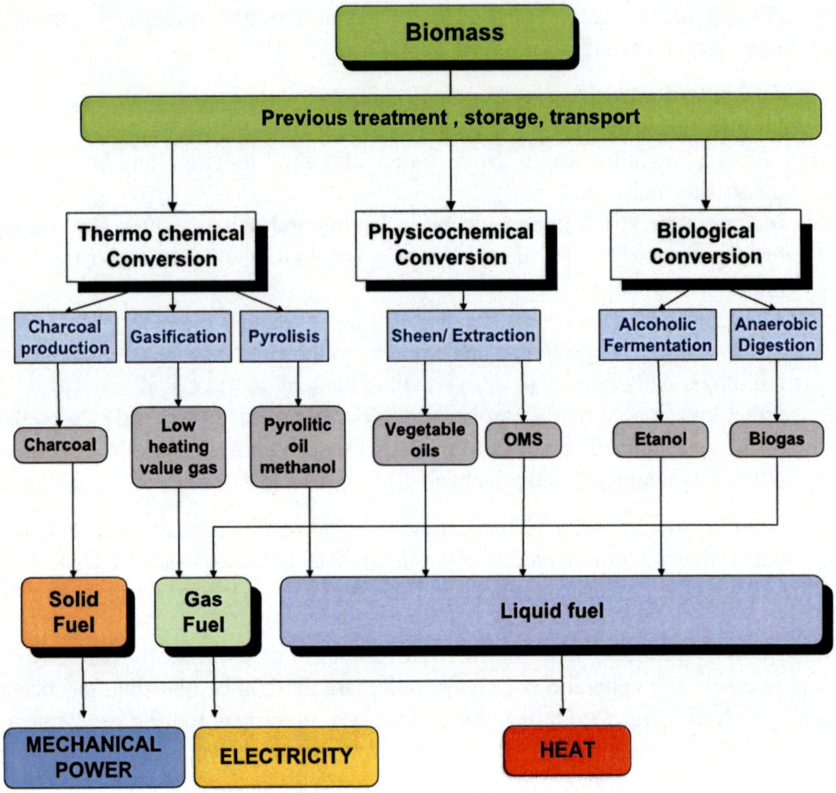

Fig. 1 Schematic of waste treatment and recycling of biomass/organic waste

gas-heating value: There is a gasification performance that depends on the type of gasifier and the oxidizing agent [7].

The main reason for requiring a certain gasification performance is that the above process (drying → pyrolysis → gasification) is endothermic, which means that it is necessary to provide energy. This can be done in two ways: through an external source or through the combustion of a portion of the gasified solid. The latter alternative, which is most often used industrially, requires careful control of the solid–air (or oxygen) ratio. In this case, a series of endothermic and exothermic reactions occurs within the gasifier. If the process is handled properly (at constant temperature), the heat generated by the exothermic reactions (combustion) compensates for the heat absorbed by endothermic reactions. Then, the process is referred to as "autothermal" [8, 9].

The gas obtained contains carbon monoxide (CO), carbon dioxide (CO_2), hydrogen (H_2), methane (CH_4), small amounts of other heavier hydrocarbons, such as ethane (C_2H_6) and ethylene (C_2H_4), water (H_2O), nitrogen (N_2) (when using air as oxidant), and several contaminants, such as small carbonaceous particles, ash, tars, and oils. Partial oxidation can be carried out using air, oxygen, steam, or a mixture of these elements. CO and H_2 confer heating value to the gas and they may react with oxygen (combustion in a boiler, engine, or turbine). CO_2 and H_2O are undesirable but unavoidable products. Although it is often formed in small proportions, methane is responsible for much of the energy content of gas.

The gas composition depends on the composition of the gasified solid, the gasifier operating conditions, and the design of the gasifier.

3 Opportunities and Future Prospects

Compared with other technologies, gasification has many positive attributes that help to stimulate its market. It is the only technology that offers both flexibility and advantages in terms of feed kinds and product uses. All raw materials containing carbon, including hazardous waste, municipal solid waste, sewage sludge, and biomass, can gasify after pretreatment to produce clean synthesis gas that can then be processed. Because of this ability to use low-cost raw materials, gasification technology is most suitable for many industrial applications, such as those used in refineries [10].

Compared with combustion systems, gasification is more efficient and causes less environmental impact to produce low-cost electricity from solid materials and gets even closer to the production of electricity from combined-cycle natural gas. Greater efficiency is achieved when it is integrated with the use of fuel cells and other advanced technologies. Increased efficiency in electricity production reduces operating costs and produces less carbon dioxide. In addition, the gasification process can be adapted to incorporate advanced technologies for CO_2 concentration with a reduced impact on cost and thermal efficiency. This feature is one of the most important factors for selecting this technology in future power plants.

Moreover, the products of gasification are much easier to clean than those of combustion, thereby reducing emissions of sulfur and nitrogen oxides. In general, the volume of gas-processed fuel in an IGCC plant to be cleaned is usually a third of that which would correspond to a conventional power plant. This reduces the cost of equipment to prevent contamination. It is also much easier to remove sulfur, nitrogen, and other pollutants from reducing gas leaving the gasifier than from combustion gases. This in turn affects the emissions of sulfur and nitrogen oxides, which are lower than those that correspond to conventional combustion processes. If necessary, gasification plants can also be configured to achieve zero emissions.

Unlike the combustion process, the ash and slag produced as a by-product of gasification are not dangerous. Therefore, ash can be deposited in landfills without any additional treatment costs, used as building materials, or further processed for value-added products, leading to a zero-discharge plant (nonproduction of solid waste).

The aforementioned features provide good prospects for gasification in an environment marked by greater competitiveness in the electricity market, by increasingly stringent regulations regarding emissions of sulfur, nitrogen oxides, air pollutants, and other particles, and by treaties to reduce emissions of greenhouse gases. To be competitive and choose the most appropriate technology in this frame of reference, Stiegel and Maxwell [5] at the US Department of Energy point out that technologies entering the market should have a thermal efficiency greater than 60% and investment costs less than US$1,000/kWe; issue little or no sulfur and nitrogen oxides, other air pollutants, and particulate matter; use noncarbon sources; produce a wide range of specialty products; and capture and sequester carbon dioxide. Of all the advanced technologies being developed, gasification-based technologies are the only ones that have the potential to meet these objectives with production costs at or less than current market costs.

4 Commercial Status

In the late twentieth century, gasification was deployed widely throughout the world. In 1999, there were 128 plants with 366 gasifiers in operation. Most of these facilities were in Western Europe, the Eastern Pacific, Africa, and North America. Combined, these plants generated 42,000 MW of synthesis gas. In the 1999–2003 period, there were plans to build 33 plants with 48 additional gasifiers, which adds another 18,000 MW of capacity to produce synthesis gas. Most of these plants belong to Asian countries, which need to expand electricity production because of economic development [11].

At present, the main raw materials used in gasification plants are coal and petroleum residues, which account for more than 70% of the synthesis gas produced, followed by natural gas, which accounts for about 20%. In the latter case, natural gas is only used as a raw material for chemicals. In the coming years, growth can be expected in the use of low-grade coal, petroleum residues, and other waste.

In the current market situation of power generation, gasification cannot compete with combined-cycle natural gas because of the high investment costs and low price of natural gas. The low costs of fossil fuels or waste materials that can be gasified, compared with the cost of natural gas, are not sufficient in most scenarios to achieve the return on capital investment in the gasification plant. An acceptable return is only possible when the raw material cost is very low, the local cost of natural gas is high, high added-value products are obtained, or improved technology is integrated into existing systems [12, 13].

The Värnamo plant in Sweden is an example of a successful exploitation. In the United States, there is a limited number of biomass gasification projects that receive government support, and most are in a demonstration phase. A recent study [12] shows that 50 manufacturers from Europe, the United States, and Canada market gasification technologies, 75% of which are moving-bed or fixed-bed designs (Lurgi, Wellmann Galusha, Woodall Duckham, Merc, Riley Morgan, Willputte, Wellmann, FW Stoic, BGC Lurgi, etc.) and 20% are fluidized bed systems (Winkler, CO_2 acceptor, Hygas, Synthane, Cogas, Eron, Batelle Union Carbide,

Westinghouse, U Gas, etc.) or entrained bed systems (Koppers Totzek, Shell-Koppers, Texaco, Ruhrgas, Combustion Engineering, Foster Wheeler, and Babcock & Wilcox).

5 Future Development

Considerable activity is underway to develop biomass gasification for efficient and environmentally acceptable energy conversion applications, as evidenced by the status of efforts made in several countries to generate clean power [14]. However, an underlying framework is required for the conceptual design of *Emerging Technologies for the Generation and Conditioning of Syngas From Waste*. This framework should be useful not only for assessing current technologies but also for incorporating novel concepts and being open to future advancements.

Such a framework should encompass a detailed treatment of the following technical and operational issues:

(a) New processes for the pretreatment of biomass wastes, through the modification of their properties prior to gasification, so as to make them more attractive for their subsequent use (see chapter Raw Materials, Selection, Preparation and Characterization).
(b) Efficient energy supply chains from biomass: design and planning of efficient multiple source—multiple product bioenergy supply chains (see chapter Raw Materials Supply).
(c) Gasification precise modeling for improved design of high energy efficiency plants considering the enlargement of the fuel panorama (see chapter Modelling Syngas Generation).
(d) Conditioning and gas cleaning of synthesis gas from biomass and other waste types requiring specific treatments for commercial use, including removal of tars, particulates, alkali, ammonia, chlorine, and sulfur (see chapter Main Purification Operations).
(e) Process intensification: development of high-temperature purification processes that avoid or reduce the loss in performance employing "important value" adsorbents (see chapter Emerging Technologies on Syngas Purification: Process Intensification).
(f) Implementation of innovative CO_2 Capture and Storage (CCS) technologies increasing CO_2 uptake and improving H_2 and CO_2 separation, as well as promoting process intensification and avoiding energy intensive options (see chapter H_2 production and CO_2 separation).
(g) Process systems engineering approach to integrated conceptual design of syngas process networks leading to the flexible and versatile representation of all possible flowsheets for its production from different raw materials using different processing units by means of a process *superstructure*. This superstructure would be subsequently optimized (see chapter Modelling Superstructure for Conceptual Design of Syngas Generation and Treatment).

(h) Process integration: heat exchange network synthesis and exergy opportunities should be explored for enhanced energy efficiency of improved designs (see chapter Process Integration: HEN synthesis, Exergy opportunities).
(i) Global clean gas process synthesis and multiobjective optimization taking into account key performance indicators associated to syngas generation (cost, environmental impact, and safety requirements) (see chapter Global Clean Gas Process Synthesis and Optimization).
(j) Selection of best designs for specific applications: techno-economical database involving many sustainable integrated processes for clean power production from solid raw materials mixtures co-gasification (focused on entrained flow and fluidized bed gasifiers), with corresponding flowsheets, detailed technical characteristics, and optimal kWh cost price (see chapter Selection of Best Designs for Specific Applications).
(k) Examples of industrial applications: successful case studies and limitations. Data collection from industry, including mining and reconciliation techniques for their exploitation in upgrading the model of the entire process of gasification and its influence on future designs. Also, this collection leads to the knowledge of the process response time to perturbations in the main variables. Each industrial design shall be finished off by the evaluation of the profitability of the overall process in question, the cornerstone of its commercial viability (see chapter Examples of Industrial Applications and Industrial Data Collection).

The aforementioned issues are examined in detail in this book.

6 About This Book

The objective of this book is to demonstrate how production and cleaning technologies applied to syngas can be improved by developing perfected models that are validated, interlinked in optimized processes, and economically assessed taking into account the expansion of the fuel panorama. The content is divided into two main areas. The first area focuses on the improvement of the conceptual design of syngas production and purification technologies at two levels: (a) development of better performing models and (b) scientific advancement of improved and innovative technologies. The second area focuses on the modeling, validation, and techno-economic evaluation of improved and innovative systems for clean syngas production, targeted to enhance the industrial design of gasification processes at a preliminary stage. A state-of-the-art systems approach is used throughout the book: A superstructure is introduced to describe the flowsheet configurations for intermediate syngas production aiming at power and H_2 production. This superstructure is based on commercial and custom-made simulation tools, which are standards compliant, interoperable, and open to future developments. The student and practicing engineer are guided in the use of the superstructure, which acts as a versatile and flexible tool for the preliminary design of novel process alternatives and redesign of existing ones and serves to improve the operating conditions of the real process plant. In all cases, several process-wide performance indicators related to the simultaneous consideration of economic and environmental aspects are contemplated for optimization of the best design.

This book contains high-quality works from leading experts in the field. It is intended as a textbook for academics (PhD, MSc), researchers, and industry practitioners in syngas production and applications who are involved in the design, retrofit design, and evaluation activities of alternative scenarios. Teachers can benefit from this book for teaching advanced courses, and industry professionals are provided with the know-how to evaluate and improve existing installations or to design a new one.

References

1. Puigjaner L (2005) Valoració de residus sòlids industrials. In: Llebot JE (ed) La Terra i el Medi. Institut d'Estudis Catalans, Barcelona, pp 201–224
2. Puigjaner L (2009) Tecnologies Emergents per al Tractament Tèrmic de Residus: La Gasificació. In: Gaya J (ed) La Gestió de Residus de Catalunya. COEIC, Barcelona, pp 39–42
3. Antares Group, Incorporation (2003) Assessment of power production at rural utilities using forest thinnings and commercially available biomass power technologies. Prepared for the US Department of Agriculture, US Department of Energy, and National Renewable Energy Laboratory
4. Belgiorno V, De Feo G, Della Rocca C, Napoli RMA (2003) Energy from gasification of solid wastes. Waste Manage 23:1–15
5. Stiegel GJ, Maxwell RC (2001) Gasification technologies: the path to clean, affordable energy in the 21st century. Fuel Process Technol 71:79–97
6. Warneche R (2000) Gasification of biomass: comparison between fixed and fluidized bed gasifier. Biomass Bioenergy 18:489–497
7. Bain R (2006) Biomass gasification presentation. Presented at the USDA thermochemical conversion workshop. NREL Biorefinery Analysis and Exploratory Research Group
8. Farriol X, Montané D, Salvadó J (1994) Tecnologies avançacades per al tractament tèrmic de residus. Eficiència Energètica 5:10
9. Higman C, van der Burgt M (2008) Gasification. Elsevier, Amsterdam
10. McGowan TF (2009) Biomass and alternate fuel systems: an engineering and economic guide. AIChE, Wiley, Hoboken, NJ
11. Kwant KW, Knoeff H (2004) Status of gasification in countries participating in the IEA and GasNet activity. IEA Bioenergy Gasification—EU Gasification Network
12. U. S. Environmental Protection Agency (EPA) Combined Heat and Power Partnership (2007) Biomass combined heat and power catalog of technologies. Report prepared by: Energy and Environmental Analysis, Inc., an ICF International Company, and Eastern Research Group, Inc. (ERG). http://www.epa.gov/chp/documents/biomass_chp_catalog.pdf
13. Resource Dynamics Corporation (2004) Combined heat and power market potential for opportunity fuels. Prepared for Oak Ridge National Laboratory
14. Smith D (2006) Entropic energy LLC community based model for forest land management and bioenergy production using distributed CHP system. Presented at World Bioenergy. Jonkoping, Sweden

Raw Materials, Selection, Preparation and Characterization

Fernando Rubiera, José Juan Pis and Covadonga Pevida

Abstract Among the different energy sources, biomass wastes hold most promise for the near future. Biomass is considered a neutral carbon fuel because the carbon dioxide released during its use is an integral part of the carbon cycle. Increasing the share of biomass in the energy supply contributes to diminishing the environmental impact of CO_2 and to meeting the targets established in the Kyoto Protocol. The use of biomass waste material as a fuel, however, has certain drawbacks related with its high-moisture content, low-energy density and the problem of reducing the size of the biomass, especially in the pulverized range of entrained flow gasifiers. Currently, there is increasing interest in developing new processes for the pre-treatment of biomass wastes, through the modification of their properties prior to gasification, so as to make them more attractive for their subsequent use. Pelletization is a proven technology for improving biomass properties, whereas torrefaction is considered a plausible alternative for decreasing the moisture content, increasing the energy density and greatly facilitating the handleability and grindability properties of the torrefied material.

Notation

CHE	Pellets from chestnut
BCOAL	Pellets from bituminous coal
COF	Pellets from coffee husks
GRA	Pellets from grape waste
HGI	Hardgrove Grindability Index

F. Rubiera (✉) · J. J. Pis · C. Pevida
Instituto Nacional del Carbón (CSIC), Apartado 73, 33080 Oviedo, Spain
e-mail: frubiera@incar.csic.es

J. J. Pis
e-mail: jjpis@incar.csic.es

C. Pevida
e-mail: cpevida@incar.csic.es

L. Puigjaner (ed.), *Syngas from Waste*, Green Energy and Technology,
DOI: 10.1007/978-0-85729-540-8_2, © Springer-Verlag London Limited 2011

PIN	Pellets from pine
RDF	Refuse derived fuel
TOP	Torrefaction and pelletization process
IGCC	Integrated gasification in combined cycle

1 Introduction

The energy systems of most of the developed countries are based on fossil fuels at a time when the energy demand is continuously increasing. The limited reserves of such fuels are leading to a heavy dependence on imported fuels. If the amount of fuel needed to produce energy (rate energy/fuel) is reduced, more energy can be generated from the same amount of fuel, and the dependence on external energy supplies can therefore be reduced. Furthermore, the burning of fossil fuels produces pollutants and has a very negative impact on the environment, especially from the point of view of CO_2 emissions and climate change. For these reasons, one of the current challenges in energy production is to reduce the dependence on fossil fuels and to create a more sustainable scenario. One of the most promising approaches for diversifying energy resources is to resort to renewables, because these have a much less harmful environmental impact than fossil fuels and help to preserve the equilibrium of ecosystems. Moreover, renewable energies are indigenous resources and an increase in their use would have positive implications for the security of energy supplies. Among the different renewable energy resources available, biomass waste is the most promising for the immediate future as it is considered a carbon neutral fuel, i.e., the carbon dioxide released when biomass is used is an integral part of the carbon cycle.

Although several waste materials, such as biomass wastes, plastic wastes, used tyres and organic municipal solid wastes (RDF fraction), can be used directly as a solid fuel because of their high-heating value, their conversion into fuel gas or chemicals is an attractive alternative. Gasification can be considered as a way to recover energy from low-grade fuels, biomass, wastes or their mixtures [1, 2]. In particular, integrated gasification in combined cycle (IGCC) is highly efficient and has great potential for reducing the amount of man-made CO_2 emissions. Various solid fuels, including coal, biomass and wastes such as petroleum coke, heavy refinery residuals and municipal sewage sludge have all been employed as feedstocks in gasification systems [3–6].

However, the use of biomass waste material as a fuel entails several problems, including its high-bulk volume, high-moisture content and relatively low-calorific value, which make raw biomass an expensive fuel to transport. For biomass to produce an equivalent amount of energy as fossil fuels such as coal, very high loads of this material would be needed. Another drawback of some types of biomass (e.g., lignocellulosic materials) is that they are difficult to grind into fine particles. This problem is especially acute when biomass is to be used in pulverised systems, such as entrained flow gasifiers. All of these drawbacks have led to the development of new technologies, the aim of which is to improve the quality of biomass fuels so that they are easier to transport, process and handle.

This chapter mainly deals with new processes, such as torrefaction, that have been proposed as a means of treating biomass wastes to overcome some of the drawbacks of these renewable sources, in particular those related with densification, handleability and grindability. These problems are especially important when biomass is gasified with coal or petcoke, in entrained flow gasifiers where fine grinding is required, making the pre-treatment of the biomass particularly difficult when fibrous materials such as wood have to be handled.

2 Materials Pre-Treatment

The origin of biomass fuels can be very wide-ranging and they may be obtained from many sources, including forestry and agriculture residues, food-processing wastes or municipal and urban wastes. Likewise, the chemical constituents and moisture content of biomass materials are also extremely varied [7]. The characteristics of the biomass feedstock, especially moisture and mineral matter content, have a significant effect on the performance of the gasifier. The gasification of biomass that has a high-moisture content produces fuel gas with lower effective-heating values and higher tar concentrations. The energy valorization of these organic wastes requires as a first-step dewatering technologies with a low-energy consumption. Various types of pressure-assisted dewatering devices, such as filter presses, belt presses and centrifuges, can be used to reduce the water content. For many applications, mechanical dewatering cannot guarantee a sufficiently low-moisture content and, in this case, thermal drying must be used instead [8]. Many types of dryers are used to dry biomass, including direct- and indirect fired rotary dryers, conveyor dryers, cascade dryers, flash or pneumatic dryers, superheated steam dryers and microwave dryers. The direct flue gas biomass-drying technologies produce exhaust gases that contain high-volatile organic compounds [9]. The drying of biomass with superheated steam provides a more uniform drying over shorter periods at high temperatures. In modern drying technologies, woody biomass is dried in recirculated gas in conditions of relatively high humidity [10].

In moving bed gasifiers, the gas is made to pass through the biomass so the feed at the bottom has to have sufficient compressive strength to withstand the weight of the feed above it. In addition, the bed can tolerate only a limited amount of fines without experiencing an excessive pressure drop. In these gasifiers, particle sizes tend to be in the 20–80 mm range [11, 12]. Most biomass gasifiers use fluidized bed technologies, and bubbling beds are more common than circulating systems. In these gasifiers, the particle size is still in the millimetre range, and size reduction is not as critical as in the case of entrained flow gasifiers, where the short residence times make it necessary to pulverize the fuel [13].

2.1 Grindability

To prepare biomass for gasification, it is first necessary to reduce the material to a smaller size. However, it is usually not practical, nor necessary, to reduce the biomass feedstock to the same size as the coal feed. For biomass fuels, a wide range of equipment is used to produce different particle sizes. Grinding mills or hammer mills are used to produce particle sizes less than 5 mm, whereas particle sizes between 5 and 50 mm are produced by drum chippers and disc chippers, and the larger particle sizes (5–25 cm) by chunkers [14].

The dependence of most biomasses on seasonal availability and the higher efficiencies attained in larger coal gasifiers are among the causes that have led to the co-gasification of coal and biomass. There are, however, some disadvantages with this approach. In the case of entrained flow gasifiers where pulverized fuel is used, the addition of biomass quickly reduces the mill capacity during fuel handling because of the fibrous nature of some feedstocks. Higher throughputs can be achieved with dedicated co-use schemes. Nevertheless, co-milling is an attractive option compared with the alternative of a separate biomass feed system acting in parallel with the coal feed system as it avoids additional maintenance and installation costs [15].

An evaluation of the grinding behaviour of blends of biomass and solid fossil fuels (coal/petcoke) was conducted at the Instituto Nacional del Carbón (CSIC), as part of a Spanish project [16] coordinated by ELCOGAS, S.A., an electricity generating company that runs an IGCC of 335 MWe, using coal and petcoke as feedstocks. One of the objectives of this project was to select the most convenient biomass for co-gasification with a 50:50 blend of coal and petcoke. A mortar and pestle was used to grind the blends to which up to 20% of biomass was added and the particle size distribution of the different blends was determined.

Three biomass samples were used: (i) almond shells, (ii) olive oil waste and (iii) olive stones. The particle size distribution of the biomasses and the 50:50 petcoke–coal blend (CP–PT) are represented in Fig. 1. It can be seen that the coal–petcoke blend produces the highest amount of fines and does not contain any material greater than 212 μm under the conditions of the study. The results achieved for the biomass samples are clearly worse than for the coal–petcoke blend, with olive oil waste showing the best behaviour, as reflected by the higher amount of fine material in the <75 μm fraction. The olive stones also have a relatively high percentage of <75 μm particles.

Regarding the grinding behaviour of blends of 50:50 CP–PT with different percentages of biomass, the performances in the case of the almond shells and olive stones were similar. The addition of low amounts of biomass to the coal–petcoke blend increased the amount of large particles in the ground material. The behaviour of olive oil waste was different to that of the other biomasses, as can be seen in Fig. 2. For percentages of biomass of up to 10%, the grinding behaviour of the ternary blend was not affected and the particle size distribution of the blend was similar to that of the original coal–petcoke blend. However, when the

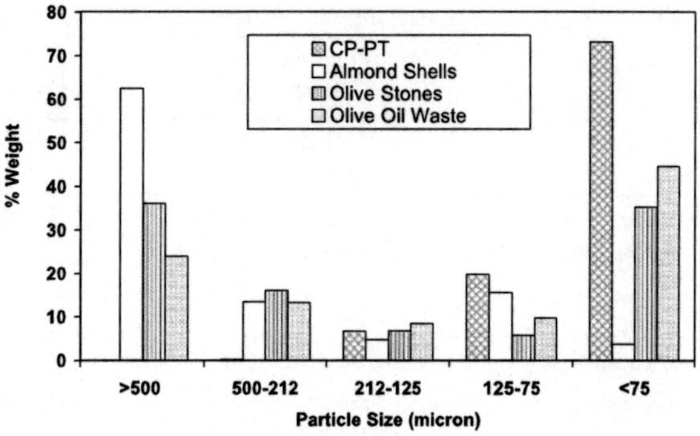

Fig. 1 Particle size distribution of samples ground in a mortar and pestle

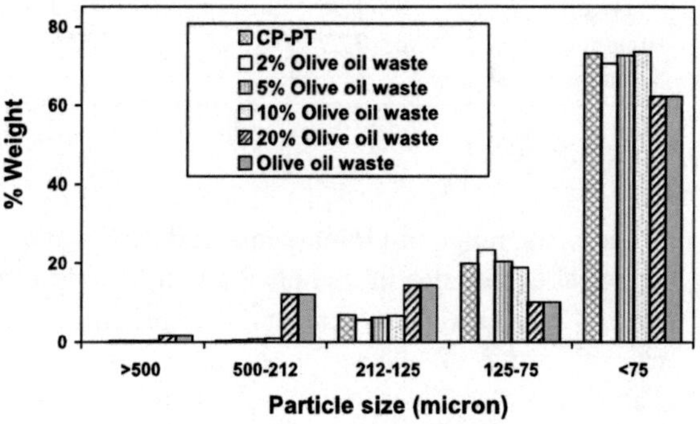

Fig. 2 Particle size distribution of the ternary blend coal–petcoke–olive oil waste ground in a mortar and pestle

percentage of olive oil waste was increased to 20%, a notable increase in the number of larger particles was observed in the milled product in comparison with the coal–petcoke blend, 7% of the particles exceeding 125 μm in the case of this particular blend, whereas after the addition of 20% of olive oil waste, the percentage of particles greater than 125 μm raised to 26%.

3 Raw Materials Characterization

The characterization of fuels normally includes the proximate and ultimate analyses, and the calorific value, which give an indication of the quality of the fuel and its suitability. Because of increasing environmental constraints, analysis of metals concentration is also becoming more common. There are, of course, other properties that could be included in a more exhaustive fuel characterization, such as the ash fusion temperatures, the fuel particles hardness as measured by the Hardgrove grindability index and fuel density.

The nature and behaviour of the mineral constituents have a significant bearing on the design, operation and performance of the gasifier. For the gasification

Table 1 Proximate and ultimate analyses and high-heating value of various samples

Sample	Proximate analysis (wt.%, dry basis)			Ultimate analysis (wt.%, dry basis)					High-heating value (MJ/kg)
	Ash	Volatile matter	Fixed carbon	C	H	N	S	O	
Pine	0.2	86.3	13.5	45.2	6.3	0.1	0	48.2	20.0
Chestnut	0.4	82.1	17.5	45.5	5.7	0.2	0	48.2	19.1
Eucalyptus	0.5	84.6	14.9	46.8	6.1	0.1	0	46.5	19.5
Cellulose residue	1.3	87.7	11.0	41.0	6.4	0.3	0	51.0	17.6
Coffee husks	4.5	79.4	16.1	43.2	6.3	2.6	0.2	43.2	20.1
Grape waste	7.5	67.9	24.6	50.0	6.0	2.0	0.1	34.4	22.1
Almond shells	1.2	79.3	19.5	49.2	6.0	0.2	0	43.4	19.7
Olive stones	0.6	81.4	18.0	50.6	6.1	0.1	0	42.6	19.0
Olive oil waste	7.1	77.3	15.7	48.9	6.2	1.4	0.2	36.2	21.6
Petcoke	0.6	12.6	86.8	87.2	4.1	1.5	5.4	1.2	35.2
High-volatile bituminous coal	7.6	37.7	54.7	77.9	5.1	1.7	1.5	6.2	32.4

process, the range of biomass materials and ash qualities, of which there is limited practical experience in real plant conditions, is a particularly acute problem in the case of entrained flow gasifiers. The proximate and ultimate analyses, and high-calorific value of selected waste materials, including a petcoke, are presented in Table 1. An analysis of a high-volatile bituminous coal is also provided for comparative purposes [17]. The ash contents of the majority of the biomass materials listed in Table 1, with the exception of the grape waste, coffee husks and olive oil waste, are quite low. In general, clean white wood materials have very low ash contents, generally less than 1%, whereas straw and grass materials, and the solid residues from the vegetable oil-producing industries, have ash contents in the 4–7% range.

The volatile matter content of the biomass materials is much higher than that of coal, ranging between 70 and 85%, as can be seen in Table 1. High-volatiles content increases the production of tars in all gasification systems except entrained flow gasifiers, which produce tar-free syngas. Nonetheless in these reactors, biomass is blended in low percentages with coal and/or petcoke [18]. In addition, the heating values of biomass fuels are lower than that of coal, as can be seen in Table 1. One of the limitations of biomass materials is their relatively low volumetric heating value or energy density (MJ/m^3), which is about five times lower than that of coal. This gives rise to concomitant problems related to the need for a large volumetric flow of biomass, even when co-gasifying small percentages of biomass with coal [19].

The drawbacks of using biomass for co-gasification, such as the difficulty of grinding to low particle sizes and the low-energy density of biomass, have prompted the search for new processes and methods of preparing biomass for co-feeding. These include densification through pelletization and/or torrefaction.

4 Improvement of Biomass Quality

4.1 Pelletization

The densification of biomass waste materials contributes to improving their behaviour as a fuel by increasing their homogeneity and allowing a wider range of lignocellulosic materials to be used as fuel [20]. Wood pellets are also easier to handle, transport and store because of their uniform size, high density and low-moisture content [21]. All over the world, wood pellet markets are experiencing exponential growth. In fact, several new pellet plants are planning to base their production on more secure and cheaper feedstocks than sawdust. Increasingly, market parties, especially large-scale users, are experimenting with pellets made of straw, coffee husks, wheat husks and other agricultural residues [22].

The use of blends of coal and biomass could produce fuel pellets with more suitable characteristics for use in industrial furnaces, because coal has a higher carbon content and calorific value than biomass [7]. Apart from their increased density, and higher homogeneity, the quality of the pellets can be evaluated by measuring their mechanical resistance. This is quantified by means of the abrasion index, I_a, lower values of I_a indicating a better pellet quality or a higher resistance [17].

The results of the abrasion index, I_a, for the pellets obtained from some of the raw materials presented in Table 1 are given in Fig. 3. The highest mechanical durability was attained with pellets from chestnut, CHE, followed by pine, PIN, and bituminous coal, BCOAL, whereas the pellets obtained from coffee husks, COF, and grape waste, GRA, were the weakest. Blending materials with high- and low-abrasion index have also been tried in an attempt to increase the mechanical resistance of some common wastes. The results of the work of Gil et al. [17] indicated that a blend of pine sawdust or chestnut sawdust with different percentages of a bituminous coal of up to 20% would be the most suitable for pellet production. In addition, a blend of this coal with eucalyptus sawdust was found to improve the mechanical durability of eucalyptus pellets considerably when the coal was added in a percentage equal to, or higher than, 10%.

As pointed out above, wood-based pellets are produced commercially around the world but there is only a limited production of agricultural biomass-based pellets. A recent study has developed a data-intensive techno-economic model for assessing the economic viability of using agricultural residue for pellet production [23]. The conclusions of this study indicate that for average and maximum yields, the cost curves are quite flat for a number of plants that produce 70,000 tonnes a year. This implies that plants smaller than the economically optimum size can be built with only a minor penalty cost. In addition, from a sensitivity analysis conducted in the same study, it was concluded that the total cost of pellet production is most sensitive to the field cost (cost of harvesting and collection, on-farm storage, nutrient replacement and farmer's premium) followed by the transportation cost.

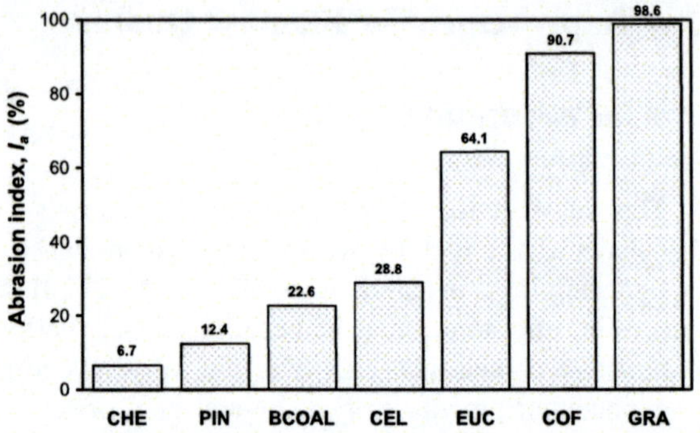

Fig. 3 Abrasion index of pellets from different materials (cf. Table 1)

4.2 Torrefaction

The drawbacks of raw biomass waste, which include a low-energy density (typically 18 MJ/kg) and high-moisture content (around 10% even after drying), make the transport of biomass relatively expensive. A drying step would be insufficient as the biomass may regain moisture and result in storage problems, such as degradation and self-heating. Increasing the energy density of biomass is, therefore, seen as a necessary step to make biomass more useful and attractive as an energy source. On the other hand, it has been shown that higher gasification efficiencies can be achieved for fuels with low O/C ratios, such as coal, than for fuels with high O/C ratios, such as wood. Therefore, rather than gasify these fuels directly, it would seem preferable to modify their properties before gasification [24].

Torrefaction is a pre-treatment technology that makes biomass more suitable for combustion or gasification applications by addressing problems such as high-bulk volume, high-moisture content and poor grindability. The improved grindability of biomass after torrefaction may lead to higher use rates in the near future. Torrefaction is performed under atmospheric pressure at a relatively low-heating rate of less than 50°C/min in the absence of oxygen. Temperatures vary from 200 to 300°C according to the type of biomass. The biomass is partly decomposed and yields a solid product with a very low-moisture content and a high-calorific value, in addition to a condensable liquid and non-condensable gases. The main advantages of torrefied biomass include the production of a hydrophobic solid with a higher calorific value than that of the raw biomass, and the enhancement of grindability, because the energy consumption for milling is reduced by three to seven times with respect to the parent biomass [25].

Biomass is completely dried during torrefaction and after torrefaction the uptake of moisture varies from 1 to 6%. The hydrophobic nature of biomass after torrefaction can be explained by the destruction of OH groups, and the subsequent impossibility of hydrogen bonding. In addition, unsaturated structures are formed which are non-polar. This property is also the main reason why torrefied biomass is almost completely preserved and why biological degradation, as often observed with untreated biomass, does not occur anymore [26]. Torrefaction removes moisture and low weight organic volatile components and depolymerizes the long polysaccharide chains, producing a hydrophobic solid product with an increased energy density (on a mass basis) and greatly increased grindability. As a result, a significantly lower amount of energy is required to process the torrefied fuel and it no longer requires separate handling facilities when it is co-used with coal in existing power stations [15].

It has been stated that applying torrefaction as a biomass pre-treatment process can be expected to contribute to the technical and economic feasibility of entrained-flow gasification of biomass using existing feeding technology. Moreover, particles can be produced that are pneumatically transportable, which would be impossible in a bed of untreated biomass particles [26].

The changes in the grindability characteristics of biomass samples subjected to mild pyrolysis treatment or torrefaction has been studied in considerable depth by several authors. Bridgeman et al. [15] conducted an experimental research on the pulverization behaviour of two torrefied energy crops, willow and Miscanthus. The untreated and torrefied fuels were subjected to standard fuel analysis techniques, including ultimate analysis, proximate analysis and calorific value determination. The grindability of these fuels was then determined using a laboratory ball mill and by adapting the Hardgrove Grindability Index (HGI) test for hard coals. As the temperature and residence time of the process increased, a trend of decreasing volatile matter content was observed. The energy content of the torrefied biomass experienced an increase, the lowest energy yield (76%) corresponding to Miscanthus, at a treatment temperature of 290°C and a residence time of 1 h. The conclusions of the study showed temperature to be the most important parameter in terms of mass loss, increase in energy content, and ease in grindability of the solid product. It was also demonstrated that residence time plays an important role in facilitating grindability. The results demonstrated that Miscanthus was easier to pulverize than willow, the particle size distribution profiles of the pulverized pre-treated Miscanthus being similar to those of coals with similar or equivalent HGI values. The authors also concluded that particle size distribution is a more satisfactory method of analysis of grinding behaviour than the modified Hardgrove Grindability Index used in their work.

Arias et al. [27] evaluated the influence of torrefaction on the grindability of eucalyptus. 10–15 g of biomass was ground and sieved to a particle size of 5 mm. Torrefaction was performed under a nitrogen atmosphere, at a heating rate of 10°C/min up to three different final temperatures (240, 260, 280°C). The samples were kept at the final temperature for different residence times (0 to 3 h). Torrefaction increased the gross calorific value with the temperature treatment and the residence time. In the most critical conditions studied in their work (280°C and 3 h), the gross calorific value of the product increased by 34% with respect to that of the raw biomass. However, during torrefaction, there was a loss of energy with respect to the original biomass, as was computed from the heating value yield. Figure 4 shows the evolution of the heating value yield of the torrefied eucalyptus. Here the heating value yield ranges from 92% for the sample treated up to 240°C (TRE-0-240) to the unacceptable value of 67% for the sample treated at 280°C for

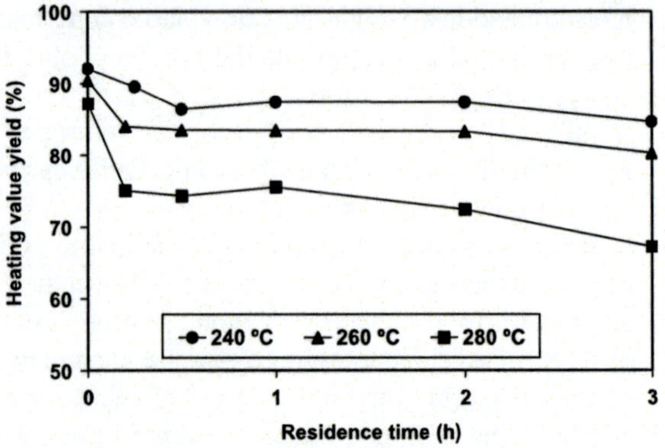

Fig. 4 Variation in the heating value yield of torrefied eucalyptus for different treatment temperatures and residence times (reprinted from Arias et al. [27] with permission from Elsevier)

3 h (TRE-3–280). It can be seen from the results in this figure that even at low residence times, treatment at 280°C produces a large decrease in the heating value yield. For the other two temperatures (240 and 260°C) the heating value yield remained practically constant from 30 min to 2 h of treatment, with a slight reduction being observed after 3 h.

The grindability characteristics of the torrefied eucalyptus were evaluated by treating the samples in a cutting mill and sieving the samples into four size fractions. The particle size distributions of the torrefied eucalyptus samples are shown in Table 2. In the case of the raw eucalyptus only 29% of the untreated biomass passed through the 425μm sieve. In all cases, there is an improvement in the grindability characteristics of the treated biomass, the percentage of particles passing to the lower size fractions greatly increasing for the samples subjected to torrefaction. The results of the work of Arias et al. [27] indicated that a mild torrefaction treatment at 240°C for 30 min could improve the grinding characteristics of the biomass with little loss of heating value yield.

A combination of the processes of pelletization and torrefaction has been developed at the Energy Research Centre of the Netherlands (ECN), and has been named as the TOP process [28]. Densification by means of pelletization is considered to be a proven technology for improving the properties of biomass for its conversion into heat and power. However, biopellets are expensive, require special treatment at the power station and can only be produced from a limited variety of biomass feedstock. The TOP process integrates the advantages of torrefaction and pelletization with respect to the quality of the biopellet. The experimental work conducted at ECN revealed that TOP pellets may have a bulk density of 750–850 kg/m^3 and a net calorific value of 19–22 MJ/kg as received. This results in an energy density of 14–18.5 GJ/m^3, which is reasonably similar to that of sub-bituminous coal (16–17 GJ/m^3). The energy density is significantly higher than that of conventional biopellets produced from softwood (sawdust: 7.8–10.5 GJ/m^3). In contrast to conventional biopellets, the TOP pellets can be produced from a wide variety of feedstock (sawdust, willow, larch, verge grass, demolition wood and straw) yielding similar physical properties.

Table 2 Particle size distribution of torrefied eucalyptus

Sample	>425 μm	425–150 μm	150–75 μm	<75 μm
RE	71.2	18.7	4.7	5.4
TRE-0–240	55.5	32.1	8.8	3.6
TRE-0.5–240	43.2	36.3	13.1	7.4
TRE-1–240	51.9	29.9	12.2	6.0
TRE-3–240	39.7	35.4	12.5	12.4
TRE-0–260	46.8	34.1	12.3	6.8
TRE-0.5-260	37.9	34.6	13.5	14.0
TRE-1–260	44.1	37.9	13.3	4.7
TRE-3–260	35.5	35.8	18.1	10.6
TRE-0–280	45.6	36.7	10.8	6.9
TRE-0.5–280	36.2	35.6	12.7	15.5
TRE-1–280	46.6	35.8	12.7	4.9
TRE-3–280	33.9	37.0	18.8	10.3

5 Conclusions

The use of biomass waste materials for gasification is increasingly seen as a means of reducing the dependence on fossil fuels and reducing the emissions of CO_2 to the atmosphere. The disadvantages associated with the high-moisture content and the low grindability properties of biomass waste materials can be greatly reduced by applying new processes such as torrefaction or a process that combines pelletization and torrefaction. The reduction in the size of the biomass is an energy consuming process that subjects the equipment to excessive wear. The torrefaction of biomass decreases energy consumption, while increasing production capacity with respect to the untreated, raw material. The higher energy efficiencies achieved in pressurized entrained flow gasifiers would make the co-gasification of coal and/or petcoke with biomass wastes, and the pre-treatment of biomass (i.e., torrefaction) more attractive for the industrial application.

References

1. Bridgwater AV, Toft AJ, Brammer JG (2002) A techno-economic comparison of power production by biomass fast pyrolysis with gasification and combustion. Renew Sust Energ Rev 6:181–248
2. Pan YG, Velo E, Roca X, Manyà JJ, Puigjaner L (2000) Fluidized bed co-gasification of residual biomass/poor coal blends for fuel gas production. Fuel 79:1317–1326
3. Fermoso J, Arias B, Plaza MG, Pevida C, Rubiera F, Pis JJ, García-Peña F, Casero P (2009) High-pressure co-gasification of coal with biomass and petroleum coke. Fuel Process Technol 90:926–932
4. Hernández JJ, Aranda-Almansa G, Serrano C (2010) Co-gasification of biomass wastes and coal–coke blends in an entrained flow gasifier: an experimental study. Energ Fuels 24:2479–2488
5. Kumabe K, Hanaoka T, Fujimoto S, Minowa T, Sakanishi K (2007) Co-gasification of woody biomass and coal with air and steam. Fuel 86:684–689

6. Min TJ, Yoshikawa K, Murakami K (2005) Distributed gasification and power generation from solid wastes. Energy 30:2219–2228
7. Heschel W, Rweyemamu L, Scheibner T, Meyer B (1999) Abatement of emissions in small-scale combustors through utilisation of blended pellet fuels. Fuel Process Technol 61:223–242
8. Mahmoud A, Arlabosse P, Fernandez A (2010) Application of a thermally assisted mechanical dewatering process to biomass. Biomass and Bioenerg. doi:10.1016/j.biombioe.2010.08.037
9. Fagernäs L, Brammer J, Wilén C, Lauer M, Verhoeff F (2010) Drying of biomass for second generation synfuel production. Biomass Bioenerg 34:1267–1277
10. Svoboda K, Martinec J, Pohořelý M, Baxter D (2009) Integration of biomass drying with combustion/gasification technologies and minimization of emissions of organic compounds. Chem Pap 63:15–25
11. Higman C, van der Burgt M (2008) Gasification, 2nd edn. Elsevier, Boston
12. McKendry P (2002) Energy production from biomass (part 3): gasification technologies. Bioresource Technol 83:55–63
13. Fernando R (2009) Co-gasification and indirect cofiring of coal and biomass. IEA Clean Coal Centre, CCC/158
14. Maciejewska A, Veringa H, Sanders J, Peteves SD (2006). In: Co-firing of biomass with coal: constraints and role of biomass pre-treatment. Edited by the European Commision, DG JRC, Institute for Energy
15. Bridgeman TG, Jones JM, Shield I, Williams PT (2008) Torrefaction of reed canary grass, wheat straw and willow to enhance solid fuel qualities and combustion properties. Fuel 87:844–856
16. PiIBE (2006) Project for the deployment of biodiesel in Spain. CDTI
17. Gil MV, Oulego P, Casal MD, Pevida C, Pis JJ, Rubiera F (2010) Mechanical durability and combustion characteristics of pellets from biomass blends. Bioresource Technol 101:8859–8867
18. García-Peña F, Coca P (2009) ELCOGAS: R&D activities towards zero emissions IGCC plants. In: Fourth international conference on clean coal technologies, CCT2009, Dresden, Germany
19. Baxter L (2005) Biomass–coal co-combustion: opportunity for affordable renewable energy. Fuel 84:1295–1302
20. Tabarés JLM, Ortiz L, Granada E, Viar FP (2000) Feasibility study of energy use for densificated lignocellulosic material (briquettes). Fuel 79:1229–1237
21. Bhattacharya SC, Sett S, Shrestha RM (1989) State of the art for biomass densification. Energy Sources 11:161–182
22. Peksa-Blanchard M, Dolzan P, Grassi A, Heinimö J, Junginger M, Ranta T, Walter A (2007) Global wood pellets markets and industry: policy drivers, market status and raw material potential. IEA Bioenergy Task 40
23. Sultana A, Kumar A, Harfield D (2010) Development of agri-pellet production cost and optimum size. Bioresource Technol 101:5609–5621
24. Prins MJ, Ptasinski KJ, Janssen FJ (2006) More efficient biomass gasification via torrefaction. Energy 31:3458–3470
25. Deng J, Wang G, Kuang J, Zhang Y, Luo Y (2009) Pretreatment of agricultural residues for co-gasification via torrefaction. J Anal Appl Pyrolysis 86:331–337
26. Bergman PCA, Boersma AR, Kiel JHA, Prins MJ, Ptasinski KJ, Janssen F (2005) Torrefaction for entrained-flow gasification of biomass. ECN Report, ECN-C-05-067, Petten
27. Arias B, Pevida C, Fermoso J, Plaza MG, Rubiera F, Pis JJ (2008) Influence of torrefaction on the grindability and reactivity of woody biomass. Fuel Process Technol 89(2):169–175
28. Bergman PCA (2005) Torrefaction in combination with pelletisation: the TOP process. ECN Report, ECN-C-05-073, Petten

Raw Materials Supply

José Miguel Laínez, Mar Pérez-Fortes, Aarón D. Bojarski and Luis Puigjaner

Abstract The pressure on reducing environmental footprint is facilitating the emergence of energy supply chains that have biomass as main feedstock. For the development of efficient energy supply chains from biomass it is required to properly integrate the various elements that comprise such systems (e.g., biomass supply, pre-treatment facilities and technologies for biomass to energy and/or fuels conversion). Additionally, it is recognised that a concerted effort is required, embracing the different supply chain entities, in order to correctly estimate environmental burdens and to propose effective environmental strategies. This chapter proposes the use of a mathematical modelling approach as an analytical tool that can support decision-making towards accomplishing the design and planning of efficient multiple source—multiple product bio-energy supply chains. The mathematical formulation of this problem becomes a multi-objective MILP (moMILP). Criteria selected for the objective function are the net present value (NPV) and the overall environmental impact, which is computed using the Impact 2002+ indicator. The main advantages of this approach are highlighted through a case study of a biomass-based supply chain that comprises different components geographically distributed over Spain. For comparison purposes, such a supply chain is contrasted to one embracing coal usage.

J. M. Laínez (✉) · M. Pérez-Fortes · A. D. Bojarski · L. Puigjaner
Universitat Politècnica de Catalunya, ETSEIB, Diagonal 647,
08028 Barcelona, Spain
e-mail: jose.miguel.lainez@upc.edu

M. Pérez-Fortes
e-mail: mar.perez-fortes@upc.edu

A. D. Bojarski
e-mail: aaron.david.bojarski@upc.edu

L. Puigjaner
e-mail: Luis.Puigjaner@upc.edu

L. Puigjaner (ed.), *Syngas from Waste*, Green Energy and Technology,
DOI: 10.1007/978-0-85729-540-8_3,

Notation

Indices

e	Suppliers
f, f'	Facility locations
i	Tasks
j	Equipment technology
s	Materials (states)
t, t'	Planning periods
a	Mid-point environmental impact categories
g	End-point environmental impact categories

Sets

A_g	Set of mid-point environmental interventions that are combined into end-point damage factors g
E_{rm}	Set of suppliers e that provide raw materials
$\hat{E}_{\mathrm{prod}}$	Set of suppliers e that provide production services
$\bar{E}_{\mathrm{tr}}$	Set of suppliers e that provide transportation services
F_e	Set of locations f where supplier e is placed
FP	Set of materials s that are final products
I_j	Set of tasks i that can be performed in technology j
$\bar{J}_e$	Technology j that is available at supplier e
$\tilde{J}_f$	Technology j that can be installed at location f
J_i	Technologies that can perform task i
Mkt	Set of market locations
NTr	Set of non-distribution tasks
RM	Set of materials s that are raw materials
Sup	Set of supplier locations
T_s	Set of tasks producing material s
$\bar{T}_s$	Set of tasks consuming material s
Tr	Set of distribution tasks

Parameters

A_{sft}	Maximum availability of raw material s in period t in location f
Dem_{sft}	Demand of product s at market f in period t
$\mathrm{distance}_{ff'}$	Distance from location f to location f'
FCFJ_{jft}	Fixed cost per unit of technology j capacity at location f in period t
I^J_{ft}	Investment required to establish a processing facility in location f in period t
$\mathrm{Norm}F_g$	Normalising factor of damage category g
Price_{sft}	Price of product s at market f in period t
Price^J_{jft}	Investment required per unit of technology j capacity increased at facility f in period t

rate	Discount rate
Water_s	Moisture for material s
$\text{Water}_{ij}^{\max}$	Maximum moisture for task i performed in equipment j
α_{sij}	Mass fraction of task i for production of material s in equipment j
$\bar{\alpha}_{sij}$	Mass fraction of task i for consumption of material s in equipment j
β_{jf}	Minimum utilisation rate of technology j capacity that is allowed at location f
ζ_{ag}	g end-point damage characterisation factor for environmental intervention a
$\theta_{ijff'}$	Capacity utilisation rate of technology j by task i whose origin is location f and destination location f'
$\rho_{eff't}^{\text{tr}}$	Unitary transportation costs from location f to location f' during period t
$\tau_{ijfet}^{\text{ut1}}$	Unitary cost associated with task i performed in equipment j from location f and payable to external supplier e during period t
τ_{sfet}^{ut2}	Unitary cost associated with handling the inventory of material s in location f and payable to external supplier e during period t
χ_{est}	Unitary cost of raw material s offered by external supplier e in period t
$\psi_{ijff'a}$	a environmental category impact CF for task i performed using technology j receiving materials from node f and delivering it at node f'
ψ_{ija}^{T}	a environmental category impact CF for the transportation of a mass unit of material over a length unit

Binary Variables

V_{jft} 1 if technology j is installed at location f in period t, 0 otherwise

Continuous Variables

DamC_{gft}	Normalised end-point damage g for location f in period t
$\text{DamC}_{g}^{\text{SC}}$	Normalised end-point damage g along the whole SC
EPurch_{et}	Economic value of purchases executed in period t to supplier e
ESales_t	Economic value of sales executed in period t
FAsset_t	Investment on fixed assets in period t
FCost_t	Fixed cost in period t
F_{jft}	Total capacity of technology j during period t at location f
FE_{jft}	Capacity increment of technology j at location f during period t
IC_{aft}	Mid-point a environmental impact associated to site f which rises from activities in period t
Impact_f^{2002}	Total environmental impact for site f
$\text{Impact}_{\text{overall}}^{2002}$	Total environmental impact for the whole SC
LHV_{si}	Lower heating value for material s in task i
NPV	Net present value
$P_{ijff't}$	Activity magnitude of task i in equipment j in period t whose origin is location f and destination location f'

$Pv_{sijff't}$	Amount of material s for flexible task i in equipment j in period t whose origin is location f and destination location f'
Profit_t	Profit achieved in period t
$\text{Purch}_{et}^{\text{pr}}$	Amount of money payable to supplier e in period t associated with production activities
$\text{Purch}_{et}^{\text{rm}}$	Amount of money payable to supplier e in period t associated with consumption of raw materials
$\text{Purch}_{et}^{\text{tr}}$	Amount of money payable to supplier e in period t associated with consumption of transport services
$\text{Sales}_{sff't}$	Amount of product s sold from location f in market f' in period t
S_{sft}	Amount of stock of material s at location f in period t

Superscripts

L	Lower bound
U	Upper bound

Acronyms

B-NET	Biomass utilisation networks
BM	Biomass
CBA	Cost.benefit analysis
CCS	Carbon capture and storage
CF	Characterisation factors
DFCF	Discounted-free-cash-flow
Eco-indicator 99	Damage environmental metric
EISA	Energy Independence and Security Act
FWR	Forest wood residues
GHG	Greenhouse gas
GrSCM	Green supply chain management
IMPACT 2002+	Mid-point and end-point (damage) environmental metric
LCA	Life cycle assessment
LCIA	Life cycle impact assessment
LCI	Life cycle inventory
STN	State task network
moMILP	Multi-objective mixed integer linear programming
NPV	Net present Value
SCM	Supply chain management
STN	State task network

1 Introduction

Global trends are promoting the utilisation of renewable sources of energy. For instance, the European Union has established a target of 20% share of renewable energy out of the total European energy consumption by 2020 [1]. Biomass energy generation is an option that is expected to play an important role in the renewable energy mix. As a matter of fact, the USA in its Energy Independence and Security Act (EISA) of 2007 states that advanced biofuels shall supply at least 21 billion gallons of US motor fuels by 2022. It is recognised that in order to achieve these targets efficient networks to supply sustainable amounts of biomass required, cost-effective technologies to convert biomass and improved distribution infrastructures to deliver the final product (i.e., energy or fuel) are to be developed [2]. Even more, the efficient integration of these three elements is equally relevant to reach the posed goals. In this chapter, we propose a supply chain modelling approach as a tool that can support decision-making towards accomplishing such integration.

The concept of supply chain (SC) refers to the network of interdependent entities (i.e., processing sites, distributors, transporters, warehouses and raw material suppliers) which is the processing and distribution channels of a product from origin of its raw materials to final delivery to the customer. Then, supply chain management (SCM) can be defined as the management of material, information and financial flows through a SC that aims at producing and delivering goods or services to consumers [3]. Note that a SC is comprised by components that may be geographically distributed. One of the main objectives of SCM is to synchronise and coordinate the flows of materials that go through the different processes so that the final product is delivered in a most efficient manner. This is especially important for biomass to energy projects which are highly geographically dependent and whose profitability can be strongly influenced by the location of the different processes and biomass sources. Commonly, biomass production and transportation account for a significant part of the whole bioenergy supply chain cost [4]. Therefore, a tool capable of evaluating the possible trade-offs between the different feedstock sources, each one with specific properties (i.e., moisture and energy density) and the location of processing sites and consumption is a requisite to develop efficient bioenergy networks.

The interest on SC biomass-based for different final purposes, ranging from energy or fuels production, has increased since the 90s, thus being a relatively new concept and also a consequence of the current energy paradigm. Typically, a biomass SC problem considers the possible use of multiple sources of biomass. It is also important to point out that biomass usually requires some pre-treatment before its use. This previous treatment is done aiming at obtaining a homogenised or densified biomass, whether in terms of mass or energy. In this sense, Hamelinck et al. [5] considers an international bioenergy SC taking into account the fact that biomass production and consumption do not need to be in the same region, i.e., that biomass residues and energy crops are in excess in some areas, but scarce in others, and energy import regions (Western Europe) do not match with the higher biomass production areas (Scandinavia, Eastern Europe and Latin America). Biomass compacted into briquettes or pellets can be used to save transport use because of their smaller volume. Nevertheless, there exists a trade-off between the type of transport, the distance and the densification methods since the cost of biomass pre-treatments as a whole should not be higher than the compensated logistic costs. This is an important consideration to take into account when long distances should be covered. They demonstrate that an international bioenergy trade has real potential; however, governments must stimulate the biomass market in terms of prices, policies and social acceptance. Panichelli and Gnansounou [4] contemplate forest wood residues (FWR) from final cuttings to produce torrified wood that supplies a gasification unit in order to produce electricity. They are able to allocate biomass quantities between predefined combinations of candidate sites to find the best set of locations for the energy units. They state fix values for the torrefaction and gasification units, and take into account cost minimisation. A mixed integer linear program (MILP) that determines the optimal sizes and locations of biomass-based methanol plants (biofuel plants) is developed by Leduc et al. [6]. The objective function considered to be optimised is the operating costs plus the investment required to establish the biomass SC. The supply is given by poplar coppice, as energy crop, and the demand is based on gasoline-methanol car blend use. The possible consumer sites are the already existing gas stations in Austria, while methanol production is considered through the use of gasification plants. The evaluation of three scenarios is performed, based on different methanol-gasoline blends. By-production of heat is also considered as economic revenue, and CO_2 emissions are accounted finally, but not introduced in their model as an environmental objective to be accomplished. Later on, Ayoub et al. [7] present a methodology for designing and evaluating the biomass utilisation networks (so called B-NETs), which are process networks aiming at producing different bioproducts, from one or more biomass resources. The idea of this methodology is to provide a framework to create the underlying superstructure that relates the biomass resources to their products via current and possible future available processes; which can be used to develop an optimisation model. Their methodology is applied at a local level and proposes better biomass uses by means of economic parameters (costs) and environmental impacts accounting. The environmental indicator that the authors use accounts for emissions to air, water pollutants and solid wastes. The work of Rentizelas et al. [8] emphasises the multi-biomass seasonal availability and combines this fact with the biomass storage problem. The stages considered before the conversion plant of the raw material include harvesting and collection, in field handling and transport, storage, loading and unloading, transport and biomass pre-treatment. This last stage can be included in any of the abovementioned stages, and optimally could precede the transportation stage. Storage can be also located in the biomass origin in an intermediate step or at the power station site. The authors state that one of the main drawbacks of the use of biomass as a source is its relatively low density and heating value when compared, for instance, with other fossil sources. On a later work, Rentizelas et al.

[9] exemplify the fact that a biomass SC can account for multiple sources, as well as for multiple final products production, such as electricity, heat and cooling. They apply the methodology in a specific region of Greece. The results provide with optimal locations and investment details for potential investors. More recently, Van Dyken et al. [10] developed a linear optimisation model for planning the capacity expansions in energy systems where several alternative biomass and technologies are considered simultaneously. The main objective of this work is to present a generic model including different components such as sources, handling, processing, storage and final usage. Heating value, moisture content and bulk density are the key parameter changes that biomass undergo along the SC. The objectives to be optimised are the operating cost and emissions of the whole SC.

Definitely, energy policies are driven by environmental considerations, more specifically by the pressure on reducing greenhouse gas (GHG) emissions. In that sense, biomass is an energy source that is expected to provide significant reductions of environmental impacts related to GHG emissions when compared to the classical fossil fuels technologies. Therefore, it is relevant the integration of environmental thinking into SCM in order to assess such expected reduced environmental impacts. The aforementioned integration may be achieved through the emerging concept regarded as 'green supply chain management' (GrSCM). This concept considers the environmental interventions associated with the raw materials sourcing and selection, manufacturing process selection, delivery of final product to the consumers as well as end of life management of the product after its useful life [11]. Traditionally, the methodologies devised to assist SC operation and design have focused on finding a solution that maximises a given economic performance indicator while satisfying a set of operational constraints imposed by the manufacturing/processing technology and the topology of the network. In recent years, however, there has been a growing awareness of the importance of including environmental aspects as objectives and not constraints associated with the SC decision support [12, 13].

The environmental science and engineering community have developed several systematic methodologies for the detailed characterisation of the environmental impacts of chemicals, products and processes. All of these methodologies have embodied the concepts of life cycle, i.e., they are based on a life cycle assessment (LCA) which is described in a series of ISO documents [14]. The LCA framework includes the entire life cycle of the product, process or activity, encompassing extraction and processing of raw materials; manufacturing, transport and distribution; re-use, maintenance recycling and final disposal. Most importantly, it takes a holistic approach, bringing the environmental impacts into one consistent framework, wherever and whenever these impacts have occurred or will occur [15]. These methodologies are based on the incorporation of an optimisation step into the four classical phases that comprise an LCA study namely, goal definition, life cycle inventory—LCI, life cycle impact assessment—LCIA and interpretation (see Fig. 1). The idea of them is to determine process conditions or topology using a multi-criteria optimisation strategy in order to evaluate the trade-off between economic and environmental issues.

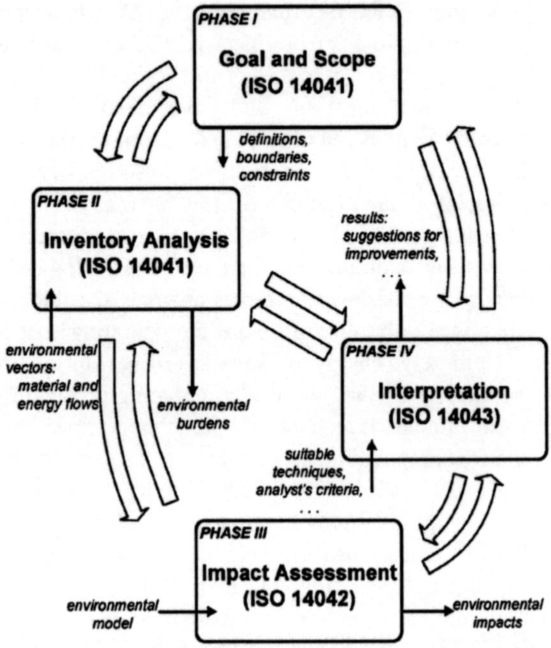

Fig. 1 Life cycle assessment steps [13]

As aforementioned, the concept of SC refers to the network of interdependent entities that constitute the processing and distribution channels of a product from the supply of its raw materials to its delivery to the final consumer. Because an LCA study ideally covers a cradle-to-grave approach, it can be clearly seen that LCA fits as a suitable tool for quantitatively assessing the environmental burdens associated with designing and operating a SC.

This chapter describes an analytical approach for the design and planning of a multiple source—multiple product interregional bioenergy SC taking into consideration not only economic issues but also environmental impact. The model accounts for different *biomass wastes* sources as well. The approach applies mixed integer modelling techniques. The model is optimised so as to select the most appropriate *pre-treatment* technologies, the most appropriate feedstock supplier for each process and the most convenient production and distribution profiles, in order to supply *electricity* and *hydrogen* to the customers. The mathematical model encompasses direct emissions, raw materials production and transport distribution emissions. LCA concepts are embedded in the approach, and going further in order to attain a comprehensive LCA application, not merely an overall environmental impact indicator is calculated but also partial environmental impact categories are studied. Furthermore, the impact associated with every SC echelon is mapped aiming at discovering possible opportunities to focus resources for environmental impact reduction.

2 Biomass Supply and Operations Planning

This chapter deals with the *strategic-tactical problem* associated to the optimal design and operation of a biomass SC network taking into account economical and environmental considerations. Generally speaking, the SC strategic level determines the network through which production and distribution serves the marketplace. The intent of the SC network design problem is typically to determine the optimal sourcing, manufacturing and distribution network for the new and existing product lines of a company (e.g., expansion or contraction of the business, introduction of new products, new strategic suppliers). The most common approach is to formulate a large-scale mixed integer linear program (MILP) that captures the relevant fixed and variable operating costs for each facility and each major product [16].

The considered biomass SC network consists of a number of potential locations where either a processing site or distribution centre or both of them can be installed, and suppliers at fixed locations which have available biomass with different characteristics. In general, each final product (energy or H_2) can be produced at several plants located at different locations using the different biomass wastes. The characteristics of the biomass can be changed by using the pre-treatment units (e.g., drying or torrefaction) so that the treated biomass meets the characteristics required to be used in further steps. Even more, such pre-treatments increase the energy content and bulk density of the biomass. By doing so the mass is reduced, thus significant savings in transportation may be achieved. The production capacity of each processing site is modelled by relating the nominal production rate per activity to the availability of the equipment technology at each plant. Distribution centres are described by upper and lower bounds on their material handling capacity and they can be supplied from more than one manufacturing plants. Given the way the problem is modelled, materials flow between any facilities may appear if selecting such flow allows improving the performance of the SC. A market demand may be served by more than one site.

The mathematical model is an analytical tool intended to support managers on planning decisions such as:

- The active SC nodes and links among them
- The facilities capacity expansion in each time period
- The product portfolio per plant, production amounts, utilisation level and transportation links to establish in the network alongside with material flows
- The amount of final products (energy or H_2) to be sold
- The environmental impact associated to each SC node or activity

A general schematic of the biomass energy SC is shown in Fig. 2. One can notice that it is comprised by four blocks: (i) sourcing, (ii) pre-treatment, (iii) generation and (iv) distribution. The sourcing block consists in collecting the different biomass that may be available from different regions and suppliers. Each type of biomass (e.g., pine waste, forest wood residues, olive pomace) has its

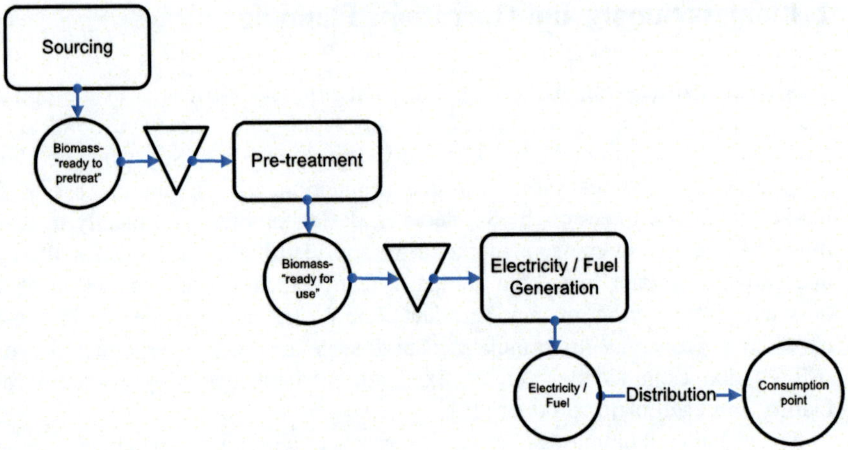

Fig. 2 General schematic of a biomass SC

own characterizing properties such as moisture content and heating value that determines its energy conversion efficiency. The pre-treatment block considers those activities that modify the quality (primarily moisture content) and/or shape of the biomass. Examples of this kind of processes are the chipping, pelletizing, drying, and torrefaction. Notice that these activities may be necessary, provided that there are some technologies which require feeding material to have a fixed maximum moisture content and/or some shape requirements in order to be processed. The generation block converts biomass into energy or any biofuel. Finally, the distribution block comprises those activities aiming at delivering the final product to the consumption points.

2.1 Supply Chain Drivers

All the aforementioned decisions will be taken such that an economic indicator, i.e., net present value (NPV), and an environmental impact metric, are optimised at the end of a predefined planning horizon.

2.1.1 Cost Benefit Analysis

The usage of NPV is the back bone of cost benefit analysis (CBA) of any project. A generic CBA consists of three steps: (i) valuation of the yearly costs and benefits of the project, (ii) discounting costs and benefits in future years to make them commensurate with present costs and benefits and (iii) calculation of the metric, (see Sect. 2.1 in chapter Global Clean Gas Process Synthesis and Optimisation).

Here, in order to compute the NPV, operational costs include those associated with production, handling of material, transportation and raw materials. Transportation costs are assumed to be linear functions of the actual flow of the product from the source echelon to the destination echelon. Revenue is obtained by the selling of products. Investments on facilities and equipment are also taken into account. The cash flows are discounted at a given return rate.

2.1.2 Environmental Metrics

Regarding the selection of environmental metrics, different methodologies have been developed; however, all of them rely on the accurate estimation of environmental interventions. Environment is compromised by industry mainly in two ways, namely, its emissions and the consumption of raw materials. See Sect. 2.3 in chapter Global Clean Gas Process Synthesis and Optimisation for further details.

Here, the environmental metrics used are the ones devised in the work of Humbert et al. [17], which presents an implementation working at both mid-point and end-point (damage) levels. For each environmental intervention two characterisation factors are proposed, which eases model implementation. Their methodology, IMPACT 2002+, is mainly a combination between IMPACT 2002 [18], Eco-indicator 99 [19] using egalitarian factors, CML [15], and the Intergovernmental Panel on Climate Change (IPCC) considerations for CO_2 emissions. IMPACT 2002+ has grouped similar category end-points into a structured set of damage categories by combining two main schools of impact model methods: classical impact assessment methods (CML/IPCC) and damage-oriented methods (Eco-indicator 99). This methodology proposes a feasible implementation of a combined mid-point/damage-oriented approach. It links all types of LCI results via 15 mid-point impacts (human toxicity, respiratory effects, ionizing radiation, ozone layer depletion, photochemical oxidation, aquatic ecotoxicity, terrestrial ecotoxicity, terrestrial acidification/nitrification, aquatic acidification, aquatic eutrophication, land occupation, global warming, non-renewable energy and mineral extraction) to four areas of protection end-point categories (human health, ecosystem quality, climate change and global warming potential and resources).

This approach contains the advantages of being able to calculate both mid- and end-point indicators. Within the presented model, and in order to avoid emission double counting, raw material emissions are not aggregated to product manufacturing, similarly transport and energy consumption are considered separately.

To conclude, Matthews et al. [20] highlight the importance of a footprint estimation that includes the total SC up to the production gate, also known as cradle-to-grave approach (tier 3). Furthermore, the authors refer to tier 4 emission estimations when the whole product life cycle is taken into account by considering emissions occurring during distribution and product end of life. This extended scope is expected to better aid effective environmental strategies since both firms and consumers have an important influence over the footprints through their 'purchase' decisions.

3 The Mathematical Formulation

This chapter describes, besides the methodology, also a tool that can be used to assist in the planning and design of a biomass SC under economical and environmental impact considerations. The resulting model is solved by using a multi-objective MILP (moMILP) algorithm, which allows observing possible environmental trade-offs between damage categories and the economic indicator.

The mathematical formulation of the LCA-SC problem is briefly described next. The variables and constraints of the model can be roughly classified into three groups. The first one concerns process operations constraints given by the SC topology. The second one deals with the environmental model used while the third refers to the economic metric applied.

3.1 Supply Chain Design: Planning Model

The design-planning approach presented is a translation of the state task network (STN) formulation [21] to SC modelling, which is adapted from the work of Laínez et al. [22]. Such a formulation is suitable to collect all SC node information through a single variable, which eases the environmental and economic metrics formulation. This way SC node characteristics are modelled with a single equation set, since manufacturing nodes and distribution centres are treated in the same way as production and distribution activities. Subsequently, it turns out that the most important model variable is $P_{ijff't}$; which represents the specific activity of task i performed using technology j receiving input materials from site f and 'delivering' output materials to site f' during period t. Indeed, to model a production activity it must receive and deliver material within the same site (P_{ijff}). In case of a distribution activity, facilities f and f' must be different. The model's equations are briefly described in the next paragraphs. The separation between tasks and technologies allows for a flexible formulation of different scenarios.

Materials mass balance must be satisfied at each one of the nodes. The expression for the mass balance for each material (state in the STN formulation) s consumed at each potential facility f in every time period t is presented next. Parameter α_{sij} is defined as the mass fraction of material s that is produced by task i performed using technology j; T_s set refers to those tasks that have material s as output, while $\bar{\alpha}_{sij}$ and $\bar{T}_s$ set, refer to tasks that consume material s (Eq. 1).

$$S_{sft} - S_{sft-1} = \sum_{f'} \sum_{i \in T_s} \sum_{j \in (J_i \cap \tilde{J}_{f'})} \alpha_{sij} P_{ijf'ft} - \sum_{f'} \sum_{i \in \bar{T}_s} \sum_{j \in (J_i \cap \tilde{J}_f)} \bar{\alpha}_{sij} P_{ijff't} \quad \forall s, f, t \qquad (1)$$

The model assumes that process parameters are fixed (such as reaction conversion, separation factors and temperatures). This assumption is acceptable for most activities. Notice that the recipe for a given activity is fixed and given by the parameters α_{sij} and $\bar{\alpha}_{sij}$; however, there are activities for which it may be desirable to let the model specify the mixture of inputs in order to achieve a given value for a specific biomass property (i.e., moisture content). For such activities the proportion of the different possible feedstocks should be variable. In order to account for those activities the mass balance shall be modified as shown in Eq. 2.

$$S_{sft} - S_{sft-1} = \sum_{f'} \sum_{i \in T_s} \sum_{j \in (J_i \cap \tilde{J}_{f'})} \alpha_{sij} P_{ijf'ft} - \sum_{f'} \sum_{i \in \bar{T}_s} \sum_{j \in (J_i \cap \tilde{J}_f)} \bar{\alpha}_{sij} P_{ijff't} + \sum_{i \in T_s} \sum_{j \in (J_i \cap \tilde{J}_{f'})} Pv_{sijft} - \sum_{i \in \bar{T}_s} \sum_{j \in (J_i \cap \tilde{J}_f)} Pv_{sijff't} \quad \forall s, f, t \tag{2}$$

For these flexible activities, it is necessary to make sure that the energy balance is achieved. This is done by introducing Eq. 3. Here, HV_{si} is the heating value for material s in activity i. Notice that heating values for feedstock are fixed. An activity changes the output heating value if (i) it is a pre-treatment task that modifies the biomass properties; or if (ii) it is a task that just changes the shape/appearance of biomass but it is receiving different kinds of biomass as input.

$$\sum_{s \in S_i} HV_{si} Pv_{sijft} = \sum_{s \in \bar{S}_i} HV_{si} Pv_{sijff't} \quad \forall i \in \bar{I}, j, f, t \tag{3}$$

Let us consider that a flexible activity must accomplish a total moisture content. In such a case, constraint (4) must be satisfied. Parameters Water_s and $\mathrm{Water}_{ij}^{\max}$ express the moisture content for material s and the maximum moisture content permitted for task i performed in equipment j.

$$\sum_{s \in S_i} \mathrm{Water}_s Pv_{sijft} \le \mathrm{Water}_{ij}^{\max} \sum_{s \in \bar{S}_i} Pv_{sijff't} \quad \forall i \in \bar{I}, j, f, t \tag{4}$$

Equation 5 models the temporal changes in facility capacities, in this sense the model allows for the simultaneous consideration of design and retrofit of SCs. Equation 6 serves for total capacity (F_{jft}) bookkeeping taking into account the amount increased during planning period $t(\mathrm{FE}_{jft})$.

$$V_{jft} \mathrm{FE}_{jft}^{L} \le \mathrm{FE}_{jft} \le V_{jft} \mathrm{FE}_{jft}^{U} \qquad \forall f, j \in \tilde{J}_f, t \tag{5}$$

$$F_{jft} = F_{jft-1} + \mathrm{FE}_{jft} \qquad \forall f, j \in \tilde{J}_f, t \tag{6}$$

Equation 7 ensures the total production rate in each plant to be greater than a minimum desired production rate and lower than the available capacity. Furthermore, parameter β_{jf} defines a minimum utilisation rate of technology j in site f, while $\theta_{ijff'}$ determines the resource utilisation factor.

$$\beta_{jf} F_{jft-1} \le \sum_{f'} \sum_{i \in I_j} \theta_{ijff'} P_{ijff't} \le F_{jft-1} \qquad \forall f, j \in \tilde{J}_f, t \tag{7}$$

$\theta_{ijff'}$ is the capacity utilisation rate of technology j by task i whose origin is location f and destination location f'. This parameter is one of the key factors to be determined when addressing aggregated planning problems, considering strategic and tactical decisions. This operational model may be applied in continuous as well as in semi-continuous processes. First, let us consider the continuous processes. For these cases, the capacity utilisation factor is a conversion factor, which allows taking into account the equipment j capacity in site f in terms of task i kg of produced material per time unit. In this way, the θ_{ijff} factor is the maximum throughput per planning period. On the other hand, this parameter is closely related to tasks operation time in the case of semi-continuous (batch) processes. Note that in this kind of production processes, the time period scale utilised in aggregated planning is usually larger than the time a task (production/distribution activity) requires to be performed. Therefore, the sequencing-timing problem of short-term scheduling is transformed into a rough capacity problem where aggregated figures are used. It is important to have in mind that capacity is expressed as equipment j available time during one planning period, then $\theta_{ijff'}$ represents the time required to perform task i in equipment j per unit of produced material. Thus, once operation times are determined this parameter can be easily estimated. Equation 8 forces the amount of raw material s purchased from site f at each time period t to be lower than an upper bound given by physical limitations $\left(A_{sft}\right)$. Also, the model assumes that part of the demand can actually be left unsatisfied because of limited production or supplier capacity. Thus, Eq. 9 forces the sales of product s carried out in market f during time period t to be less than or equal to demand.

$$\sum_{f'}\sum_{i\in\bar{T}_s}\sum_{j\in J_i} P_{ijff't} \leq A_{sft} \qquad \forall s\in \mathrm{RM}, f\in \mathrm{Sup}, t \tag{8}$$

$$\sum_{f'}\sum_{i\in T_s}\sum_{j\in J_i} P_{ijf'ft} \leq \mathrm{Dem}_{sft} \qquad \forall s\in \mathrm{FP}, f\in \mathrm{Mkt}, t \tag{9}$$

For further model details the reader should refer to Laínez et al. [22].

3.2 Supply Chain: Environmental Model

The application of the LCA methodology to a SC requires four steps, namely (i) goal setting, (ii) life cycle inventory (LCI), (iii) life cycle impact assessment (LCIA) and (iv) results interpretation towards improvement.

Regarding goal setting, it is important to define the boundaries of the system under study, and which is the functional unit (FU) that the SC will provide. Boundaries in the case of the chemical industry in general and in the case of Biomass SCs are usually drawn from cradle-to-grave, this is due to the fact that most of these SC products (chemicals and electricity) are used in different ways and the use phase of products made from these products is too difficult to model appropriately. Consequently raw material extraction, its processing and shipment to a market are considered as part of the SC system. Regarding the FU, commonly a certain amount of product produced is considered. In this sense it is advisable to compare different SCs in terms of the fulfilled amount of sales or portion of demand satisfied [12].

The LCI step requires the estimation of SC environmental interventions (emissions or natural raw material consumptions) which requires assessment of raw material producers, transportation and product processing impacts. This step is the most time consuming within a LCA due to the large amount of information that is required to be gathered, however the usage of LCI databases eases this issue. More importantly the use of a mathematical model helps in calculating the appropriate LCI for the optimal SC configuration.

The results of the LCI step of the LCA can be interpreted by means of different environmental metrics. Environmental interventions are translated into metrics related to environmental impact as end-points or mid-points metrics by the usage of characterisation factors (CFs). This translation is the Life Cycle Impact Assessment (LCIA) step. The metrics used differ in their position along the environmental damage chain (environmental mechanism).

The equations of the environmental model are briefly described next. Equation 10 models IC_{aft} which represents the mid-point a environmental impact associated to site f which rises from activities in period t; $\psi_{ijff'a}$ is the a environmental category impact CF for task i performed using technology j, receiving materials from node f and delivering them at node f'.

$$\mathrm{IC}_{aft} = \sum_{j \in \tilde{J}_f} \sum_{i \in I_j} \sum_{f'} \psi_{ijff'a} P_{ijff't} \qquad \forall a, f, t \tag{10}$$

Similarly to the case of α_{sij} and $\bar{\alpha}_{sij}$, the value of $\psi_{ijff'a}$ is fixed and constant, provided that all environmental impacts are directly proportional to the activity performed in that node $\left(P_{ijff't}\right)$. This issue is a common practice in LCA, where all direct environmental impacts are considered linear with respect to the FU [23]. In the case of transportation the FU commonly considered is the amount of material (kg) transported a given distance (kg km). Consequently the value of $\psi_{ijff'a}$ can be calculated by Eq. 11 in the case of transportation, which considers the distance between sites $\left(\mathrm{distance}_{ff'}\right)$ and where ψ^{T}_{ija} represents the a environmental category impact CF for the transportation of a mass unit of material over a length unit. The study of environmental impacts associated to transport or production can be performed by setting the indices summation over the corresponding tasks (i.e., $i \in Tr$ or $i \in NTr$). It should be noted that environmental impacts associated to materials transport are assigned to their origin node.

$$\psi_{ijff'a} = \psi^{T}_{ija} \mathrm{distance}_{ff'} \qquad \forall i \in Tr, j \in J_i, a, f, f' \tag{11}$$

Equation 12 introduces DamC_{gft} which are a weighted sum of all mid-point environmental interventions combined using g end-point damage factors ζ_{ag} and then further normalised with NormF_g factors. Equation 13 is used to compute the g normalised end-point damage along the whole SC $\left(\mathrm{DamC}_g^{\mathrm{SC}}\right)$.

$$\mathrm{DamC}_{gft} = \sum_{a \in A_g} \mathrm{NormF}_g \zeta_{ag} \mathrm{IC}_{aft} \qquad \forall g, f, t \tag{12}$$

$$\mathrm{DamC}_g^{\mathrm{SC}} = \sum_f \sum_t \mathrm{DamC}_{gft} \qquad \forall g \tag{13}$$

Equations 14 and 15 sum up the environmental damage category results for each site f and for the whole SC, respectively.

$$\mathrm{Impact}_f^{2002} = \sum_g \sum_t \mathrm{DamC}_{gft} \qquad \forall f \tag{14}$$

$$\mathrm{Impact}_{\mathrm{overall}}^{2002} = \sum_f \sum_g \sum_t \mathrm{DamC}_{gft} \qquad \forall f \tag{15}$$

$\mathrm{DamC}_g^{\mathrm{SC}}$or $\mathrm{Impact}_{\mathrm{overall}}^{2002}$ are both used as objective functions in the moMILP formulation. In this sense the use of damage categories is sometimes preferred given that they are easier to comprehend compared to mid-point values.

3.3 *Economic Model*

Many economic indicators have been proposed to assess the performance of a SC network design. The most traditional indicators are profit, NPV and total cost. Other more holistic measures have been recently proposed. Laínez et al. [24] proposed a model that pursues the maximisation of a financial key performance indicator, the corporate value of the firm at the end of the time horizon. The corporate value is computed by a discounted-free-cash-flow (DFCF) method which can be introduced as part of the mathematical formulation. Next, expressions to calculate (i) operating revenue, (ii) operating cost and (iii) capital investment are presented which would eventually permit integration with detailed financial models. Here, NPV will be used for the sake of simplicity and comprehensiveness. The application of other kind of metrics is out of the scope of this chapter, given the specific characteristics of the problem addressed in this work, but they are discussed under chapter Global Clean Gas Process Synthesis and Optimisation in Sect. 2.1

Operating revenue is calculated by means of net sales which are the income source related to the normal SC activities. Thus, the total revenue incurred in any period t can be easily computed from products sales executed in period t as stated in Eq. 16.

$$\mathrm{ESales}_t = \sum_{s \in \mathrm{FP}} \sum_{f \in \mathrm{Mkt}} \sum_{f' \notin (\mathrm{Mkt} \cup \mathrm{Sup})} \mathrm{Sales}_{sf'ft} \mathrm{Price}_{sft} \qquad \forall t \tag{16}$$

In order to calculate overall operating cost an estimation of fixed costs and variable costs is required. The total fixed cost of operating a given SC structure can be computed using Eq. 17. Where FCFJ_{jft} is the fixed unitary capacity cost of using technology j at site f.

$$\mathrm{FCost}_t = \sum_{f \notin (\mathrm{Mkt} \cup \mathrm{Sup})} \sum_{j \in \tilde{J}_f} \mathrm{FCFJ}_{jft} F_{jft} \qquad \forall t \tag{17}$$

The cost of purchases from supplier e, which is computed through Eq. 18, includes raw materials purchases, transport and production resources.

$$\mathrm{EPurch}_{et} = \mathrm{Purch}^{\mathrm{rm}}_{et} + \mathrm{Purch}^{\mathrm{tr}}_{et} + \mathrm{Purch}^{\mathrm{prod}}_{et} \qquad \forall e, t \tag{18}$$

The purchases $\left(\mathrm{Purch}^{\mathrm{rm}}_{et}\right)$ associated to raw materials made to supplier e can be computed through Eq. 19. Variable χ_{est} represents the cost associated to raw material s purchased from supplier e.

$$\mathrm{Purch}^{\mathrm{rm}}_{et} = \sum_{s \in \mathrm{RM}} \sum_{f \in F_e} \sum_{i \in \bar{T}_s} \sum_{j \in J_i} P_{ijfft} \chi_{est} \qquad \forall e \in E_{\mathrm{rm}}, t \tag{19}$$

The costs of transportation and production are determined by Eqs. 20 and 21, respectively. Here, $\rho^{tr}_{eff't}$ denotes the e provider unitary transportation cost associated to material movement from location f to location f' during period t. $\tau^{\mathrm{ut1}}_{ijfet}$ represents the unitary production cost associated to perform task i using technology j, whereas $\tau^{\mathrm{ut2}}_{sfet}$ represents the unitary inventory costs of material s storage at site f, both of them using provider e during period t.

$$\mathrm{Purch}^{\mathrm{tr}}_{et} = \sum_{i \in Tr} \sum_{j \in J_i \cap \bar{J}_e} \sum_{f} \sum_{f'} P_{ijff't} \rho^{\mathrm{tr}}_{eff't} \qquad \forall e \in \bar{E}_{tr}, t \tag{20}$$

$$\mathrm{Purch}^{\mathrm{prod}}_{et} = \sum_{f} \sum_{i \notin Tr} \sum_{j \in (J_i \cap \hat{J}_f)} P_{ijfft} \tau^{\mathrm{ut1}}_{ijfet} + \sum_{s} \sum_{f \notin (\mathrm{Sup} \cup \mathrm{Mkt})} S_{sft} \tau^{\mathrm{ut2}}_{sfet} \qquad \forall e \in \tilde{E}_{\mathrm{prod}}, t \tag{21}$$

In the case of $\tau^{\mathrm{ut1}}_{ijfet}$, this parameter entails restrictions associated with α_{sij} and $\bar{\alpha}_{sij}$, which forces the plant to operate at the same fixed conditions, meaning that the amount of utilities and labour spent is proportional to the amount of raw material processed. However the utilities and labour unitary cost may change over time. Moreover, possible cost decrease associated to economies of scale are disregarded by using the former assumption, higher production rates are associated linearly to higher production costs. Finally, the total investment on fixed assets is computed through Eq. 22. This equation includes the investment made to expand the technology's capacity j in facility site f in period t $\left(\mathrm{Price}^{\mathrm{FJ}}_{jft} \mathrm{FE}_{jft}\right)$.

$$\text{FAsset}_t = \sum_f \sum_j \text{Price}^J_{jft} \text{FE}_{jft} + I^J_{ft} \text{JB}_{ft} \qquad \forall t \tag{22}$$

Equation 23 represents the calculation of profit at period t. To conclude, NPV is computed by means of Eq. 24.

$$\text{Profit}_t = \text{ESales}_t - \left(\text{FCost}_t + \sum_e \text{EPurch}_{et} \right) \qquad \forall t \tag{23}$$

$$\text{NPV} = \sum_t \left(\frac{\text{Profit}_t - \text{FAsset}_t}{(1 + \text{rate})^t} \right) \tag{24}$$

The selection of the discount rate (*rate*) for any time discounted metric is subject to controversy, given that it represents the trade-off between the enjoyment of present and future benefits and affects directly intergenerational aspects of sustainability. Higher values of *rate* devaluate future impacts and consequently they count little on long time horizon projects, which could be perceived as contrary to the interest of future generations. Identically to the case of a weighting set for a composite environmental index, the selection of a given discount rate is highly subjective and should represent the decision makers belief in terms of intergenerational aspects.

Finally, the SC network design-planning problem whose objective is to optimise a given set of objective functions can be mathematically posed as follows:

$$\underset{x,y}{\text{Min}} \left\{ -\text{NPV}, \text{DamC}^{\text{SC}}_g, \text{Impact}^{2002}_{\text{overall}} \right\}$$

subject to Eqs. 2 – 24

$$\mathcal{X} \in \{0, 1\}; \mathcal{Y} \in \mathbb{R}^+$$

Here X denotes the binary variables set, while Y corresponds to the continuous variable set.

4 Case Study: A Biomass Supply Chain Geographically Located in Spain

The case study used to illustrate the concepts behind the presented design strategy addresses a Biomass SC problem comparing the generation of electricity and H_2 from two different kinds of feedstock: (i) different biomass wastes and (ii) coal, which is a wide extended fossil fuel resource. The SC under study comprises biomass collection sites and processing sites where biomass is pre-treated and used, in distribution centres and marketplaces.

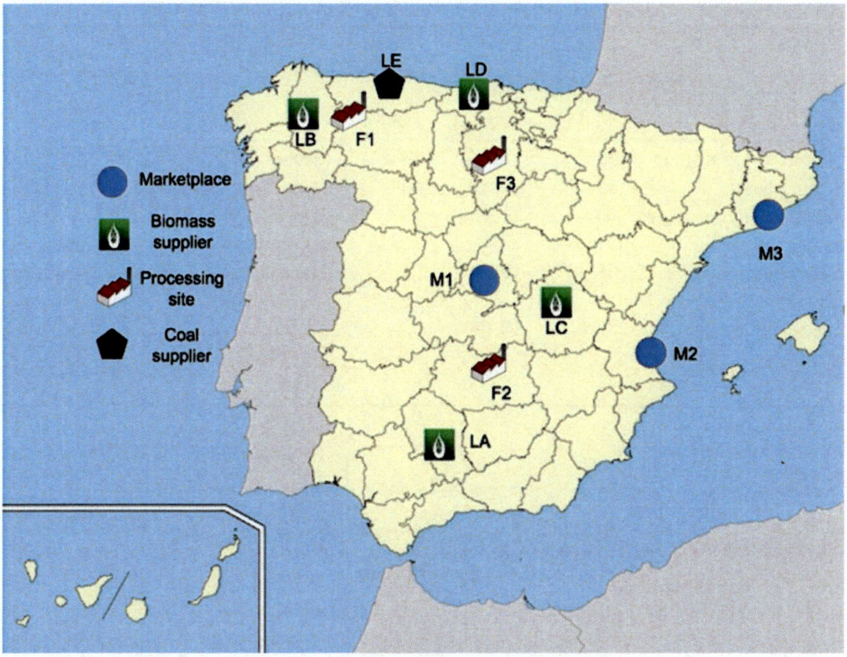

Fig. 3 Location map for the potential SC network

A simplified potential network is proposed and restricted to Spain (see Fig. 3). Lugo (*F*1), Ciudad Real (*F*2) and Burgos (*F*3) are considered to be possible facilities location nodes. The feedstock is supposed to be available at Cordoba (*LA*), Lugo (*LB*), Cuenca (*LC*), Santander (*LD*) and Oviedo (*LE*). This last site is the one supplying coal. Hydrogen is supposed to be sold at three market places located at Madrid (*M*1), Valencia (*M*2) and Barcelona (*M*3), while electricity is fed to the Spanish electricity network at their respective generation places.

Different biomasses are modelled considering that each of them possesses different energy content and moisture. Table 1 shows the characteristics of the considered feedstocks, biomass wastes with useable energy content (see chapter Raw Materials, Selection, Preparation and Characterization) that can be found in Spain. Biomass types, approximate availability, costs and moisture contents are based on data published in Gomez et al. [25], Van Dyken et al. [10], Rentizelas et al. [8] and Panichelli and Gnansounou [4]. Lower heating values (LHV) for biomasses are taken from Phyllis database [26]. Hypothetically, coal has been considered as a dry material which does not require any pre-treatment. It is interesting to observe that higher moisture contents are present in forest and wood residues. Olive residues have low moisture content and high heating value which is comparable to the one of the sub-bituminous coal considered here.

The biomass may be *pre-treated* before being finally processed. Figure 4 depicts the different pre-treatment processes that may be applied to the biomass

Table 1 Feedstock properties. LHV is in ar (as received) basis; Wt: Weight

Biomass	Cost (€ ton^{-1})	Lower Heating Value, (MJ/kg)	Moisture content (%wt)	Seasonality	Monthly availability (tons)	Source location
Forest wood residue	25	8.597	48.9	None	40,500	LA
Pine waste	35	10.450	40	None	3,350	LB
Almond tree prunings	40	11.313	40	December to February	12,400	LC
Olive pomace	65	19.098	7.6	January to March	73,400	LD
Olive pit	35	18.778	6.1	February to May	27,900	LD
Coal	45	15.000	–	None	65,000	LE

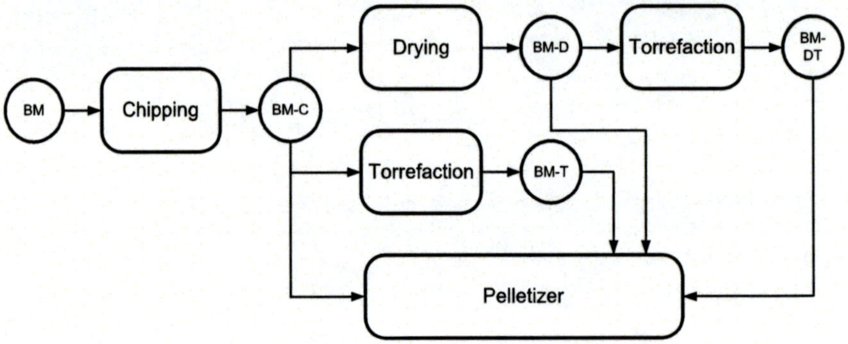

Fig. 4 STN representing the pre-treatment activities for a generic biomass

(BM) so that it achieves the adequate shape and properties (energy content and humidity) to follow up later processes. Several conditions have been assumed to select the specific path of pre-treatment units for each feedstock. They depend basically on the moisture content, shape and LHV of the input required for the treatment plant. Energy densification and matter densification through drying as well as dry matter loss in the different processes affecting the heating value have been considered. Nevertheless, in this case study BM bulk density has not been considered for the sake of simplicity and also given the planning horizon required for the strategic decisions that are being addressed. Note that coal does not need any pre-treatment. The characteristics for the pre-treatment processes are presented in Table 2 The data are from Panichelli and Gnansounou [4] and from Hamelinck et al. [5]. The pre-treatment options considered here are:

- *Chipper*: Transformation of the biomass as received, into chips. It is the first biomass size reduction step, and mandatory for all the biomass sources considered. It is considered that a first step of shape homogenisation is crucial for an integrated Biomass SC. Moisture content is not modified, but there exists a loss of dry matter, which is considered.

Table 2 Pre-treatment processes and their main modelling assumptions

Activity/ equipment	Moisture losses (%)	Dry matter losses (%)	Operating cost (€ ton^{-1})	Capacity (tons h^{-1})	Investment (1×10^6 €)	Electricity consumption (MWh ton^{-1})
Chipper	0	0.17	2.5	30	0.370	5
Dryer	88	0.08	55	100	5.000	20
Torrefactor	55	19	40	20	0.100	37
Pelletizer	0	0	3.5	6	0.485	30

Table 3 Parameters for the processes for electricity and H_2 generation

Technology	Operating cost (€)	Capacity	Investment (1×10^6 €)	Product price (€)	Total monthly demand
Electricity	34.2 MWh^{-1}	300 MW	860	0.151 kWh^{-1}	75,000 MWh
H_2	1,880 ton^{-1}	33.6 tons h^{-1}	1,500	3 kg^{-1}	650 ton

- *Dryer*: Active dryer is needed when the source has too high moisture contents. The condition to enter this step is to have a humidity higher than 7%. Passive drying is assumed not to be significant according to the unit of time that the model considers. The dry matter loss is insignificant if compared with the moisture loss.
- *Torrefactor*: Its main objective is the LHV increase through volatiles release in an inert atmosphere. The biggest dry matter loss is obtained here, and the drying process also takes place. The condition to pass through this unit is to have a LHV lesser than 15 MJ/kg.
- *Pelletizer*: In this case study we assume that the biomass must be pelletized before its final transformation, thus a high homogeneous raw material is arriving to the final plants. Note that the pelletizer requires input to have humidity equal to or lower than 7%. This requirement may be fulfilled by mixing different types of biomasses or pre-treated material. It is the main characteristic of this unit (note that no moisture or dry matter losses are considered here).

The technology that is employed to provide the final product is gasification. An integrated gasification combined cycle (IGCC) power plant is assumed for the electricity generation, and a gasification plant with carbon capture and storage (CCS) for the H_2 generation. Efficiencies of 40 and 30% are assumed for each plant, respectively, the adoption of such figures will be extensively discussed in chapter Modelling Superstructure for Conceptual Design of Syngas Generation and Treatment. It is assumed that the lower capacity that can be installed for the energy generator is 85 MWh. Other relevant information concerning these technologies is presented in Table 3. These data are from Pérez-Fortes et al. [27] and IEA-GHG [28]. Biomass and H_2 transportation prices were estimated at 0.03 and 0.05 € ton^{-1} km^{-1} from current economical trends, respectively. The demand of H_2 is evenly distributed along the three markets (*M*1, *M*2 and *M*3, while the demand of electricity is supplied to the grid from any facility location). It is assumed that the demand must be completely satisfied. For the case of NPV optimisation return rate is assumed to be 8% which is a common value for this kind of projects.

In order to assess the environmental impact associated to the biomass-based energy SC, the available LCI values were retrieved from the LCI database EcoinventV1.3 [29] and using SimaPro 7.1.6 [30] and converted directly to the IMPACT 2002+ mid-point indicators. For those activities which were not available, the impacts were assumed based on similar products or activities. The environmental impacts associated to energy generation, H_2 production and pre-treatment processes without consideration of feedstock and transportation consumption are found in Table 4. The environmental impact for transportation activities is presented in that table as well. The environmental impact for feedstock can be found in Table 5 which does not consider impacts associated to transportation.

The project is evaluated along a planning horizon of 25 years, considering monthly planning decisions. The model has been implemented in GAMS which is an algebraic modelling software. The formulation of the SC-LCA model leads to a MILP with 4,159 equations, 41,221 continuous variables and 96 discrete variables. It takes 61 CPU seconds to reach a solution with a 0.1% integrality gap on a 2.0 GHz Intel Core 2 Duo computer using the MIP solver of CPLEX.

Figure 5 shows the obtained dominant biomass-based SC that maximises NPV. It is found that the three potential locations are considered and on each one of them a facility is opened. All pre-treatment technologies are installed in location *F*1 besides the required equipment to produce H_2. From this site H_2 is delivered to all markets. Note that *F1* is collecting all the forest wood residues (FWR) for which larger mass flows are required due to their low LHV. By establishing *F1*, which is near to the FWR collection site, significant savings in transportation are obtained. The electricity is generated in site *F2*. In this site; equipments to perform chipping, drying and pelletising are installed. The electricity demand of each market is satisfied from site *F2*. Site *F3* is used just as a distribution centre for pre-treated biomass. Equipment for chipping and drying is installed in such a site.

Table 6 shows the proposed capacity to be installed for each of the equipments at every site to obtain the maximum NPV configuration. Notice that for this configuration there are some inter-site flows, clearly showing the capabilities of the model to tackle with inter-site distribution tasks. Forest wood residues which have been dried and torrefied are being sent from site *F1* to *F2*, while *F3* is transferring dried pine waste and dried almond tree pruning to location *F2* in order to be converted to energy later on. By having material flows of pre-treated biomass the transportation cost is reduced due to the mass decrease that is achieved through the utilisation of such processes.

The optimal configuration for the environmental impact has also been obtained. Figure 6 shows the minimum IMPACT 2002+ configuration for the biomass-based SC. Please note that this supply chain fulfills with the same demand as the one obtained by optimising NPV. The capacity proposed to be installed for the equipment of this configuration is presented in Table 7. Note that for this case the location *F3* is not considered, and all biomass is sent from the collection sites to locations *F1* and *F2*. This configuration is satisfying the demand of electricity from both locations *F1* and *F2*, whereas H_2 is delivered from site *F2*. There is one

Table 4 Environmental impacts associated to SC tasks

Mid-point categories	Chipper (points ton^{-1})	Dryer (points ton^{-1})	Torrefactor (points ton^{-1})	Pelletizer (points ton^{-1})	Electricity generator (points GJ^{-1})	H_2 generator (points ton^{-1})	Transport. (points ton^{-1} km^{-1})
Carcinogens	2.16E−06	8.64E−06	1.60E−05	1.30E−05	0.00E+00	0.00E+00	6.93E−08
Non-carcinogens	9.28E−06	3.71E−05	6.86E−05	5.57E−05	0.00E+00	0.00E+00	4.93E−07
Respiratory inorganics	4.18E−04	1.67E−03	3.10E−03	2.51E−03	2.09E−03	2.49E−01	2.38E−05
Ionizing radiation	6.03E−06	2.41E−05	4.46E−05	3.62E−05	0.00E+00	0.00E+00	5.17E−09
Ozone layer depletion	1.23E−08	4.93E−08	9.11E−08	7.39E−08	0.00E+00	0.00E+00	2.58E−09
Respiratory organics	6.01E−08	2.40E−07	4.45E−07	3.61E−07	0.00E+00	0.00E+00	3.43E−08
Aquatic ecotoxicity	3.40E−07	1.36E−06	2.51E−06	2.04E−06	0.00E+00	0.00E+00	2.23E−08
Terrestrial ecotoxicity	6.50E−06	2.60E−05	4.81E−05	3.90E−05	0.00E+00	0.00E+00	5.34E−06
Terrestrial acid/nutri	5.75E−06	2.30E−05	4.26E−05	3.45E−05	6.91E−05	8.23E−03	5.13E−07
Land occupation	1.30E−07	5.20E−07	9.63E−07	7.80E−07	0.00E+00	0.00E+00	1.61E−10
Aquatic acidification	0.00E+00	0.00E+00	0.00E+00	0.00E+00	0.00E+00	0.00E+00	0.00E+00
Aquatic eutrophication	0.00E+00	0.00E+00	0.00E+00	0.00E+00	0.00E+00	0.00E+00	0.00E+00
Global warming	2.57E−04	1.03E−03	1.90E−03	1.54E−03	2.06E−02	2.45E+00	1.37E−05
Non-renewable energy	3.22E−04	1.29E−03	2.39E−03	1.93E−03	0.00E+00	0.00E+00	1.34E−05
Mineral extraction	2.96E−09	1.19E−08	2.19E−08	1.78E−08	0.00E+00	0.00E+00	4.54E−12

Table 5 Environmental impacts associated to feedstock production (points ton^{-1})

Mid-point categories	Forest wood residue	Pine waste	Almond tree prunings	Olive pomace	Olive pit	Coal
Carcinogens	2.12E−05	2.18E−05	1.96E−05	2.84E−05	9.40E−06	1.51E−04
Non-carcinogens	1.37E−05	1.31E−05	1.53E−05	3.28E−05	1.76E−06	2.52E−04
Respiratory inorganics	1.16E−03	1.19E−03	1.08E−03	1.11E−03	4.36E−04	2.89E−02
Ionizing radiation	2.62E−05	2.42E−05	3.12E−05	4.77E−06	2.64E−07	9.37E−05
Ozone layer depletion	1.05E−07	1.07E−07	1.00E−07	8.00E−08	3.70E−08	1.42E−06
Respiratory organics	5.52E−06	6.22E−06	3.71E−06	2.25E−06	4.95E−06	2.28E−05
Aquatic ecotoxicity	1.24E−06	1.15E−06	1.44E−06	1.89E−06	6.60E−08	2.13E−04
Terrestrial ecotoxicity	5.05E−05	4.91E−05	5.42E−05	1.10E−04	6.64E−06	8.41E−03
Terrestrial acid/nutri	1.89E−05	1.97E−05	1.67E−05	1.68E−05	8.28E−06	6.78E−04
Land occupation	1.90E−03	1.62E−03	2.62E−03	4.80E−03	1.38E−03	1.63E−04
Aquatic acidification	0.00E+00	0.00E+00	0.00E+00	0.00E+00	0.00E+00	0.00E+00
Aquatic eutrophication	0.00E+00	0.00E+00	0.00E+00	0.00E+00	0.00E+00	0.00E+00
Global warming	1.08E−03	1.05E−03	1.16E−03	4.50E−04	1.88E−04	1.57E−02
Non-renewable energy	1.49E−03	1.43E−03	1.65E−03	5.72E−04	1.97E−04	1.84E−01
Mineral extraction	6.52E−08	6.69E−08	6.05E−08	6.75E−07	3.59E−08	7.72E−07

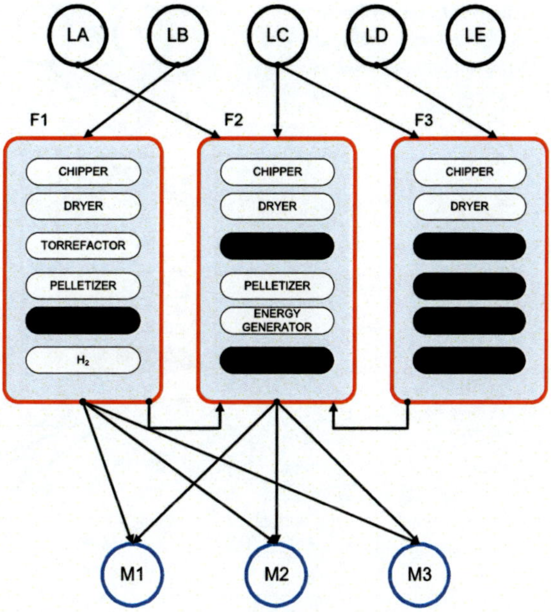

Fig. 5 Optimum NPV network configuration for the biomass-based SC

Table 6 Equipment capacity for the optimum NPV Biomass SC

Equipment	Facility		
	F1	F2	F3
Chipper (tons h^{-1})	67.5	60.4	40.0
Dryer (tons h^{-1})	47.4	20.7	20.0
Torrefactor (tons h^{-1})	55.8	–	–
Pelletizer (tons h^{-1})	23.2	60.3	–
Electricity (GJ h^{-1})	–	450.0	–
H_2 (tons h^{-1})	1.1	–	–

inter-site flow from site *F2* to *F1*, which corresponds to a flow of dried olive pomace, in this environmental friendly configuration.

Recall that we introduced a 'flexible' task to account for those tasks for which we would like the model to decide how to better mix different biomasses so as to achieve a given specified biomass property. We have assumed that the pelletizer is one of such tasks for this case study. To give an example, there are periods in which the model proposes to make the following mix: 1.4% forest wood residues, 30.3% dried and torrefied forest wood residues, 10.5% dried pine waste, 14.4% dried almond tree prunings and 43.5% chipped olive pomace (mass basis). This mixture is then fed to the syngas production plant. The values of humidity corresponding to these materials are 10.0, 6.0, 7.0, 7.0 and 7.5%, respectively. It can be proved that the humidity of this mix is 7.0% which is the maximum humidity allowed for the pelletizer input.

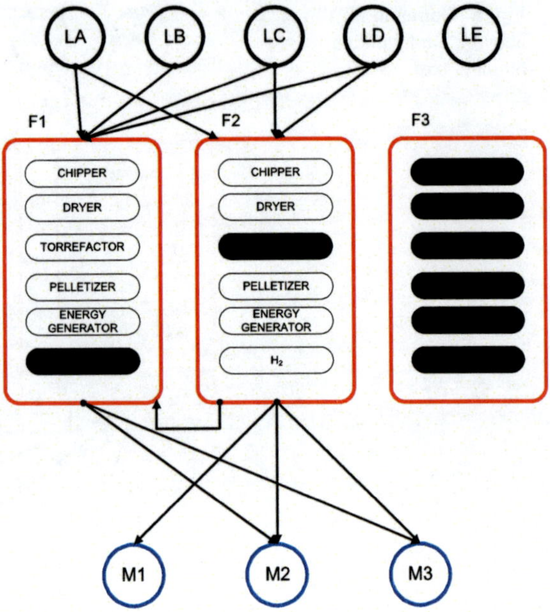

Fig. 6 Optimum Impact 2002+ network configuration for the biomass-based SC

Table 7 Equipment capacity for the optimum Impact 2002+ Biomass SC

Equipment	Facility		
	F1	F2	F3
Chipper (tons h^{-1})	73.1	70.3	–
Dryer (tons h^{-1})	54.4	20.7	–
Torrefactor (tons h^{-1})	51.7	–	–
Pelletizer (tons h^{-1})	40.0	62.7	–
Electricity (GJ h^{-1})	300.0	300.0	–
H_2 (tons h^{-1})	–	1.1	–

Tables 8 and 9 summarize the most significant values corresponding to both solutions regarding environmental and economic aspects. By deploying the SC configuration corresponding to the more profitable SC configuration, a NPV equal to 228.51 M€ is obtained. This value is reduced by 3% when the environmental friendly configuration is established. Note that the main difference between these two configurations is the investment required for installing the proposed capacity in the different sites. For the case of NPV maximisation the installation of an electricity generation plant at location *F2* makes it necessary to send larger amount of biomass to this site. In order to reduce this additional cost a distribution centre at location *F3* is established where biomass is treated before being transferred to the other sites. The reduction in capacity investment offsets, the increase in the transportation cost and the extra costs are associated to the pre-treatment of biomass. Table 8 also shows the payback periods and internal rate of return for both solutions.

Table 8 Economic aspects arising from single objective optimisation (NPV and Impact 2002+) (M€)

	Impact 2002+ optimisation	NPV optimisation
Investment	170.21	148.13
Biomass cost	820.16	819.87
Transportation cost	92.73	122.65
Production cost	2,222.77	2,224.02
Sales	3,987.00	3,987.00
Profit	851.34	820.46
NPV[a]	220.59	228.51
IRR (%)[b]	24.02	27.38
Payback period (years)	5.00	4.51

[a] Based on a return rate of 8%

[b] Based on a planning horizon of 25 years

Table 9 Environmental impacts arising from single economic and overall environmental objective function optimisation results (Impact 2002+ points per year)

End-point impact category	Impact 2002+ optimisation	NPV optimisation
Human health	16,255.29	17,267.21
Ecosystem quality	3,375.79	3,610.96
Climate change	90,383.37	90,950.66
Resources	5,292.64	5,852.73
Impact 2002+	115,307.09	117,681.56

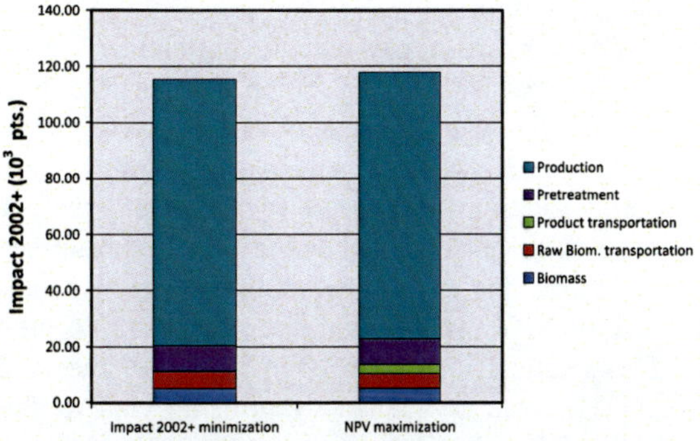

Fig. 7 Distribution of annual environmental impacts for single objective optimisation solutions, according to different SC activities

With regard to environmental interventions, the NPV optimum solution renders an environmental impact of 117681 pts. The environmental friendly configuration can slightly reduce the impact by 2%. Notice that the impacts for each damage category (see Table 9) are very similar for both solutions. These configurations have a major impact on the climate change category which represents 78% of the overall impact. Figure 7 presents the distribution of environmental impacts according to different SC activities. Notice that as for the NPV comparison, the difference between the environmental impacts is mainly due to transportation.

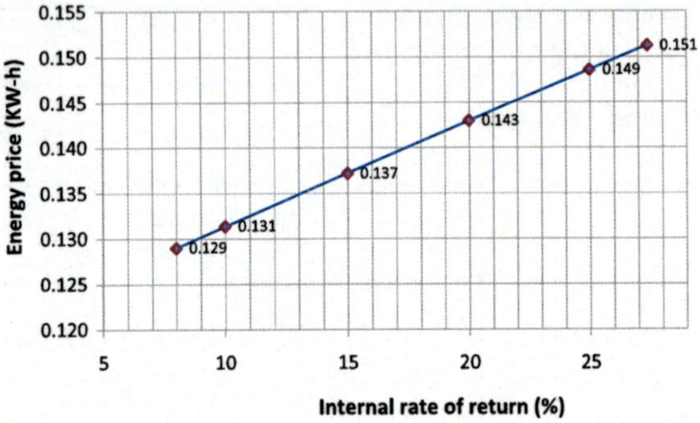

Fig. 8 Energy price versus internal rate of return for the biomass-based SC obtained by maximising NPV

In the presented case study electricity generation and H_2 production are the most important factors contributing to the overall environmental impact in both single objective optimisation cases; while biomass sourcing is the least impacting aspect. This clearly shows that activities to reduce environmental impact should be focused on improving the technologies used to produce energy and H_2.

We would like to highlight how sensitive the solutions for the biomass-based SC are to the final product prices. For instance if we assume that the energy price is reduced to 50% (i.e., 0.075 € kWh^{-1}) the NPV at the end of the planning horizon (i.e., 25 years) would be negative and equal to -553.25×10^6 €. The price that renders an internal rate of return of 8% considering a project life time of 25 years is 0.129 € kWh^{-1}. Any price below this value would require some sort of subsidy. Figure 8 shows the energy prices that are required for different internal rates of return.

For comparison purposes the optimal SC based on coal has been also obtained. The optimal NPV configuration for this case proposes to deliver electricity from location *F1*, where a capacity of 450 GJ h^{-1} is installed, while the H_2 is produced at location *F3* where a capacity of 1.1 tons h^{-1} is installed. The minimum Impact 2002+ solution for this case (where coal is the only raw material available) is the same as the one obtained by the NPV optimisation. The reason is that the only way to improve the environmental impacts is by means of transportation, which is the case of NPV optimisation since there are no pre-treatment associated with coal. This solution is summarised in Tables 10 and 11. Note that an NPV improvement of 219% can be gained by utilising coal as feedstock when compared to the biomass-based SC. Notice that the main difference is from the production cost which is due to the pre-treatment activities that are required in the biomass-based SC. This fact also makes the investment increase in the Biomass SC.

Regarding the environmental impacts, the Impact 2002+ is increased in 203% compared with the biomass-based SC. It is noteworthy that the impact associated with the climate change category is very similar for both cases. We have to bear in mind that CO_2 is still emitted when using a Biomass SC, however this biomass is regenerated faster than fossil fuels. Nevertheless, the other categories are significantly increased specially for the case of resources and human health. The resource regeneration issue is the reason why the impact related to the resources category is

Table 10 Economic aspects arising from single objective optimisation for the coal-based SC (M€)

	NPV optimisation
Investment	111.74
Coal cost	839.55
Transportation cost	181.70
Production cost	1,136.10
Sales	3,987.00
Profit	1,829.65
NPV[a]	729.40
IRR (%)[b]	112.75
Payback period (years)	1.53

[a] Based on a return rate of 8%
[b] Based on a planning horizon of 25 years

Table 11 Environmental impacts arising from single economic and overall environmental objective function optimisation results for the coal-based SC (Impact 2002+ points per year)

End-point impact category	NPV optimisation
Human health	109,640.53
Ecosystem quality	11,077.16
Climate change	95,334.48
Resources	140,605.89
Impact 2002+	356,658.06

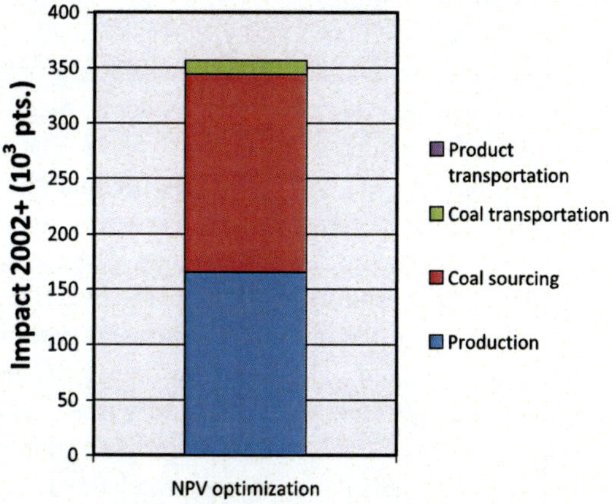

Fig. 9 Distribution of annual environmental impacts for the coal-based SC

larger in the coal-based case. This fact also emphasises the importance of having an overall impact indicator instead of a partial indicator such as CO_2 kg. Notice that if we compare the coal and biomass-based SCs based on CO_2 kg, there would not be an important reduction in the emissions to environmentally justify the project. Figure 9 shows the distribution of environmental impacts according to different SC activities for the coal-based SC. Notice that for the coal-based case the impact of production accounts for 46% of the overall impact. The other activity generating environmental impacts is the coal sourcing which accounts for the 50%

of the overall impact. Moreover, from Fig. 9, it can also be concluded that if the sourcing of raw materials would be disregarded then different solutions would be obtained. If we disregard the raw materials sourcing from the analysis, the biomass-based SC and the coal-based SC would result in a environmental impact of 112,701 and 178,225 points, respectively. This highlights the significance of a SC approach to the problem, rather than focusing only in the processing sites. This also points out the relevance of 'purchase' decisions on the environmental impact of an SC.

5 Conclusions

This chapter presents an approach for designing and planning efficient Biomass SCs. The model consists of a multi-period MILP that accounts for the multi-objective optimisation of economic and environmental interventions. The model considered the long-term strategic decisions (e.g., establishing of pre-treatment trains of units and their respective location, selection of biomass sources, location of processing sites and distribution centres). The problem has been formulated as a multi-objective optimisation, where two objective functions are considered, the net present value and the Impact 2002+ metric was adopted as a measure of overall environmental impact.

A biomass SC case study, geographically located in Spain, is presented here considering the availability of five different potential biomass waste sources. It has been considered that two final products, electricity and H_2, are delivered to marketplaces by means of gasification plants. The most environmental friendly and the most profitable SC configurations for this case study have been shown and their differences have been discussed. However, the indicator values were very similar. It was shown that the more impacting damage category was climate change which accounts for approximately 80% of the overall environmental impact. Moreover, the sensitiveness of the optimal solutions to change in prices and rate of return values was demonstrated. The most profitable SC results in a positive NPV, however reductions of product prices can easily lead to economic losses. The prices to achieve different internal rates over a project life time of 25 years were presented. For this specific case study it was demonstrated that if a return rate of 8%, which is typical for this type of projects, is required, the price cannot be lower than 0.129 € kWh^{-1}. Otherwise the project should be subject to a subsidy if it is meant to be sustainable. This demonstrates how this type of models can be used to determine subsidy policies in order to actually drive industry towards more environmental practices.

The problem was also solved for a coal-based SC. This comparison emphasised the relevance of (i) using an overall environmental impact rather than a partial metric such as kg of CO_2, and (ii) a supply chain approach to the problem so that the entire life cycle of the product and/or process is included and analysed. Narrower approaches (i.e., merely analyzing the processing/manufacturing sites) may lead to biased solutions when the magnitude of environmental impacts associated with producing, collecting or extracting the raw materials are considerably high.

References

1. Capros P, Mantzos L, Tasios N, De Vita A, Kouvaritakis N (2010) EU trends to 2030: update 2009. Technical report, Directorate-General for Energy and the Directorate-General for Climate Action: European Commission
2. US Department of Energy (2010) Energy efficiency and renewable energy: biomass program Technical report, US Department of Energy. http://www1.eere.energy.gov/library/pdfs/biomass_green_jobs_factsheet_2010_01.pdf. Accessed 10 Sept 2010
3. Tang CS (2006) Perspectives in supply chain risk management. Int J Prod Econ 103:451–488
4. Panichelli L, Gnansounou E (2008) GIS-based approach for defining bioenergy facilities location: a case study in Northern Spain based on marginal delivery costs and resources competition between facilities. Biomass Bioenerg 32:289–300
5. Hamelinck CN, Suurs RAA, Faaij APC (2005) International bioenergy transport costs and energy balance. Biomass Bioenerg 29:114–134
6. Leduc S, Schwab D, Dotzauer E, Schmid E, Obersteiner M (2008) Optimal location of wood gasification plants for methanol production with heat recovery. Int J Energy Res 32: 1080–1091
7. Ayoub N, Seki H, Naka Y (2009) Superstructure-based design and operation for bimoass utilization networks. Comput Chem Eng 33:1770–1780
8. Rentizelas AA, Tolis AJ, Tatsiopoulos IP (2009) Logistics issues of biomass: the storage problem and the multi-biomass supply chain. Renew Sustain Energy Rev 13:887–894
9. Rentizelas AA, Tatsiopoulos IP, Tolis A (2009) An optimization model for multi-biomass tri-generation energy supply. Biomass Bioenerg 33:223–233
10. Van Dyken S, Bakken BH, Skjelbred HI (2010) Linear mixed-integer models for biomass supply chains with transport, storage and processing. Energy 35:1338–1350
11. Srivastava SK (2007) Green supply chain management: a state of the art literature review. Int J Manag Rev 9:53–80
12. Bojarski AD, Laínez JM, Espuña A, Puigjaner L (2009) Incorporating environmental impacts and regulations in a holistic supply chains modeling: an LCA approach. Comput Chem Eng 33:1747–1759
13. Puigjaner L, Guillén G (2008) Towards an integrated framework for supply chain management in the batch chemical process industry. Comput Chem Eng 32:650–670
14. ISO14040 (1997) Environmental management: life cycle assessment: principles and framework. Technical report. ISO. Geneva
15. Guinee J, Gorree M, Heijungs R, Huppes G, Kleijn R, de Konig A, van Oers L, Sleeswijk A, Suh S, de Haes HU, de Brujin H, van Duin R, Huijbregts M, Lindeijer E, Roorda A, van-der Ven B, Weidema B (2001). Life cycle assessment. An operational guide to the ISO standards Part 3: scientific background. Ministry of Housing, Spatial Planning and the Environment (VROM) and Centre of Environmental Science, Leiden University (CML)
16. Graves SC, Willems SP (2003) Handbooks in operations research and management science. Supply chain management: design, coordination and operation,vol 11. Elsevier, Amsterdam
17. Humbert S, Margni M, Jolliet O (2005) Impact 2002+: user guide draft for version 2.1. Technical report. Industrial Ecology & Life Cycle Systems Group, GECOS. Swiss Federal Institute of Technology Lausanne (EPFL), Lausanne, Switzerland
18. Pennington D, Margni M, Amman C, Jolliet O (2005) Multimedia fate and human intake modeling: spatial versus non-spatial insights for chemical emissions in Western Europe. Environ Sci Technol 39:1119–1128

19. Goedkoop M, Spriensma R (2001) The eco-indicator 99: a damage oriented methods for life cycle impact assessment methodology report. Technical report. Pré Consultants B.V., Amersfoort, The Netherlands
20. Matthews HS, Hendrickson C, Weber C (2008) The importance of carbon footprint estimation boundaries. Environ Sci Technol 42:5839–5842
21. Kondili E, Pantelides CC, Sargent RW (1993) A general algorithm for short term scheduling of batch operations. Comp Chem Eng 17:211–227
22. Laínez J, Kopanos G, Espuña A, Puigjaner L (2009) Flexible design-planning of supply chain networks. AIChE J 55:1736–1753
23. Heijungs R, Suh S (2002) The computational structure of life cycle assessment. Kluwer, Dordrecht, The Netherlands
24. Laínez JM, Guillén-Gozálbez G, Badell M, Espuña A, Puigjaner L (2007) Enhancing corporate value in the optimal design of chemical supply chains. Ind Eng Chem Res 46: 7739–7757
25. Gomez A, Zubizarreta J, Rodrigues M, Dopazo C, Fueyo N (2010) An estimation of the energy potential of agro-industrial residues in Spain. Resour Conserv Recycling 54:972–984
26. ECN-Biomass (2010) Phyllis database for biomass and waste. Energy Research Centre of the Netherlands (ECN). http://www.ecn.nl/phyllis. Accessed 12 Aug 2010
27. Pérez-Fortes M, Bojarski AD, Velo E, Puigjaner L (2010) IGCC power plants: conceptual design and techno-economic optimization. In: Clean energy: resources production and developments. Energy science engineering and technology. NOVA, New York
28. IEA-GHG (2008) Co-production of hydrogen and electricity by coal gasification with CO_2 capture: updated economic analysis. Technical report, UK
29. EcoinventV1.3 (2008) The ecoinvent database v1.3. Technical report, Swiss Centre for Life Cycle Inventories
30. PRe-Consultants-bv (2008) Simapro 7.1.6. Technical report, PRe-Consultants-bv The Netherlands

Modelling Syngas Generation

Mar Pérez-Fortes and Aarón D. Bojarski

Abstract Syngas generation refers to the production of a synthetic or synthesis gas that is mainly composed of CO and H_2, in different proportions according to the generation process used and the raw material composition. Gasification is the referred technique to produce syngas. It can be used for different purposes, such as power and/or heat generation or for chemicals and fuels production. This chapter describes, we comment the generalities of syngas and its main characteristics and properties, also discuss its possible sources and focus on biomass waste and its co-gasification with coal and petcoke. Then, gasification modelling most common approaches are mentioned. A thermochemical equilibrium model is presented here as the model used for gasification plant conceptual design. Through sensitivity analysis technique, the effects of the reactor temperature and pressure are seen in syngas composition. This chapter enumerates the major hypothesis assumed in this syngas generation step, which must be inevitably applied in modelling and optimizing the entire gasification plant.

Notation

ASU	Air separation unit
CC	Combined cycle
CGE	Cold gas efficiency
EOS	Equation of state
ER	Equivalence ratio
FT	Fischer–Tropsch process
GT	Gas turbine

M. Pérez-Fortes (✉) · A. D. Bojarski
Universitat Politècnica de Catalunya, ETSEIB, Diagonal 647, 08028 Barcelona, Spain
e-mail: mar.perez-fortes@upc.edu

A. D. Bojarski
e-mail: aaron.david.bojarski@upc.edu

L. Puigjaner (ed.), *Syngas from Waste*, Green Energy and Technology,
DOI: 10.1007/978-0-85729-540-8_4,

HHV	Higher heating value
HP	High pressure
HRSG	Heat recovery steam generator
IGCC	Integrated gasification combined cycle
IP	Intermediate pressure
LHV	Lower heating value
P_{gasif}	Pressure of gasification
PRENFLO	Pressurized entrained flow gasifier
PSD	Particle size distribution
SA	Sensitivity analysis
SG	Solid–gas
ST	Steam turbine
T_{gasif}	Temperature of gasification
WHB	Waste heat boiler

1 Introduction

Synthesis gas or producer gas (syngas) is a mixture of H_2 and CO, with different proportions of CH_4 and CO_2. Flexibility is one of its main characteristics, because it is not restricted to a single source of fuel; but it can be obtained from natural gas, coal, petroleum refinery fractions, biomass and organic wastes. Traditionally, natural gas and petroleum fractions have been the largest syngas sources worldwide because of the trade-off between costs and availability; however, because of global economic, energetic and environmental contexts, coal and biomass/waste are of growing interest and use. Syngas is the worldwide most used source of H_2 and CO. The proportion of hydrogen and carbon monoxide depends on the source and the syngas generation process [1]. Further details of the syngas use processes are given in chapter "Main Purification Operations".

Two main routes are currently available for syngas generation; both are traditionally and highly used for hydrogen generation from fossil fuels, specifically natural gas and coal. Syngas from natural gas mainly refers to partial oxidation with oxygen, oxidation with steam or oxidation with steam and oxygen; being the principal *steam reforming*. Syngas from coal involves *gasification*. From these two possible ways, a second process is used for the hydrogen generation; the so-called water–gas shift (WGS) reaction, where the conversion of CO into CO_2 takes place.

Gasification can be defined as the partial combustion of organic matter, with oxygen, air and/or steam as *gasifying agents*, thus less oxygen is used than that required for the raw material complete combustion. Steam is used as the customary temperature moderator; but we can also find the CO_2 and N_2 in this role. The usual gasification process refers to solid organic matter as feedstock, where gas–solid and gas phase reactions take place. Nevertheless, several applications demonstrate

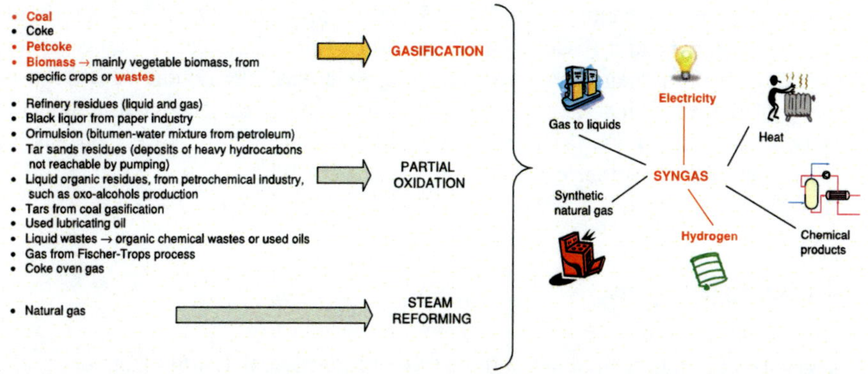

Fig. 1 Syngas generation pathways, contemplating the mixture of coal, petcoke and solid biomass wastes, to produce electricity and hydrogen in this specific case

that the concept of gasification is also applicable to liquid and gas feedstocks, being referred therefore as a 'partial oxidation' [2]. In this book, we consider the gasification process referred to solid feedstock. In Fig. 1, a summary of the possible pathways to obtain syngas is shown, and the pathway of concern in this work is specified in bold; as solids, coal, coke and petcoke (wastes from the refinery industry) are traditionally and widely used materials, whereas biomass and other wastes are nowadays being considered with more interest. It is worth mentioning that when we talk about biomass gasification, we are referring to vegetable biomass gasification, which is the most extended practice. Nonetheless, as declared in chapter "Examples of Industrial Applications", animal biomass can be also gasified with encouraging results. It is appreciated in Fig. 1 that all organic matter, even waste from a main process, are susceptible to be used in a partial oxidation process to produce syngas.[1]

2 Gasification: Principles

In this section, the most relevant gasification issues for process design are revised: reactor and feedstock types and characteristics. Concerning feedstock types, we are focused on the feedstock used in the case studies of this book.

A general gasification picture is given in [3], which provides gasification data for year 2007, revealing that the global marketplace has coal as dominant feedstock and that Sasol Lurgi, GE energy and Shell are the main gasifier providers.

[1] In this book, solid wastes from vegetable sources, forestry or agricultural wastes are of concern, as described in chapter "Raw Materials, Selection, Preparation and Characterisation". From now, when we speak about biomass, biomass waste or simply waste, we are referring to these forest or agricultural wastes, treated in this book.

It is important to mention that China is developing its own technology, as the current project Tianjin IGCC power plant exemplifies. Gasification plants are now operating in more than 27 countries, leading Asia/Australia region, because of the rapid growth of China. The preferred products are mainly chemicals production (such as fertilizers) followed by Fischer–Trops (FT) liquids, power and gaseous fuels, that will be further discussed in the next chapter.

2.1 Gasifier Types

The gasification is carried out in three types of gasifiers that differ from each other by the type of bed that consists of the raw material itself: moving bed or fixed bed, fluidised bed and entrained bed gasifiers. Only a few processes do not fit into these three categories, namely, in situ gasification of coal (underground gasification), molten iron bath gasification, plasma gasification or hydrogasification, as described in Highman and van der Burght [2].

Moving bed or fixed bed gasifier is the most widely used gasifier, and is the one that allows the feeding of the largest particle sizes. In these reactors, the bed moves downwards as the feedstock is consumed, and the residence time can be in the order of hours. The feedstock is introduced at the top of the reactor; and gasifying agents can be introduced at the bottom (counter-current configuration or updraft) or also at the top (co-current or downdraft). Oxidant requirements are low, and gasification temperature is also relatively low if compared with the other types of beds. The main drawbacks of these types of beds are their production of *tars* (see Sect. 2.2.1), their temperature profiles that do not allow for a considerable ash *slagging phenomena*,[2] which results in high amount of fines (fly-ash), and the large quantity of pyrolysis products (mainly CH_4). Looking at the temperature profile along the bed, a peak temperature can be observed in the combustion zone. Nevertheless, it is stated by Reed and Das [4] that the downdraft configuration might achieve higher conversion rates with relatively low tar formation; thus, it is the most widely applied configuration for power generation. This type of bed is not suitable for large-scale syngas production, because of its scale-up problems caused by agglomeration. Agglomeration avoids obtaining a good bed because of penetration and mixing problems for the gasifying agent. Nowadays, it is the most widely used gasifier for biomass gasification, particularly for rural potential profit and thus, suitable for rural development. Specifically, India countries are well positioned using and extending this technology when paying special attention to the supply chain management of the small-scale bioenergy chain, in an analogous way to the features described in chapter "Raw Materials Supply" [5]. We can find both batch and continuous moving bed gasifiers in the market, even if the batch mode is the most usual one.

---

[2] Ash slagging is important given that it prevents the formation of fly-ash.

Fluidised bed gasifier is based on the principle of fluidisation of the solid particles inside the reactor. When fluidisation occurs, particles and the mixture of gasifying agent, moderator and other recirculation flow velocities are the same, thus, driving the particles to a state of levitation. Consequently, this type of bed offers a good mixing between feedstock and gasifying agent, therefore, allowing better heat and mass transfer between gas and solid phases. As a consequence, it approaches the behaviour of an ideal continuously stirred tank reactor (CSTR). This fluidisation phenomenon allows for a stable temperature along the bed. In this case, and differently from what occurs in the previous bed, where the temperature is limited by the bed itself moving downwards as being consumed, the temperature must be controlled to avoid ash slagging, because it would disturb the fluidisation phenomena [2]. As a result of the temperature limitation, the solid conversion is restricted. As the particles become lighter because of the conversion rate, it is very common to entrain them out of the reactor. That is the reason why this type of beds has a recirculation stream. The residence time is in the order of seconds to minutes. This type of beds is well extended too for biomass gasification, as well as for coal.

Entrained bed gasifier profits the particles property of entrainment with oxidant and moderator flows, in co-current way. The residence time is the shortest one, being in order of several seconds. Particle size is the most restrictive one, because particles should acquire the level of microns to be transported in the gas phase. To assure the highest solid conversion, temperatures should be higher if compared with the other two types of beds. As high temperatures are achieved, this reactor is in slagging mode. The highest oxygen demand as well as the highest solid conversion are obtained here. As the slag should be liquid enough not to block the gasifier, and because of the different types of feeds, it is usual to use an additive, such as limestone, to lower the ash melting point till an appropriate operating gasification temperature. Consequently, in this type of beds, it is assumed that the solid is completely converted and that no tar is produced, when gasifying biomass, because 'all' inorganic matter is susceptible to be converted into slag, and due to the high temperature reached, tars destruction is assured by thermal cracking [2, 6]. Integrated gasification combined cycle power plants (IGCC) usually use entrained bed gasifiers.

Even if the gasification mechanism itself is the same in each type of reactor (see Sect. 3), a typical syngas composition can be suited for each type of gasifier by means of their most common conditions of pressure and temperature. Note that in Table 1, apart from these characteristics, particle size and other important features are summed up. Note also that these table values are tentative and only serve as a general overview. Any type of bed can work at high or at low pressure, and they can be fed by different feedstock sources, and use by different gasifying agents. In summary, the syngas composition on a dry mole basis ranges from 15 to 21% for H_2 and from 18 to 60% for CO. At the end, the choice of the gasifier type and its working conditions depends on the feed material and the syngas final use.

Downstream the gasifier, different processes can provide with electricity, heat, chemicals or liquid fuels. The gasifier pressure is usually given by the final application requirement. For instance, in a combined cycle (CC) final application, the gasifier pressure is given by the gas turbine pressure. The feedstock pressure is

Table 1 Comparison between the three types of gasifiers by their most typical characteristics

	Moving bed [2, 32–34]	Fluidised bed [2]	Entrained bed [2, 6]
T (°C)	400–875	800–1,100	1,250–1,600
P (bar)	1	1	25
Particle size (mm)	6–50	6–10	<0.1
Moisture content (%)	15–20	10	2
Feedstock	Wood	Biomass	Coal–petcoke
Oxidant	Air	Air	Oxygen
Ash slagging	No	No	Yes
Carbon conversion (%)	99	97	99
Syngas (dry mol%)	–	–	–
CO	18	31	60
H_2	15	18.9	21
CO_2	10	6.7	4
CH_4	1.5	2.1	0

usually matched with the syngas application because a solid compression is less energy consuming than a gas compression. The temperature is always controlled by the melting point of the ashes. The effect of the pressure and the temperature is further discussed in the following Sect. 3. Other gasifier possibilities to consider include the moisture content of the feedstock, that is, a dry feed or a slurry feed. This last approach is beyond the scope of this chapter.

2.2 Syngas Sources: Bioenergy Sector

As already mentioned, coal, coke and petcoke are the most widely used materials. Biomass is nowadays attracting more attention. Biomass is considered as the renewable energy source with the highest potential to contribute to energy needs of modern society in developed and developing areas. According to [7], biomass is a very important source of energy especially in remote areas where centralised systems cannot arrive. Indeed, biomass is assumed to be a zero-emissions source, if disregarding the supply chain, where fossil fuels should be consumed [5]. If waste is of concern, no controversy appears in relation with the land use. For these three reasons, biomass waste, and even more, any type of organic waste (mainly produced in an urban area) are of concern when talking about renewable sources.

The bioenergy sector is nowadays accepted as having the potential to provide the major part of the projected renewable energy provisions of the future. Nevertheless, it has to overcome several barriers, not only technological but also economic and of social perception, to obtain good political support and to be attractive as investment project [8]. Bioenergy in a global market, which covers the generation and use of biomass for heat, power and biofuels production, should still develop its specific regulation to control international and national trade, land use and the increase and development of this sector in general. It can be seen from different organisation reports, from IEA and EPRI for instance, that different countries need *common strategies* to mitigate greenhouse gases emissions and to secure the energy supply by lowering the dependence on fossil fuels. Furthermore, it is estimated that the energy demand will be doubled by 2030, and the bioenergy sector plays a key role in the new energy generation paradigm, where the share of clean and renewable sources should be increased. It is then in this context where biomass is increasing its importance.

2.2.1 Biomass: Tars Production and Reactor Suitability

When regarding to biomass gasification, it is inevitable to tackle with *tars* and to think about small-scale gasification. As defined in [9], 'Tars are the organics produced under thermal or partial oxidation regimes of any organic material and are generally assumed to be largely aromatics'. Even though the tars tolerance of gasifier downstream units is subject of research, it can be stated through the experience that tars constitute a problem when the syngas is not simply burnt in a combustor. The handicap is mainly because of tars condensation before syngas use; because of their carcinogenic effects, they imply health damage to humans and generate environmental issues because of their disposal. A European *tar protocol* has been developed not only to assess a standard methodology for their measure, but also for characterising the quality of clean syngas, and the determination of gasifier downstream technology contamination [10].

Tar avoiding passes through two methodologies. First, tar formation reduction in the gasifier itself; *primary methods* include adequate selection of main operating parameters (pressure and temperature), the use of a catalyst and specific design modifications (shape, dimensions, etc.). Second, tar removal from syngas; *secondary methods* imply hot gas cleaning downstream the gasifier by means of thermal or catalytic tar cracking, and as well as wet-scrubbing or mechanical methods using cyclones and filters. Sutton et al. [11] review the use of dolomite, alkali metals and nickel as catalysts, showing their suitability for hydrocarbons removal or production reduction through experimental analysis, whereas the work of Arena et al. [12] demonstrate that olivine is a good candidate for catalysed gasification, where the tar amount is highly decreased in the case of plastic wastes. An example of current research work in ECN (Energy research Center of the Netherlands) is the OLGA (OiL GAs scrubber) technology that contemplates the use of special scrubbing oils to remove tars from syngas. On the other hand, the challenge of all the actual small gasification pilot plants, that use biomass waste from rural areas, is to find an adequate design to produce a syngas free of tars, avoiding the syngas-cleaning process before the final application, thus gaining compactness and avoiding waste water treatment before final disposal. In short, the formation of carbonaceous materials (char or particle fines) and that of heavy compounds (tars) are strictly correlated to the fuel structure and composition.

Waste gasification is not a wide commercial process because of the multiple composition options of the waste itself, and consequently because of the different amount of fly-ashes, tars, chloride and fluoride, ammonia and sulphur compounds that can be present in the syngas. According to Mastellone et al. [12], among all gasification technologies applied for waste processing, *fluidised beds* are the most promising ones because of their operation flexibility for different oxidants (thus, for different fluidising agents), different temperature and residence time range. They also allow for catalyst addition. According to Highman and van der Burgt [2], low-rank coals and biomass are more suitable for fluidised beds, because of their *ashes reactivity*. However, biomass ashes have lower melting point, and in molten state have an aggressive behaviour with refractory material. The temperature required for complete gasification of both coal and biomass is of the same order.

Concerning the gasification itself and the tars removal method, it is found that the most common type of process is the gasification with secondary methods of tar avoiding. Apparently, because of the low ashes melting point, an entrained bed gasifier looks very attractive to obtain a tar-free syngas, with less oxidant consumption. Nevertheless, because of the aggressive behaviour of ashes, a non-slagging process is recommended (except if the biomass is mixed with high amounts of other feeds, such as coal or petcoke). Moreover, entrained bed gasifiers require small particle diameter, however until now, there is no effective method for size reduction of fibrous biomass. Fixed beds, with no highly restrictive particles size, are extensively used for small-scale gasification of biomass waste applied successfully in rural areas [4].

In summary, ashes reactivity and tars formation are the main drawbacks in biomass waste gasification. The most extended bed for large-scale application is the fluidised one, whereas the most extended bed for small-scale is the fixed one. Entrained bed gasifiers are normally used when biomass is mixed with coal, for instance, and for large-scale applications.

2.2.2 Co-Gasification

Co-gasification can be defined as the 'joint conversion of two carbonaceous fuels (one of them of fossil origin) into a gas with a useable heating value' [13]. Therefore, renewable sources, such as biomass waste, allow reducing environmental and disposal costs, with no relevant efficiency reduction in the gasification process. As a practical example, as seen in chapter "Examples of Industrial Applications", ELCOGAS power plant in Spain has demonstrated it by replacing a 10% in mass of its main feedstock (a mixture of coal and petcoke) with biomass waste (olive pomace).

It has been seen that a *synergic effect* is reached by the co-feeding, leading to better feeding properties that derive into carbon losses reduction and into syngas energy content increase. Even if these synergetic effects are far from being well understood, they can be defined as positive. Several authors have investigated different mixtures of fuels into pressurised entrained bed gasifiers. Fermoso et al. [14] demonstrate a synergic effect found in ternary blends of coal, petcoke and biomass (almond shells, olive stones and eucalyptus), where the ratio H_2/CO decreases with the addition of biomass, and the carbon conversion as well as the cold gas efficiency (CGE) increase. Binary blends of coal with different biomass show an analogous performance. Hernández et al. [13] demonstrate experimentally that an increase in the proportion of biomass in the fuel blends (dealcoholised grape marc with coal–petcoke) results in an improvement of the gasification parameters (syngas composition and CGE). Mastellone et al. [12] at their turn, work with co-gasification in a fluidised bed using coal, plastic and wood, remark that it is crucial to choose the correct proportions of raw materials to achieve the desired requirements for the final syngas application.

In conclusion, biomass waste use in co-gasification is not only an option for a final waste disposal or an option as a renewable source but also it implies better gasification results by improving gasifier efficiency and carbon conversion.

3 Gasification: Modelling

Even if a lot of work has been developed concerning gasification modelling, it is still a challenge. The model's level of detail depends on the final purpose of the model itself and on the plant state (transient or stationary): The different modelling approaches range from the macroscopic level that encompasses thermochemical equilibrium approaches to the microscopic level incorporating chemical kinetics. In the first case, two different models are available: one where reactions are predefined and the other where only the chemical compounds considered in equilibrium are considered. In both cases, Gibbs's free energy is minimised, but in the first it is restricted to the predefined chemicals reactions, whereas in the second, all possible reactions incorporating those compounds are considered. In addition to the kinetics, other transport phenomena can be included, particularly boundary layer or pore diffusion are of concern. In all cases, mass and energy balances must be followed. In this section, we describe the two approaches. Despite the fact that both of them have been implemented and validated, the equilibrium approach is the one selected for its incorporation into the *superstructure* in chapter "Modelling Superstructure for Conceptual Design of Syngas Generation and Treatment", mainly because of its calculation time.

Table 2 provides with a summary of possible literature gasifier models, showing the wide variety of models proposed, ranging from equilibrium to kinetics. As discussed earlier, the two possible equilibrium calculations are proposed (i) restricted by a predefined set of reactions or (ii) by a set of chemical compounds. Kinetics has fewer consensuses, and specific parameters are highly dependant on the reactor and feedstock type. The three types of gasifiers are well represented for both approaches. When the process dynamics is of concern, the kinetic model is the only option, given that equilibrium models cannot provide with a time-dependant response. An important aspect to consider for determining the suitability of a model is the gasifier residence time. For large residence times, equilibrium models can be selected without hesitation, but in the cases of short residence times, the selection of a kinetic model must be considered. For the

Table 2 Gasifier modelling review

Source	Raw material	Gasifier type	P (bar), T[a] (°C)	Scale[b]	Time dependence	Approach
Gautam et al. [19]	Biomass	Fixed	1, 800	Pilot-Small	Steady	Equilibrium chemical reactions
Loeser and Redfern [35]	Biomass	Fixed	1, n.s.	Small	Steady	Equilibrium chemical compounds
Melgar et al. [36]	Biomass	Fixed	1, 700	Pilot-Small	Steady	Equilibrium chemical reactions
Giltrap et al. [37]	Biomass	Fixed	1, 927	Small	Steady	Kinetic model
Altafini et al. [38]	Biomass	Fixed	1, 800	Pilot-Small	Steady	Equilibrium chemical compounds
Di Blasi [20]	Biomass	Fixed	1, 850	Small	Dynamic	Kinetic model
Wang and Kinoshita [39]	Biomass	Fixed	1, 800	Small	Dynamic-Steady	Kinetic model
Nikoo and Mahinpey [40]	Biomass	Fluidised	1, 800	Lab	Steady	Kinetic model for heterogeneous reactions; equilibrium chemical compounds for homogeneous reactions
Jand and Foscolo [21]	Biomass	Fluidised	1, 675	Lab	Dynamic	Kinetic model
Ruggiero and Manfrida [17]	Biomass	Fluidised Entrained	1 and pressurised; n.s.[c]	Pilot-Large	Steady	Equilibrium chemical compounds
Emun et al. [41]	Coal slurry	Entrained	8; 1,250	Large	Steady	Equilibrium chemical compounds
Robinson and Luyben [42]	Coal slurry	Entrained	55.17, 1,370	Large	Dynamic-Steady	Kinetic model
Nathen et al. [43]	Coal	Entrained	n.s.[c], 1,500	Large	Steady	Equilibrium chemical compounds

(continued)

Table 2 (continued)

Source	Raw material	Gasifier type	P (bar), T[a] (°C)	Scale[b]	Time dependence	Approach
Frey and Akunuri [44]	Coal slurry	Entrained	42, 1,260	Large	Steady	Equilibrium chemical compounds
Wen and Chaung [15]	Coal slurry and coal liquefaction residues	Entrained	22, 1,800	Large (pilot plant)	Dynamic-Steady	Kinetic model
Govind and Shah [22]	Coal slurry and coal liquefaction residues	Entrained	22, 1,800	Large (pilot plant)	Dynamic-Steady	Kinetic model
Ullmann's [16]	General	General	General	General	Steady	Equilibrium chemical reactions with K_{eq}[d] values at a temperature other than the one in the reactor

[a] The temperature is an average value (more sources give a sensitivity analysis or a profile of temperatures)
[b] Large- and small-scales belong to industrial scale
[c] *n.s.* non-specified
[d] K_{eq} is the equilibrium constant for a specific reaction

kinetic approach, the particle level is of concern. The most extended approach is the unreacted core-shrinking model [15].

Another important parameter in both cases is the gasification temperature. Generally in all gasifiers, the controllable temperature is the gasifier outlet temperature, thus in lumped parameters models, it is generally adapted as the gasifier bulk temperature, even if the gasifier does not have one unique temperature (but a profile according to the different gasification zones). The approach of Ullmann's [16] adjusts the gasification composition to real data, using the thermochemical equilibrium at a temperature that is not the actual gasification temperature.

Ruggiero and Manfrida [17] stated that although the syngas composition is not perfectly predicted with the equilibrium approach, this approach allows for a fast and reliable method of syngas composition estimation when varying the gasifier inputs. Different sources, for instance, Usón et al. [18] or Gautam et al. [19], point out that at higher temperatures (more than 1,200°C approximately), the equilibrium approach is more accurate; given that at low temperatures the reaction rate is smaller, and the residence time should be higher to reach the equilibrium. In spite of all these arguments, it is difficult to know at which residence time 'the equilibrium is reached': To answer such question, the only options available are to compare with kinetic model results or with experimental results. Only two of the studies revised tackles with the formation of tars during biomass gasification, the papers by Di Blasi [20] and Jan and Foscolo [21].

Two reactor models are discussed in this chapter; the first one divides the gasification process into several stages, studies gas–particle interaction and simulates the final gas phase equilibrium. Thus, this model considers char formation dependant on the reactor conditions, and chemical kinetics. The second one considers the total conversion of the fuel and the final gas equilibrium through the minimisation of Gibbs free energy, at the gasifier temperature, restricted to a set of chemical compounds. The temperature in this case is determined by the heat of formation of the raw material and the cooling of ashes. The residence time of particles was calculated for our case study of a gasifier to infer the proximity of the system to equilibrium.

The implementation of kinetic and equilibrium models using different process simulation environments is discussed in the next sections.

3.1 Physicochemical Representation of the System

One important aspect of modelling coal, petcoke, biomass and wastes is the difficulty in finding appropriate models to represent and estimate their physicochemical behaviour. Most simulation environments provide with the possibility of representing non-conventional chemical compounds, understood as those that do not have an integer molecular representation.

Typically, any of the above-mentioned raw materials can be represented with a molecular formula, such as $C_aH_bO_cN_dS_e(H_2O)_wA$, whereas any treated-raw material would convert into $C_\alpha H_\beta O_\gamma N_\delta S_\varepsilon A$, where A represents the material's ashes content.

The materials enthalpy and density can be extrapolated from the ultimate and proximate analyses, for example, Aspen Plus© commercial simulator provides with such possibility by using the ULTANAL and PROXANAL properties for defining a non-conventional compound.

Regarding the behaviour of the gas phase, an equation of state (EOS) is typically selected from the wide variety of EOS that the Aspen One© software package provides. This selection is mainly driven by the possibility of estimating properties accurately near critical points. Typically, the Peng–Robinson EOS is selected.

3.2 Chemical Kinetics Approach

This approach is mainly based on the models developed by Wen and Chaung [15] and Govind and Shah [22] for the gasification of coal. Aspen Hysys© is the software selected for modelling the gasification of coal following the chemical kinetics approach. The described model has been developed in Pérez-Fortes et al. [23], where not only the gasifier but also the entire IGCC *superstructure* is implemented in Aspen Hysys©.

The proposed gasifier model encompasses a sequence of four main steps: pyrolysis and volatiles combustion, oxidation, gasification and gas equilibrium. The main assumptions are:

- Steady state and one dimension.
- Oxidation and gasification zones have uniform temperatures with adiabatic behaviour; consequently, the considered overall reactor is a non-isothermal reactor.
- The fuel particle is of concern. The solid–gas reactions consider the *unreacted core-shrinking model*. Thus, chemical reactions occur on a spherical surface and are considered to advance from the surface till the particle's core, without any release of mineral matter; which is considered as inert. The transport considerations comprise the chemical reaction itself, ashes layer diffusion and surface convection (from the inner to the outside face of the particle). Kinetic constants for solid–gas reactions have the following expression:

$$\frac{1}{k_t} = \frac{1}{h_{\text{gas}}} + \frac{Y-1}{Y} \times \frac{1}{k_{\text{ash}}} + \frac{1}{Y^2 k_q}, \tag{1}$$

where k_t is the specific reaction rate; h_{gas} is the external convection coefficient; k_{ash} is the ash film diffusion and finally, k_{q} is the chemical rate constant (that follows the Arrhenius law). Y is a parameter function of the ratio between the particle and the core radius.

- Particles interactions are not considered. The temperature along the particle is uniform.
- Solid and gas phases are assumed to be completely intermixed. Consequently, the approach contemplates the simulation of each reactor section using CSTRs in series, one tank for each step.
- The chosen equation of state is Peng–Robinson (fairly used for hydrocarbons and light gases).

Pyrolysis is the decomposition of the feedstock into solid (char) and volatiles, in absence of oxygen and because of the effect of the temperature. The raw material is represented with its molecular formula as previously described: $C_aH_bO_cN_dS_e(H_2O)_wA$, whereas the produced char is represented as $C_\alpha H_\beta O_\gamma N_\delta S_\varepsilon A$. Both solids are modelled in the software as Hypo-Components, because the default options of Aspen Hysys© do not allow to handle with solids with such complex stoichiometry. The molecular formula of char is calculated based on experimental correlations from Balzioc and Hawksley [24], which state the total amount of volatiles released based on the reactor temperature and raw material volatile matter (proximate analysis). According to these authors, coal devolatilisation corresponds to a kinetic process of first-order. Loison and Chauvin [25] define volatiles composition based on experimental experiences. To ease the estimation effort and simplify the model, tars formation is considered as benzene production. However, other approaches could provide with a set of typical aromatic compounds to be produced if data were made available. The generation of the acid and basic species (H_2S, COS, NH_3 and HCN) is modelled with experimental correlations taken from [26, 27]. Equation 2 represents the pyrolysis step in general terms; however, a set of those reactions is coded for the production of benzene and acid and basic species. This above-mentioned set of reactions is introduced into Aspen Hysys© as a unit extension programmed in MS Visual Basic. The volatile species obtained at the end of this step are: CO, H_2, CH_4, CO_2, C_6H_6, H_2S, COS, NH_3 and HCN.

$$C_aH_bO_cN_dS_e(H_2O)_wA \rightarrow C_\alpha H_\beta O_\gamma N_\delta S_\varepsilon A + \text{volatiles} + wH_2O \quad (2)$$

Volatiles combustion is produced as the volatiles released in the pyrolysis step are placed in contact with oxygen. It is assumed that they are completely transformed to produce CO_2 and H_2O. Equations 3–7 represent the main volatiles combustion.

$$CO + 0.5O_2 \rightarrow CO_2 \quad (3)$$

$$H_2 + 0.5O_2 \rightarrow H_2O \quad (4)$$

$$CH_4 + 2O_2 \rightarrow CO_2 + 2H_2O \quad (5)$$

$$C_6H_6 + 7.5O_2 \rightarrow 6CO_2 + 3H_2O \quad (6)$$

$$HCN + 1.25O_2 \rightarrow CO_2 + 0.5H_2O + 0.5N_2 \quad (7)$$

Oxidation or *combustion* refers to char combustion, and is considered to take place after the combustion of volatile compounds. It comprises a set of heterogeneous reactions. This step comprises three main reactions of char with oxygen, steam and carbon dioxide (Eqs. 8–10), and it is considered to be complete when all the introduced oxygen is consumed.

$$C_\alpha H_\beta O_\gamma N_\delta S_\varepsilon A + \left(\frac{\alpha}{2} - \frac{\gamma}{2} + \frac{\beta}{4} - \frac{\varepsilon}{2}\right) O_2 \rightarrow \alpha CO + \left(\frac{\beta}{2} - \varepsilon\right) H_2O + \varepsilon H_2S + \frac{\delta}{2} N_2 + A \tag{8}$$

$$C_\alpha H_\beta O_\gamma N_\delta S_\varepsilon A + (\alpha - \gamma) H_2O \rightarrow \alpha CO + \left(\alpha - \gamma + \frac{\beta}{2} - \varepsilon\right) H_2 + \varepsilon H_2S + \frac{\delta}{2} N_2 + A \tag{9}$$

$$C_\alpha H_\beta O_\gamma N_\delta S_\varepsilon A + \alpha CO_2 \rightarrow 2\alpha CO + \gamma H_2O + \left(\frac{\beta}{2} - \varepsilon - \gamma\right) H_2 + \varepsilon H_2S + \frac{\delta}{2} N_2 + A \tag{10}$$

These reactions are modelled in Aspen Hysys© using chemical extensions written in MS Visual Basic. The char that reacts in each Eq. 8, 9, 10 and 11 is calculated considering the variation of a parameter that depends on the reactor volume.

Gasification follows the oxidation section and is considered to start with the depletion of oxygen in the gasifier. In this section, reactions (9) and (10) are also occurring and the produced hydrogen reacts with char as in Eq. 11.

$$C_\alpha H_\beta O_\gamma N_\delta S_\varepsilon A + \left(2\alpha + \gamma + \varepsilon - \frac{\beta}{2}\right) H_2 \rightarrow \alpha CH_4 + \gamma H_2O + \varepsilon H_2S + \frac{\delta}{2} N_2 + A \tag{11}$$

This step is assumed to be finished when all char is consumed.

As a final step, the final produced gas from the heterogeneous reactions, enters into an equilibrium reactor. This gas equilibrium is performed with three main reactions, Eqs. 12, 13 and 14. The equilibrium constants are extracted from Aspen Hysys (c) library.

$$CO + H_2O \longleftrightarrow CO_2 + H_2 \tag{12}$$

$$CH_4 + H_2O \longleftrightarrow CO + 3H_2 \tag{13}$$

$$COS + H_2O \longleftrightarrow H_2S + CO_2 \tag{14}$$

Combustion of volatiles and combustion of char are modelled with a continuous stirred tank reactor (CSTR) with custom-made kinetics equations for the heterogeneous solid-gas (SG) reactions (Eqs. 8, 9 and 10). The same is done for the gasification step (Eqs. 9, 10 and 11). One isothermal zone comprises volatiles and

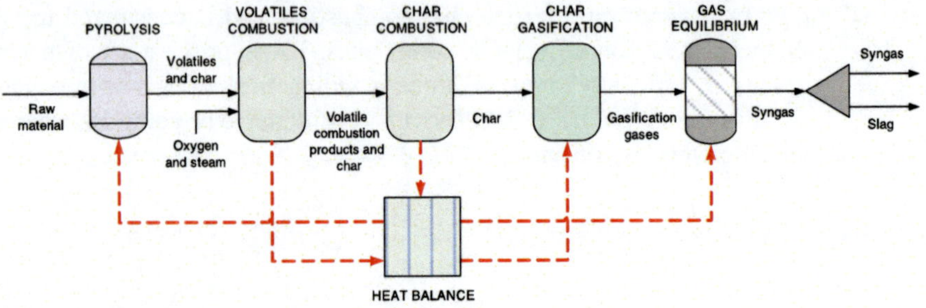

Fig. 2 Modelling blocks for the proposed chemical kinetics approach

char combustion; the other one is for gasification. As an adiabatic reactor is considered, for this autothermal gasification process, all the heat released in the combustion is used for the gasification step. In Fig. 2, a simple outline of the different steps modelled and the heat integration consderation are represented. All heat streams are included in a balance unit, thus the heat is distributed from exothermic to endothermic reactors.

Before leaving the gasification reactor, the syngas is sent to an ashes distribution model, which splits the solid stream into slag and fly ash, based on industrial data. This component splitter takes into account a base ashes composition built on the same industrial data from ELCOGAS power plant. It has been considered that ashes are composed by metal oxide and heavy metals present at very low concentrations and as pure compounds. The most common type of ashes in a mixture of coal and petcoke is composed of Al_2O_3, SiO_2, Ar, Cd, Pb and Hg. This model is introduced in Aspen Hysys© as a customer model, and the data come from the work developed in Jaume Almera Institute (CSIC, Barcelona).

3.3 Equilibrium Approach

As described in Borel and Favrat [28], a thermochemical equilibrium is a stable state that can be determined by thermodynamic methods, not being necessary a detailed knowledge of the reaction mechanisms. As seen in Table 2, one of the most extended approaches is the equilibrium *minimising Gibbs's free energy,* by taking into account the final species composition.

The equilibrium constant K_{eq} is given by the following expression, for a generic reaction and assuming ideal-gas behaviour:

$$aA + bB \leftrightarrow cC + dD \tag{15}$$

$$K_{eq} = \frac{p_C^c \times p_D^d}{p_A^a \times p_B^b} = \frac{x_C^c \times x_D^d}{x_A^a \times x_B^b} \times P^{(c+d-a-b)} \tag{16}$$

$$f(T) = \log(K_{eq}) = \frac{k_1 + k_2}{T + k_3} \times \log T + k_4 \times T + k_5 \times T^2 \tag{17}$$

For the case of a reaction such as Eq. 15, K_{eq} can be expressed as a function of partial pressures or mole fractions as shown in Eq. 16, where P is the pressure at which the reaction takes place and T is the temperature. Typically, K_{eq} can be calculated as in Eq. 17 as a function of the absolute temperature in Kelvin; k_n are particular constants for each reaction.

Equation 16 expresses the equilibrium composition dependence on the systems pressure and on the reactants and products mole number. If the numbers of moles on the right-hand side and on the left-hand side are identical, then the total pressure does not have influence on the reaction final composition. The *Le Châtelier's principle* provides with a general overview of the gasification reactions, as equilibrium reactions, where if some factor of the system (pressure, temperature, volume or partial pressures) changes, the system progresses to counteract this change, and a new equilibrium is established at other conditions. For example, in an endothermic reaction, if the temperature is increased, the reaction tends to react to the products side. Or if the total pressure increases, the partial pressures of the individual gas species increase, and the global system tends to shift the reaction to the side with fewer number of moles.

Gasification reactions can be described by many different equations but they may be limited to a certain number of representative expressions since the reaction enthalpy of some of them can be derived by combining those representative ones. To calculate the heat streams associated with them, that is the integration between endothermic and exothermic reactions, a limitation to certain characteristic reaction groups is enough, because the reaction enthalpy of all other reactions can be derived by combining these basic equations. The most characteristic reactions are reactions with oxygen, reactions with steam, reactions with carbon dioxide and hydrocarbon decomposition, as in the kinetics approach. The most characteristic gasification equations are given in Eqs. 18–27. Next to each reaction, the gas mole difference between products and reactants is indicated. Note that even though raw material composition contains C, H, O, N, S and Cl, main reactions only involves C as reactant (as a big difference with the previous approach, which counted with the molecular formulation of the solid feedstock). Equations 18–20 correspond to combustion; Eq. 21 is the Boudouard reaction, Eq. 22 is the primary and Eq. 23 is the secondary water–gas shift; Eq. 24 is the water-gas shift reaction with only gas species, and Eq. 25 is the methanation reaction. Equations 26 and 27 correspond to steam and dry reforming. As reactions with oxygen are complete under gasification conditions, the main final composition influence comes from Eqs. 21–25 considering equilibrium [14].

$$C + 0.5O_2 \leftrightarrow CO \quad \Delta n = 0.5 \tag{18}$$

$$CO + 0.5O_2 \leftrightarrow CO_2 \quad \Delta n = -0.5 \tag{19}$$

$$H_2 + 0.5O_2 \leftrightarrow H_2O \quad \Delta n = -0.5 \tag{20}$$

$$C + CO_2 \leftrightarrow 2CO \quad \Delta n = 1 \tag{21}$$

$$C + H_2O \leftrightarrow CO + H_2 \quad \Delta n = 1 \tag{22}$$

$$C + 2H_2O \leftrightarrow CO_2 + 2H_2 \quad \Delta n = 1 \tag{23}$$

$$CO + H_2O \leftrightarrow CO_2 + H_2 \quad \Delta n = 0 \tag{24}$$

$$C + 2H_2 \leftrightarrow CH_4 \quad \Delta n = -1 \tag{25}$$

$$CH_4 + H_2O \leftrightarrow CO + 3H_2 \quad \Delta n = 2 \tag{26}$$

$$CH_4 + CO_2 \leftrightarrow 2CO + 2H_2 \quad \Delta n = 2 \tag{27}$$

For these different equilibrium reactions, a compilation of the equilibrium constant values as well as their dependency with temperature at 25 bar is provided in Table 3. Gasification temperatures of 800–1,800°C are considered. The values are extracted from Aspen Plus© through a REquil reactor.

Equations 18, 19 and 20 that have oxygen as a reactant are exothermic. Equations 24 and 25 are also exothermic. As mentioned earlier, and according to the Le Châtelier's principle, K_{eq} decreases because the reaction tends to move towards the reactants side. The contrary behaviour is seen in endothermic reactions, where as the temperature increases, they tend to react towards the product side, thus diminishing K_{eq}. Higher values of K_{eq}, for instance, in Eq. 18, implies a higher predilection for the reactants side than lower values of K_{eq}, as in Eq. 24, which are highly displaced to the products side.

3.3.1 Residence Time Considerations

As mentioned earlier, one important assumption here is that the *residence time* is high enough to reach the equilibrium. As it is reached, the highest possible conversion is obtained. Nevertheless, if time is too short, this lack of time can be compensated by higher gasification temperatures.

One possible way of estimating the residence time in a reactor is the methodology based on Chap. 4 of Kunzing [29], which follows the principles of gas–solid pneumatic transfer. The residence time (t_r) calculation is the result of a particle force balance taking into account the effect of weight (gravity force g), the drag force (the drag component is the aerodynamic force component parallel to the gas flow) and the solid friction loss. In vertical transport of solids, like in an entrained bed gasifier, the gas velocity is reduced to the velocity of the transport of solids, decreasing consequently the pressure drop. The overall pressure drop is calculated as the contribution of the static, frictional and acceleration contributions. The particle velocity (U_p) is of concern, and can be assumed as the velocity of transport fluid without particles (U_g) minus the terminal velocity (U_t, which is the reached velocity when the drag force is equal to the weight of the particle minus the buoyant force). Finally, the residence time is given by the length of the

Table 3 Equilibrium constants for the gasification equations at different temperatures

Equations	K_{eq} units	800°C	1,000°C	1,200°C	1,400°C	1,600°C	1,800°C
18	$bar^{0.5}$	1.057×10^{10}	1.339×10^{9}	2.799×10^{8}	8.019×10^{7}	2.838×10^{7}	1.164×10^{7}
19	$bar^{-0.5}$	1.643×10^{9}	1.147×10^{7}	3.118×10^{5}	2.027×10^{4}	2.382×10^{3}	4.261×10^{2}
20	$bar^{-0.5}$	1.512×10^{9}	1.893×10^{7}	7.693×10^{5}	6.676×10^{4}	9.725×10^{3}	2.050×10^{3}
21	bar	6.430	1.168×10^{2}	8.976×10^{2}	3.956×10^{3}	1.191×10^{4}	2.732×10^{4}
22	bar	6.986	7.074×10^{1}	3.638×10^{2}	1.201×10^{3}	2.918×10^{3}	5.679×10^{3}
23	bar	7.591	4.285×10^{1}	1.474×10^{2}	3.647×10^{2}	7.147×10^{2}	1.181×10^{3}
24	–	1.087	6.057×10^{-1}	4.053×10^{-1}	3.036×10^{-1}	2.449×10^{-1}	2.079×10^{-1}
25	bar^{-1}	4.248×10^{-2}	8.027×10^{-3}	2.263×10^{-3}	8.252×10^{-4}	3.583×10^{-4}	1.759×10^{-4}
26	bar^{2}	1.645×10^{2}	8.813×10^{3}	1.608×10^{5}	1.455×10^{6}	8.143×10^{6}	3.229×10^{7}
27	bar^{2}	1.513×10^{2}	1.455×10^{4}	3.967×10^{5}	4.794×10^{6}	3.325×10^{7}	1.554×10^{8}

Table 4 Input data to calculate the residence time and the reactor pressure drop

d_p (m)	5.5×10^{-5}
$D_{reactor}$ (m)	3.8
L (m)	60
U_g (m/s)*	8
Particles inlet mass flow (kg/s)	21.15
ρ_p (kg/m^3)*	1,200
ρ_f (kg/m^3)*	1.2
Inlet gas P (N/m^2)	3.333×10^6
μ_f (Pa × s)*	1×10^{-5}

Specific data from the model* are taken from Chap. 4 of Kunzing [29]

reactor (L) divided by the particle's velocity (U_p). The necessary inputs are shown in Table 4, where ρ_p and ρ_f are the solid and the fluid densities, D_p is the particle diameter and μ_f is the fluid dynamic viscosity. The representative equations for the residence time calculation for this particular case of study are 28, 29 and 30.

$$U_t = \frac{0.153 \times g^{0.71} \times D_p^{1.14} \times (\rho_p - \rho_f)^{0.71}}{\rho_f^{0.29} \times \mu_f^{0.43}} \tag{28}$$

$$U_p = U_t - U_g \tag{29}$$

$$T_r = \frac{L}{U_p} * k_a sh = h_g as * \epsilon_p^2.5 \tag{30}$$

Otherwise, in the calculation following Chap. 4 of Kunzing [29] methodology, the residence time becomes the objective. The superficial gas velocity is assumed to be 8 m/s, being in dilute phase regime. For the sake of simplicity, the solid is considered as coal and the fluid as air. In that case, the *residence time is about* 7.7 s. On the other hand, the reactor pressure drop is evaluated. Here, the voidage and solid and gas friction factors are of concern (see Chap. 4 of Kunzing [29] for further detail). The result is a *pressure drop of about* 1×10^{-2} bar, thus, negligible. In the validation section, this residence time obtained at the specific assumed gasifier temperature is checked if sufficient for the equilibrium approach. At this point, we can conclude, as have been discussed before, that gasification equilibrium approaches are more suitable when higher temperatures are of concern.

4 PRENFLO Gasifier Model

We follow the syngas generation section with the description of the two approaches, but focussing on the second one, because it is the one used in the *superstructure*. The first one has been modelled in Aspen Hysys©, whereas the second one in Aspen Plus©. Related to this second approach, there are other *hypotheses* to be considered:

- The equilibrium approach through minimisation of free Gibbs's energy is chosen here to model the pressurised entrained flow (PRENFLO) gasifier.
- Tars and synergetic effects are not considered.
- There is no presence of char in the outlet of the gasifier.
- When processing biomass, no pre-treatment limits exist because of its fibrous nature.

In our modelling approach, the syngas is treated to produce electricity in an IGCC power plant. Aspen Plus© is the chosen commercial software to perform not only the gasifier but also all the power plant model. The gasification block comprises the gasifier itself and the waste heat boiler (WHB), which is the syngas cooling system before gas-cleaning units (see chapter "Main Purification Operations"). Before the gasification process itself, the feedstock must be conditioned to meet the gasifier requirements. The physical property method chosen to calculate thermodynamic and transport properties of the streams is the Peng–Robinson EOS with the Boston–Mathias alpha function (PR–BM) for the above-mentioned units. This equation of state is recommended to model gas phase systems at high-medium pressure.

As already mentioned in Sect. 2.1, the particle size of an entrained bed gasifier should be in the order of microns. The *feedstock dust preparation* comprises principally a drying and a grinding step. Other pre-treatment options account for raw material properties enhancement, such as pyrolysis and torrefaction that improve the lower heating value (LHV) of the mixture, which are not taken into consideration here. Solid inlet stream is modelled in Aspen Plus© as an addition of non-conventional streams, which is the specific manner how the software handles with solids (see Sect. 3.1). The composition of the *base case*, to calibrate and to validate the model, is the design composition of ELCOGAS power plant: 50% of coal and 50% of petcoke on a mass basis. Coal and petcoke comes from local industries close to the plant: ENCASUR mines and Puertollano REPSOL refinery, respectively (see chapter " Examples of Industrial Applications"). The coal is of sub-bituminous type, with high ash content. The petcoke is obtained as a by-product in the refinery, with high sulphur content. The biomass wastes considered here are olive pomace or orujillo and forest wood residues, which are abundant waste resources in Spain as already mentioned in chapter "Raw Materials Supply". Table 5 shows the main data concerning feedstock composition. The higher heating value (HHV) is considered as the calorific power to be used in the modelling because it is the value considered by the simulator. Proximate and ultimate analyses are reported on a dry basis, except for the moisture content. In addition to the main feed composition, limestone is added to the gasifier as a catalyst to decrease the ash fusion temperature. The composition of the limestone is about a 95% of $CaCO_3$ and a 5% of ashes. It is approximately 2–3% in weight of the total feedstock stream introduced in the system. The model allows for ternary blends (coal–petcoke–waste) mass composition changes through a FORTRAN code, introduced in the model as a calculator block. It calculates the ultimate and proximate analyses of the mixture, as well as its HHV, based on each feed proportion. The maximum flowrate is 2,600 ton/day that corresponds to a 100% of the gasifier load. The already-mentioned FORTRAN code allows for load variations.

Table 5 Raw materials ultimate and proximate analyses [6, 45]

Percentage mass basis (dry)	Coal	Petcoke	Orujillo
Ultimate analysis			
C	41.07	88.40	50.00
H	2.81	3.34	6.50
O	7.51	0.02	36.30
N	0.92	2.04	0.80
S	1.05	5.91	0.10
Cl	0.04	0	0.2
Proximate analysis			
Moisture	11.80	7.00	7.60
Ashes	46.60	0.28	6.30
Fixed carbon	32.05	85.74	21.3
Volatiles	21.35	13.98	72.4
HHV_{dry} (MJ/kg)	13.58	32.65	20.38

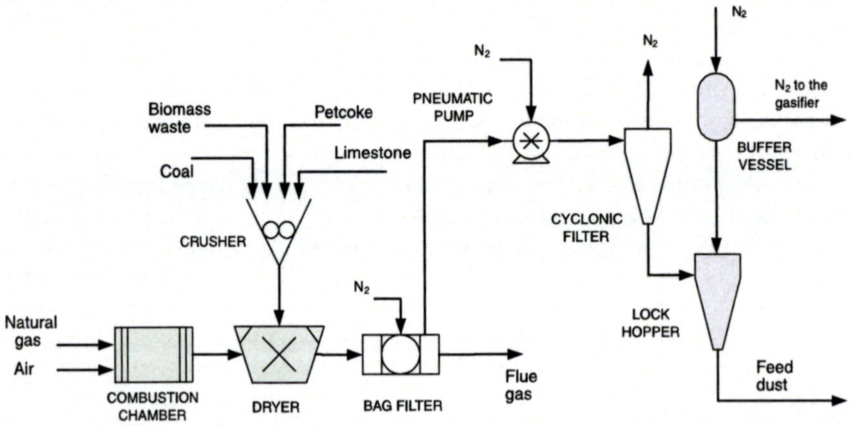

Fig. 3 Feedstock conditioning step

The three main steps in the feedstock dust preparation block are summarised in Fig. 3 and described as follows:

- *Dust preparation*. It takes place in a crusher unit that allows for particle fineness change from 100 mm to about 50–60 μm. Additional information to be introduced is the mixture's grindability, characterised by the Bond work index. This is calculated in a calculator block using a FORTRAN code and taking into account the mass proportion of each component. Bond index values are 73.80 kWh/ton for petcoke and 11.37 kWh/ton for coal. For the orujillo and the forest wood residues, the same value than the one for coal is considered. For the limestone, because it does not appear in the consulted database, the dolomite index has been chosen, being 11.31 kWh/ton [30].
- *Fuel drying*. It considers two reactors: the combustion chamber, where hot gases are generated, and the dryer itself, where the feedstock stream is dried till

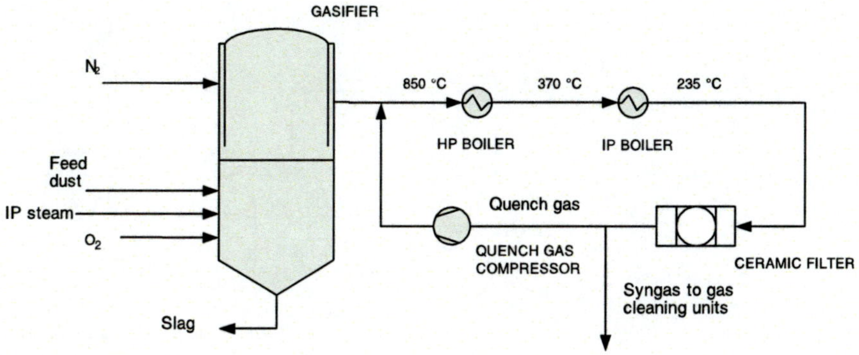

Fig. 4 Gasifier and WHB

the desired humidity by gas–solid contact. The combustor is simulated using a Gibbs's reactor, where natural gas is burnt with excesses of air. The stoichiometric value of air is calculated using the property set COMB-O2 from Aspen Plus©, and introduced in a calculator block. It operates at 0.9396 bar, with a temperature estimated as a fraction of the flame temperature. The dryer is modelled with a stoichiometric reactor with the feedstock as the limiting reactant to achieve a final moisture content of 2% on a mass basis.

– *pressurisation and feeding*. Once the feed is dried, it goes to the bag filters to be separated from the inert gas before entering the gasifier. These filters are simulated as a flash separator. Then, the feed is pressurised till 30 bar inside the lock hopper system. This system is simulated in a simple way as a mixer, considering the mixture of the powdered fuel with pure N_2.

The *gasifier* is based on the gasifier from ELCOGAS IGCC power plant, from Krupp–Koppers. Its input and output data have been used to calibrate the model. Gasifier operating conditions, pressure and temperature (P_{gasif} and T_{gasif}) are mainly driven by the gas turbine (GT) and ashes melting point, respectively. In this case, the GT pressure settles the gasifier pressure into 25 bar. According to ELCOGAS operating conditions, T_{gasif} is approximately between 1,400 and 1,500°C for coals, depending on the limestone content. In the model, this temperature interval is established as a condition in gasification when the raw material composition is changed. Specifically, a temperature of 1450 °C is fixed as a design specification in Aspen Plus©. As seen in Fig. 4, inputs to the gasifier are the feedstock powder, oxygen from the air separation unit (ASU) as main gasifying agent, intermediate pressure (IP) steam (as moderator and gasifying agent) and N_2 (as moderator).

The WHB is the group of heat exchangers that profit the syngas heat that is released before the gas-cleaning units (see Fig. 4). This syngas cooling process is modelled in Aspen Plus© taking place into two main steps: First, gasification gases are cooled down with the quench gas, which is a fraction of syngas recycled for

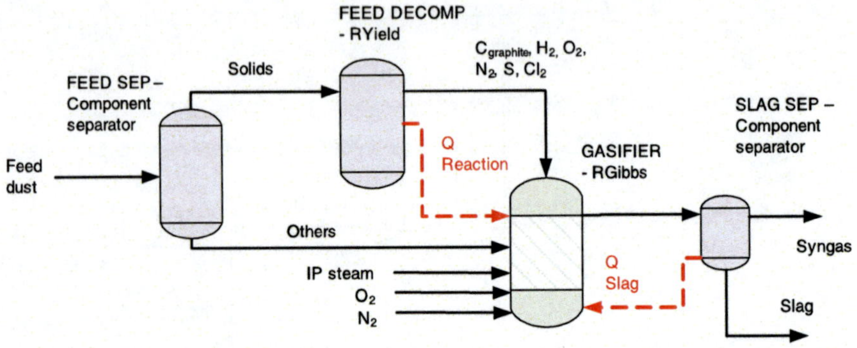

Fig. 5 Gasifier modelling approach

this purpose. The temperature is reduced till 850°C, which is the limit of the heat exchangers material [6]. The syngas goes out of the gasifier at around 1,450°C, and the quench gas is at 235°C. Second, the WHB profits the gas heat in a high pressure (HP) heat exchanger, cooling the gas down till 370°C and generating HP steam. Then, the gas moves to an intermediate pressure (IP) heat exchanger that cools the gas till 235°C, generating in this way IP steam. After this step, the quench gas is recycled to be mixed with the syngas at the outlet of the gasifier. The quench gas flowrate is determined by a FORTRAN block (a design specification block) that adjusts the temperate to 850°C.

The gasifier equilibrium approach as modelled in Aspen Plus© can be seen in Fig. 5. The gasification process itself (the equilibrium approach) only takes place in the RGibbs reactor. As the feedstock stream is introduced as a non-conventional component, before being introduced into the Gibbs reactor, it should be divided into its elements, because there is no Gibbs's energy information for a non-conventional component. Therefore, before the gasifier equilibrium, a yield reactor is used to decompose the raw material into $C_{graphite}$, H_2, O_2, N_2, S, Cl_2, H_2O and ashes according to its ultimate analysis and moisture content from the proximate analysis. As the ultimate and proximate analysis of the feedstock dust are of concern, previously to the RYield, a separator component is introduced to separate the feedstock dust from the N_2 and the limestone introduced in the feedstock system, which is latter introduced into the Gibbs reactor. The T_{gasif} is estimated by introducing into the Gibbs reactor the heat of reaction associated with the feedstock stream decomposition and the heat released into a component separator that aims at purifying the raw syngas produced from solids (ashes and limestone, named Slag in Fig. 5), named as Q Reaction and Q Slag in the same figure. According to the gasifier Gibbs's model, all the inlet carbon is converted, thus, for practical purposes, all the char is converted.

Oxygen, steam and nitrogen are also introduced into the Gibbs's reactor. Oxygen and steam flowrates are estimated by specific ratios that depend on raw material composition according to their ultimate analysis characteristics. The oxygen ratio is defined as the relation between the oxygen flowrate divided by the stoichiometric oxygen needed to completely oxidize the raw material mixture after the decomposition into its components on a mole basis (calculated through the property set COMB-O2 [31]). This ratio is commonly known as equivalence ratio (ER). The steam ratio is defined on a mass basis and has as denominator the mass flow of $C_{graphite}$. The ER value for the base case is 0.42 and the value for the steam ratio is 0.16. They have been estimated by adjusting the gasification temperature to 1,450°C, and according to ELCOGAS input conditions, respectively.

4.1 Calibration and Validation

For the gasifier model *calibration*, as well as for calibrating the entire superstructure, real ELCOGAS power plant working conditions have been considered. The objective is to reach the closest model outputs to real data, maintaining certain model independency from the ELCOGAS conditions, and thus maintaining the sensitivity to raw material and operating conditions variations. This problem is further discussed in chapter "Industrial Data Collection" under the data reconciliation section. Different model variables are suitable for calibration, such as raw material, proportions and proximate and ultimate analyses, HHV, grindability and solid's particle size distribution (PSD) that can be adapted to other specific case studies.

In this model's feedstock dryer, the required natural gas and air are settled as constant values depending on the gasifier load, and as 1.2 times of the necessary stoichiometric oxygen, respectively. Global mass flowrate, gasifier load and limestone proportion can be changed freely in a calculator block. The ER and the steam ratio can be adapted to different specific feedstock requirements. Throughout all the cases proposed and studied in this book, the gasifier operating temperature is maintained constant (see chapter "Selection of Best Designs for Specific Applications").

Validation is done through syngas composition comparison with the actual composition obtained by ELCOGAS, and the Gibbs's equilibrium approach. The required information is gathered through sensitivity analysis and compared to bibliographic and plant-operating data.

First, the entrained bed gasifier being modelled produces a syngas that consists of 21% of H_2, 60% of CO, 4% of CO_2, 3.5% of H_2O, 1% of H_2S, 10% of N_2 and the rest of COS, on a mole basis at the above-mentioned conditions. The model syngas flow counts on a 20% of H_2, 58% of CO, 4.5% of CO_2, 5% of H_2O, 1% of H_2S, 11% of N_2 and the rest of COS, on a mole basis. It can be seen that compositions are close to each other and that the equilibrium model closely predicts the plant composition. Concerning the residence time, it can be said that according to the syngas composition, for this type of gasifier and these specific conditions of temperature and pressure, 7.7 s allows for a syngas close to the equilibrium composition.

Second, the results of the work of Fermoso et al. [13] and Hernández et al. [14] are used to validate the Gibbs's equilibrium approach. The first work is used to

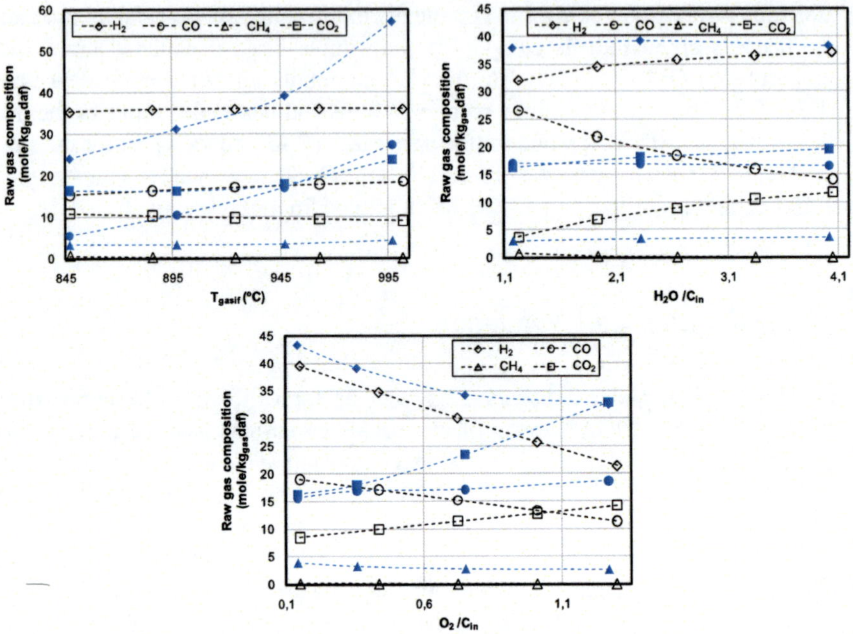

Fig. 6 Effect of T_{gasif} (°C), O_2/C_{in}, H_2O/C_{in} in the syngas composition (mole/mass basis); *daf* dry and ash-free basis. *Filled symbols* represent the data from [14]

compare the gasifier behaviour as input parameters variations, whereas the second one is used to evaluate the effect of biomass addition to the base case blend.

Validation with Fermoso et al. [14] implies the adaptation of the gasifier model conditions to their working conditions, i.e., the gasification of coal is considered and the steam and oxygen flowrates have also been adapted, as well as the possibility of reactor heating by dropping the adiabatic assumption. Simulation results as the gasification temperature, the oxygen ratio and the steam ratio vary are represented in Fig. 6. Analogous trends to the ones reported in Fermoso et al. [14] are found. Nevertheless, absolute composition ratios for the syngas composition are approximate, but this was expected according to Fermoso et al. [14] because temperatures are not high enough for the equilibrium approach to be used. Related with the gasifier temperature effect, H_2 and CO formation is favoured when the temperature increases, whereas CH_4 remains quite constant and CO_2 concentration decreases. The former syngas behaviour might be the result of first, a temperature increase that leads to K_{eq} decrease in all exothermic reactions, diminishing the composition of CO_2, and second, endothermic reactions are enhanced, increasing the composition of H_2 and CO. Oxygen ratio increase at constant temperature leads to the decrease in H_2 and CO formation and increase in CO_2 formation (as the oxidation reactions are shifted towards products). In the same way, as the steam ratio increases, the H_2 formation is enhanced (according to Eq. 22) as well as the CO_2 one, whereas occur the contrary to the formation of CO. The most affected equation is then Eq. 23.

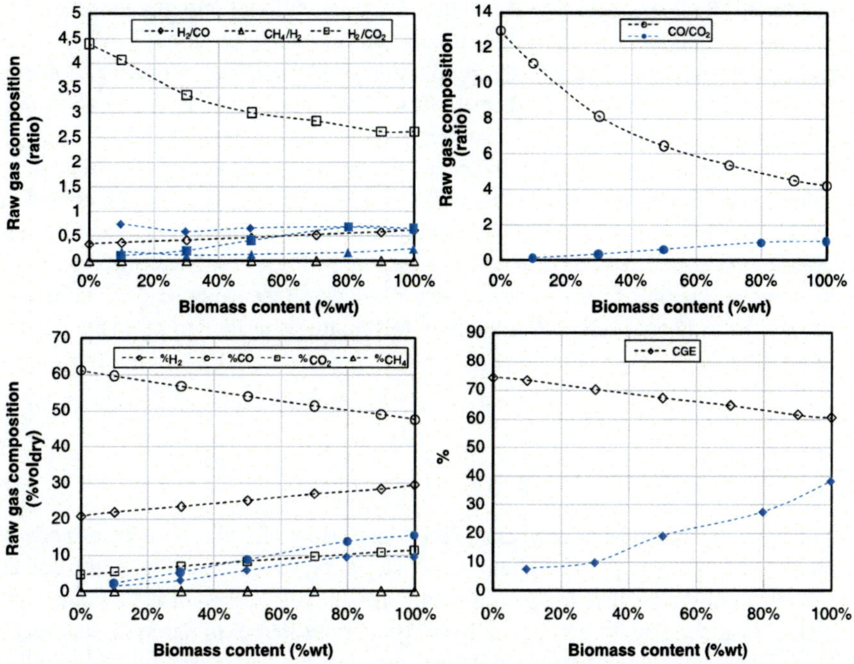

Fig. 7 Effect of biomass addition in syngas composition and CGE. *Filled symbols* represent the data from [13]

Figure 6 shows that at lower reactor temperatures, gasification reactions occur at a too slow velocity (because endothermic reactions should be motivated) with little practical value, given that for syngas applications CO and H_2 production is preferred. Nevertheless, for power production purposes, CH_4 is the one that offers a higher calorific value. Regarding the latter, high values of methane are related to low gasification temperatures, and reforming reactions (Eqs. 26 and 27) take place at relatively low temperatures as seen in Fig. 6 with CH_4 the most stable component along the performed analysis, but showing higher values at lower ratios. Overall, at lower temperatures of gasification, the LHV of the gas is higher but, for syngas applications, it is not the optimal composition. Even more, higher char conversions are linked with higher temperatures. At too high temperatures, gasification products start to combust.

According to the work of Hernández et al. [13], more sensitivity analyses have been performed with the aim of evaluating the behaviour of biomass blends, with regard to representative syngas composition ratios, H_2/CO, CH_4/H_2, H_2/CO_2 and CO/CO_2, absolute composition in mole basis, and the CGE (Eq. 31). In this case, the ELCOGAS PRENFLO gasifier conditions have been maintained (thus, the gasification temperature is around 1,400°C). See in Fig. 7 the obtained plots. It is worth noting that our results show different tendencies with the temperature increase in syngas composition and CGE, except for H_2 composition. The main reason for the discrepancies should be given to the fact that *co-gasification synergies* are not quantified. Besides, regarding the values of our model, not only the tendencies but they are also higher than the values reported in Hernández et al. [13]. All in all, the addition of biomass enhances the production of H_2, but leads to lower CGE (that means that the LHV contained in the raw gas is smaller). Regarding the syngas composition, the increase of biomass proportion implies less CO and more CO_2.

This calibration and validation procedure shows that an equilibrium reactor is a suitable approximation for the PRENFLO gasifier behaviour under the considered working conditions. It is also a good approximation for the biomass gasification at high temperature, even though tars are assumed to be zero. Nevertheless, the co-gasification scenarios are not well described, because the synergetic effect between the species is not modelled. Research on this matter is needed to try to characterise these effects.

4.2 Results

We have considered the performance of the gasifier for the base case feedstock composition carrying out different sensitivity analyses (SA) that take into account variations in the gasifier temperature and pressure, as well as in the ER and steam ratio. Also feedstock variations have been considered to analyse the reactor equilibrium behaviour. General output parameters that have been plotted for all the SA are the syngas composition and the cold gas efficiency (CGE) defined as per Eq. 31. As it is seen from the equation, if the feed does not vary, the CGE varies according to the LHV_{rg} obtained.

$$\mathrm{CGE} = \frac{m_{\mathrm{rg}} \times \mathrm{LHV_{rg}}}{m_{\mathrm{feed}} \times \mathrm{LHV_{feed}}} \tag{31}$$

$$\mathrm{LHV_{fuel}} = \sum w_{\mathrm{i}} \times \mathrm{LHV_i} \tag{32}$$

The CGE is calculated as the ratio between the chemical energy contained in the raw gas (rg), as the syngas is called as it comes out from the gasifier, and the total chemical energy contained in the fuel, both of them measured through their LHV. In the case of fuel blends, the LHV_{fuel} is calculated as the contribution of each *i*th fuel to the final blend, on a mass basis (w_i) (Eq. 32). The estimation of LHV_{rg} is performed in Aspen Plus© using the QVALNET property set.

The considered flowsheet contemplates the feeding system with the gasifier itself as modelled in our approach calibrated with ELCOGAS data. First (case SA 1), variation of T_{gasif} effects is shown in Fig. 8. As stated in Highman and van der Burgt [2], the gasification temperature range is between 800 and 1,800°C. It is seen from the first graph that syngas composition varies considerably till around 1,000°C, and then, the proportions of H_2, CO, CH_4 and CO_2 are noticeably constant. Nonetheless, CO_2 and CO show certain variability, the first one decreasing and the second one increasing. As seen in the plot of the H_2/CO value, variation of T_{gasif} leads to a variation of the two main syngas

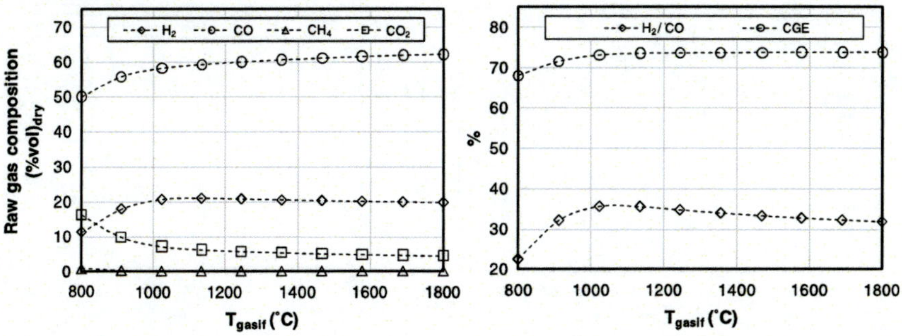

Fig. 8 SA 1: Effects of T_{gasif} variation

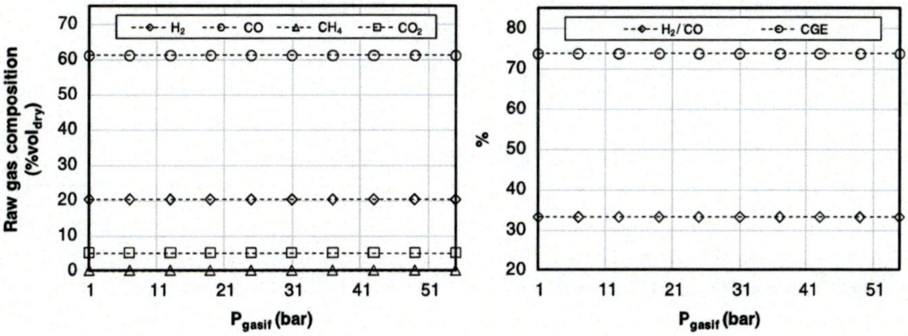

Fig. 9 SA 2: Effects of P_{gasif} variation

components that favour the decreasing of H_2 proportion and the increasing of the CO, as stated before; thus, the ration decreases. Because from around 1,000°C again, the CGE shows a positive trend (from 68 to 74%), even if it is fairly constant. It means that as the LHV_{feed} is constant, the LHV_{rg} is the one that has the positive trend. It is directly seen too the influence of the higher calorific value of the CO than that of the H_2.

SA 2 corresponds to P_{gasif} variation. It leads to a T_{gasif} variation of around 2°C (from 1,489°C at the lowest pressure considered, 1 bar; till 1,487°C at the highest pressure taken into account, 55 bar), not plotted in Fig. 9. The range of values considered is based on literature gasification plants (independently of the type of reactor) working pressure, which at its turn and as commented in previous sections is normally based on requirements of the process downstream the gasifier.

The effect of the pressure is small if compared with the effect of the temperature; it is possible to distinguish a trend of the values, but changes in composition and in H_2/CO and CGE are negligible. According to Sect. 3.3, only Eq. 24 has a zero in mole increment. However, it is possible to say that in the pressure range established the pressure effect is not noticeable, even for the reactions whose mole increment is different from zero.

SA 3 corresponds to ER variation (see Fig. 10). The ER range has been defined according to the base value determined for the base case. This scenario has an ER

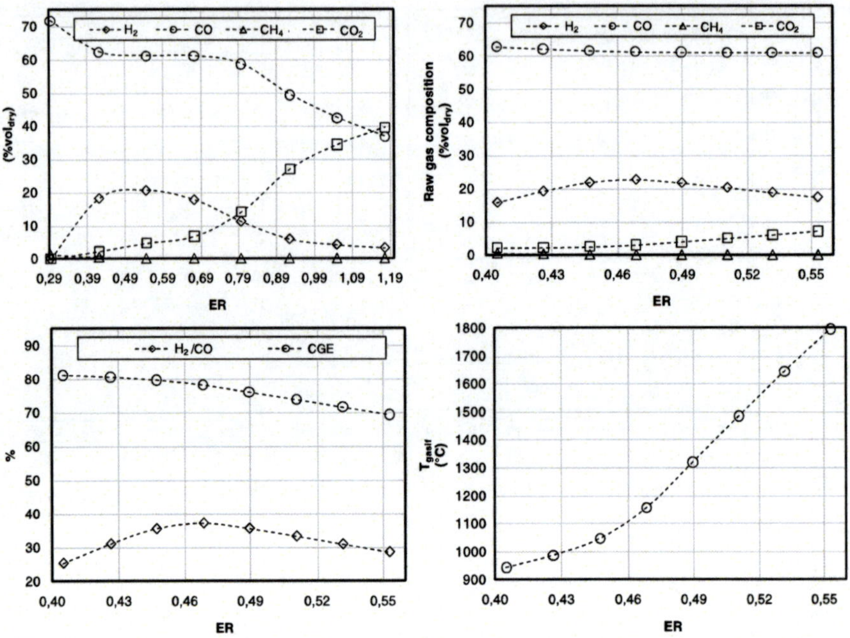

Fig. 10 SA 3: Effects of ER variation

of 0.42. To start with, the oxygen ratio has been varied in an interval of ±30%. Then, this ratio has been reduced to be focused in the *gasification area* that has been enclosed to the interval where methane production is low (high methane composition correspond to pyrolysis) and sulphur oxides production does not appear (oxides different from carbon leads to think about combustion; ER = 1 is a complete combustion). It fairly corresponds to the gasification temperature range commented before, between around 800 and 1,800°C. Finally, the ER range oscillates between 0.4 and 0.55. If representing the whole range of results (±30%) for ER between 0.29 and 0.55, it is seen that there is a significant change in syngas composition around ER = 0.33.

As seen in Fig. 10, and in the H_2/CO behaviour, there is a maximum in H_2 production around ER = 0.46. The CO_2 production tends to increase with an increase in inlet oxygen (thus, Eqs. 15, 16 and 17 are of concern). As in all the other SAs, methane is quite constant. The CGE tends to decrease as the ER increases, because of the CO_2 increment, and CO decreases.

SA 4 corresponds to steam ratio variation. It varies between 0.15 and 0.4, according to ELCOGAS real experiences. The main role of the steam is as a temperature moderator. It is seen in Fig. 11 that the gasification temperature diminishes as the ratio increases. With the decrease of temperature, contrary at what has been seen in Fig. 8, CO diminishes and CO_2 increases. Hydrogen also increases, as the temperature is lower. The CGE is quite constant, but the H_2/CO

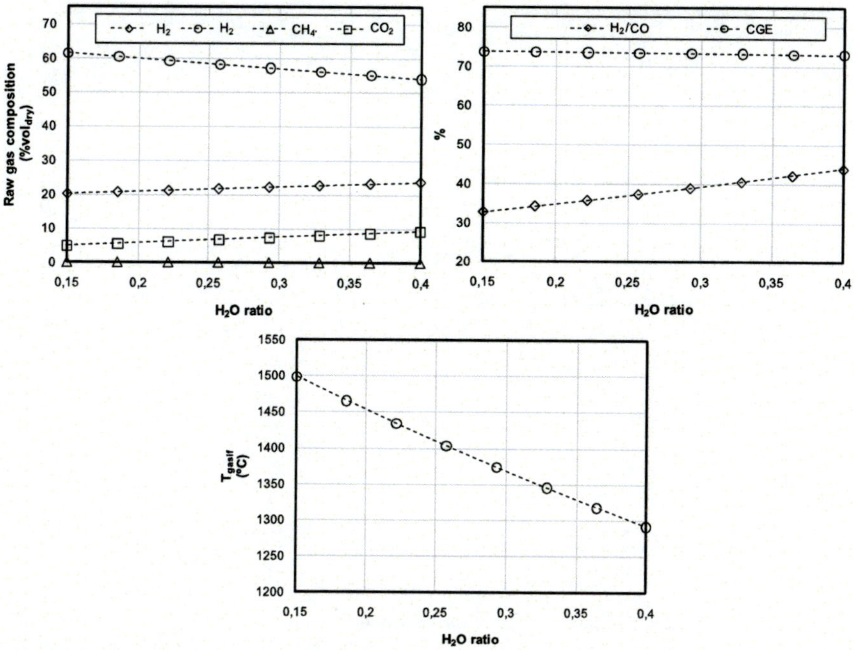

Fig. 11 SA 4: Effects of steam ratio variation

ratio increases. In endothermic reactions, according to what it has been discussed in Sect. 3.3, as the temperature decreases, tends to react to the reactants side, thus being Eqs. 22 and 25 of concern, where as the temperature increases, they tend to react to the reactant side.

SA 5 leads with variations in feedstock composition. A certain amount of petcoke is added to the coal according to the percentages seen in Fig. 12. As the proportion of petcoke increases, as it has a LHV higher than coal, the T_{gasif} increases. It implies a higher LHV_{rg}, therefore, the CGE increases with the percentage of petcoke. CO composition is the one that experiments the highest change in tendency, increasing a 20% in the interval of study. H_2 increases, as well as CO_2. O/C and H/O are the ratios derived from the feedstock inlet composition. Higher values of C (lower O/C) imply higher values of calorific power. The same as H/O increases.

In summary, as seen in this results analysis, the P_{gasif} has a minimal influence in the final syngas parameters. Assuming a gasification reactor where only T_{gasif} is changed (when it is heated externally), it seems that the syngas composition leads to a more or less 'fixed composition', where increasing the temperature does not imply better results. On the contrary, if the temperature is increased because more oxygen is introduced, in the gasification zone, H_2 tends to decrease, whereas CO_2 tends to increase. Steam ratio increase leads to a gasification temperature decrease. H_2/CO is favoured. The final gas composition depends on the relations between

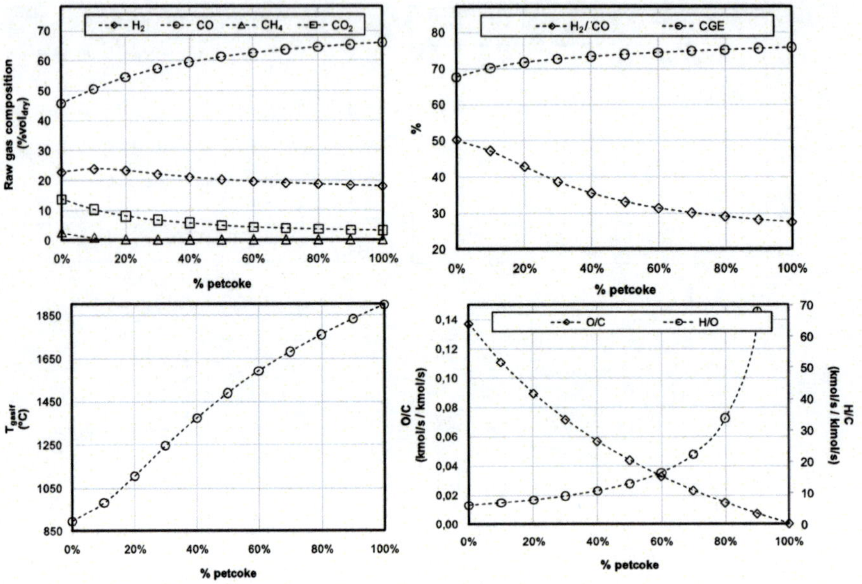

Fig. 12 SA 5: Effects of feed composition variation

carbon, oxygen and hydrogen from the feedstock composition, because T_{gasif} and the mass balance are influenced. Equations 18–24 are the most important, being the equations that imply CH_4 less significant because in the gasification zone, the methane value is very low.

5 Conclusions

In this chapter, the equilibrium model minimising Gibbs free energy is used to simulate the performance of a PRENFLO gasifier with T_{gasif} 1450 °C, P_{gasif} 25 bar and residence time 7.7 s. The model has been validated with ELCOGAS plant data, showing acceptable agreement. This chapter highlights that the model fitting is highly dependant on the specific gasification conditions, better at higher temperatures and/or longer residence times, as well as dependant on the raw materials introduced in the reactor. In this view, if co-gasification with biomass is of concern, tars and synergect effect should be known to adequately represent the gasification behaviour of temperature, pressure, ER, steam ratio and feedstock composition are studied. Feedstock composition is the most influential parameter. Reactor pressure and temperature are fixed by the final syngas application and the ashes melting point, respectively. That is why in the following chapters, the SAs are performed by varying the ER and the steam ratio. The desired gasifier temperature is obtained by adjusting the ER ratio in each case (thus, playing with exothermic and endothermic reactions).

References

1. Wender I (1996) Reactions of synthesis gas. Fuel Process Technol 48:189–297
2. Highman C, van der Burght M (2003) Gasification. Elsevier Science, New York
3. NETL (2007) Gasification world database 2007. Current industry status: robust growth forecast. Technical report, National Energy Technology Laboratory, Pittsburgh, PA
4. Reed TB, Das A (1988) Handbook of biomass downdraft gasifier engine systems. Solar Energy Research Institute (SERI), US Department of Energy, Colorado, US. Residence time 199X
5. Sharma AK (2009) Experimental study on 75 kWth downdraft (biomass) gasifier system. Renew Energy 34:1726–1733
6. Coca MT (2003) Tecnología de gasificación integrada en ciclo combinado: GICC. Aplicación real en España: ELCOGAS. Puertollano. Monterreina, S.A. Spain, ENERCLUB
7. Zainal ZA, Rifau A, Quadir GA, Seetharamu KN (2002) Experimental investigation of a downdraft biomass gasifier. Biomass Bioenergy 23:283–289
8. Bridgwater AV (2003) Renewable fuels and chemicals by thermal processing of biomass. Chem Eng J 91:87–102
9. Milne TA, Evans RJ, Abatzoglou N (1998) Biomass gasifier "tars": their nature, formation, and conversion. NREL/TP-570-25357, Technical report. EEUU, Springfield, VA
10. Van de Kamp W, Wild P, Zielke U, Suomalainen M, Knoef H, Good J, Liliedahl T, Unger C, Whitehouse M, Neeft J, van de Hoek H, Kiel J (2005) Tar measurement standard for sampling and analysis of tars and particles in biomass gasification product gas. ECN Biomass, ECN-RX-05-185
11. Sutton D, Kelleher B, Ross JRH (2001) Review of literature on catalysts for biomass gasification. Fuel Process Technol 73:155–173
12. Mastellone ML, Zaccariello L, Arena U (2010) Co-gasification of coal, plastic waste and wood in a bubbling fluidized bed reactor. Fuel 89:2991–3000
13. Hernández JJ, Aranda-Almansa G, Serrano C (2010) Co-gasification of biomass wastes and coal coke blends in an entrained flow gasifier: an experimental study. Energy Fuels 24:2479–2488
14. Fermoso J, Arias B, Plaza MG, Pevida C, Rubiera F, Pis JJ, García-Peña F, Casero P (2009) High-pressure co-gasification of coal with biomass and petroleum coke. Fuel Process Technol 90:926–932
15. Wen CY, Chaung TZ (1973) Entrainment coal gasification modelling. Industrial Eng Chem Process Des Dev 18:684–695
16. Gas Production/Gas Treating (2003) Ullmann's encyclopedia of industrial chem. Online edition
17. Ruggiero M, Manfrida G (1999) An equilibrium model for biomass gasification processes. Renew Energy 16:1106–1109
18. Usón S, Valero A, Correas L, Martínez A (2004) Co-gasification of coal and biomass in an IGCC power palnt: gasifier modeling. Int J Thermodynamics 7:165–174
19. Gautam G, Adhikari S, Bhavnani S (2010) Estimation of biomass synthesis gas composition using equilibrium models. Energy Fuels 24:2692–2698
20. Di Blasi C (2000) Dynamic behaviour of stratified downdraft gasifiers. Chem Eng Sci 55:2931–2944
21. Jand N, Foscolo PU (2005) Decomposition of wood particles in fluidized beds. Industrial Eng Chem Res 44:5079–5089
22. Govind R, Shah J (1984) Modeling and simulation of an entrained flow coal gasifier. AIChe J 30:79–92
23. Pérez-Fortes M, Bojarski AD, Velo E, Nougués JM, Puigjaner L (2009) Conceptual model and evaluation of generated power and emissions in an IGCC plant. Energy 34:1721–1732
24. Balzioc S, Hawksley PGW (1970) Kinetics of thermal decomposition of pulverized coal particles. Industrial Eng Chem Process Des Dev 9:521–530

25. Loison R, Chauvin R (1964) Fast pyrolysis of coal pyrolyse rapide du charbon. Chimie Industrie 91:269–275
26. García-Labiano F, Adánez J, Hampartsoumian E, Williams A (1996) Sulfur release during the devolatilization of large coal particles. Fuel 75:585–590
27. Kambara S, Takarada T, Toyoshimab M, Katob K (1993) Relation between functional forms olf coal nitrogen and formation of NOx precursors during rapid pyrolysis. Energy fuels 7:1013–1020
28. Lucien B, Favrat D (2010) Thermodynamics and energy systems analysis; From energy to exergy., Engineering Sciences, Mechanical EngineeringEPFL Press, Lausanne, Switzerland
29. Kunzing GE (1981) Gas–solid transport. McGraw-Hill, New York
30. Perry R, Green D (eds) (1997) Size reduction and size enlargement. In: Perry's chemical engineers' handbook. McGraw-Hill, New York
31. AspenTech Incorporation (2010) Aspen Plus user's guide. Online
32. Dogru M, Howarth CR, Akay G et al (2002) Gasification of hazelnut shells in a drowndraft gasifier. Energy 27:415–427
33. His C-L, Wang T-Y, Tsai C-H, Chang C-Y, Liu C-H, Chang Y-C, Kuo JT (2008) Characteristics of an air-blown fixed-bed downdraft biomass gasifier. Energy Fuels 22: 4196–4205
34. Jayah TH, Aye L, Fuller RJ, Stewart DF (2003) Computer simulation of a downdraft wood gasifier for tea drying. Biomass Bioenergy 25:459–469
35. Loeser M, Redfern MA (2009) Modelling and simulation of a novel micro-scale combined feedstock biomass generation plant for grid-independent power supply. Int J Energy Res 34(4):303–320. doi:10.1002/er.1556
36. Melgar A, Pérez JF, Laget H, Horillo A (2007) Thermochemical equilibrium modelling of a gasifying process. Energy Convers Manag 48:59–67
37. Giltrap DL, McKibbin R, Barnes GRG (2003) A steady state model of gas-char reactions in a downdraft biomass gasifier. Sol Energy 74:85–91
38. Altafini CR, Wander PR, Barreto RM (2003) Prediction of the working parameters of a wood waste gasifier through an equilibrium model. Energy Convers Manag 44:2763–2777
39. Wang Y, Kinoshita CM (1993) Kinetic model of biomass gasification. Sol Energy 51:19–25
40. Nikoo MB, Mahinpey N (2008) Simulation of biomass gasification in fluidized bed reactor using Aspen plus. Biomass Bioenergy 32:1245–1254
41. Emun F, Gadalla M, Majozi T, Boer D (2009) Integrated gasification combined cycle (IGCC) process simulation and optimization. Comput Chem Eng 34:331–338
42. Robinson PJ, Luyben WL (2008) Simple dynamic gasifier model that runs in Aspen dynamics. Industrial Eng Chem Res 47:7784–7792
43. Nathen SV, Kirkpatrick RD, Young BR (2008) Gasification of New Zealand coals: a comparative simulation study. Energy Fuels 22:2687–2692
44. Frey CH, Akunuri N (2001) Probabilistic modeling and evaluation of the performance, emissions, and cost of texaco gasifier-based integrated gasification combined cycle systems using ASPEN. Technical report. Prepared by North Carolina State University for Carnegie Mellon University and US Department of Energy, Pittsburgh. http://www4.ncsu.edu/~frey/reports/Frey_Akunuri_2001.pdf. Accessed 10 Oct 2010
45. ECN-Biomass (2010) Phyllis, database for biomass and waste. Energy research Centre of the Netherlands (ECN). http://www.ecn.nl/phyllis. Accessed 12 Aug 2010

Main Purification Operations

Mar Pérez-Fortes and Aarón D. Bojarski

Abstract Syngas final usage requires a previous step of cleaning and conditioning to meet with the requirements of its final use which might range from chemicals and fuels production to power and/or heat. This chapter deals with the description and the modelling of the required syngas treatment units before electricity production or before hydrogen generation, specifically in an IGCC power plant. In the case of electricity generation application, the pursued objective is to avoid as much as possible nitrogen and sulphur oxide emissions to the atmosphere. In a first step, the gas is cleaned from solids. Secondly the gas before its combustion goes through an acid and basic species removal train of units. In the case of hydrogen generation, besides syngas cleaning from other species, the main pursued objective is to separate CO from H_2. In order to accomplish the former, CO should be converted into CO_2 and then separated from the main stream. Hydrogen can be further purified to be sold to the market, or used in a combined cycle, in an analogous way as the syngas. Modelling calibration and validation are shown, and the chapter finishes with a model utilisation to evaluate the behaviour of the already built up superstructure to produce hydrogen or syngas for electricity generation section, or hydrogen for other applications.

Notation

ASU	Air separation unit
CC	Combined cycle
CCS	Carbon capture and storage
CGE	Cold gas efficiency

M. Pérez-Fortes (✉) · A. D. Bojarski
Universitat Politècnica de Catalunya, ETSEIB,
Diagonal 647, 08028 Barcelona, Spain
e-mail: mar.perez-fortes@upc.edu

A. D. Bojarski
e-mail: aaron.david.bojarski@upc.edu

L. Puigjaner (ed.), *Syngas from Waste*, Green Energy and Technology,
DOI: 10.1007/978-0-85729-540-8_5, © Springer-Verlag London Limited 2011

COE	Cost of energy
DMPEG	Dimethyl ethers of polyethylene glycol
ELECNRTL	Non-random two liquid with electrolytes formation
EOS	Equation of state
FT	Fischer–Tropsch process
GHG	Greenhouse gases
GT	Gas turbine
GTL	Gas to liquid fuels
HP	High pressure
HRSG	Heat recovery steam generator
IGCC	Integrated gasification combined cycle
IP	Intermediate pressure
LHV	Lower heating value
MCFC	Molten carbonate fuel cells
MDEA	Methyl diethyl ethanol amine
NRTL	Non-random two liquid
PEM	Proton exchange membranes
PSA	Pressure swing adsorption
SOFC	Solid oxide fuel cell
ST	Steam turbine
TIT	Turbine inlet temperature
VLE	Vapor–liquid equilibrium
WGS	Water gas shift
WHB	Waste heat boiler

1 Introduction

From a practical point of view, the syngas can be considered as an intermediate between a wide range of solid feedstocks and a wide range of products, derived from a gas (see in Fig. 1). One plant can produce different products, being called *polygeneration.* When two products are produced, it is called *co-production.* As already commented in "Modelling Syngas Generation", the syngas can be produced from a mixture of feedstocks (*co-gasification*). The concept of polygeneration and co-use of different types of raw materials is the essence of the *biorefineries* that aim at mimicking the energy efficiency of oil refineries through the production of fuels, power and chemicals from biomass. There are four types of biorefineries, and one of them is the biosyngas-based refinery. The other are pyrolysis, hydrothermal and fermentation-based [1].

On the one hand, *syngas final application* downstream the gasification process decides the syngas cleaning processes. On the other hand, the option of using biomass or waste for gasification or co-gasification with coal and/or petcoke depends on the syngas quality technologies [2]. It means that even though, the final

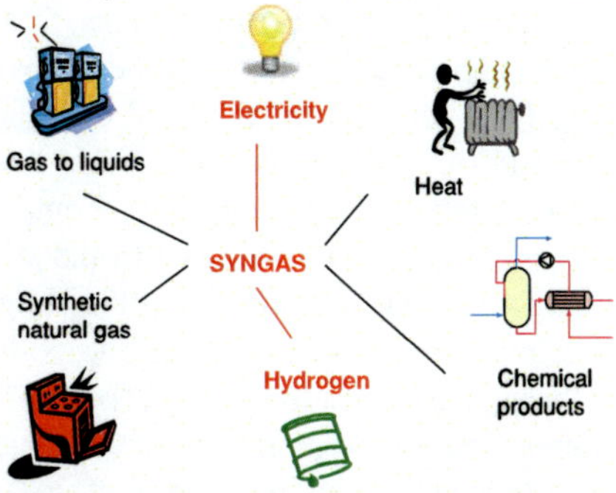

Fig. 1 Syngas uses

syngas requirements dictate the syngas cleaning processes, *raw material* used and *syngas generation conditions* (mainly pressure and temperature, and oxygen purity) are key factors to be taken into account at the moment of choosing a cleaning gas train. It implies that when designing a specific gasification plant for a specific feedstock composition, if this composition changes, it is important to guarantee that the abovementioned key parameters are maintained between certain ranges to assure the capabilities of the purification units before the final syngas application. Or from a different point of view, the purification unit train has to be designed allowing for a wide range of operating options which might rise from the use of different feedstocks. Different feedstock mixtures have different ultimate and proximate analyses, heating value and different physical properties; they lead to obtain different syngas compositions in terms of H_2 and CO partial pressures, different H_2/CO ratio, and different amounts of sulphur, nitrogen, chlorine and phosphorous. Therefore, this fact leads to different gas cleaning processes, or at least, to different gas cleaning units operation conditions [3].

At its turn, particulate dust and tar are of concern; the latter is particularly important in biomass or waste gasification. In this way, removal of syngas pollutants can be divided into two categories: during gasification (generally for solid removal) and after gasification (fluid pollutants removal), being called respectively *primary* and *secondary cleaning methods*. Overall, regarding the application, the type of gasifier and the level of pollutants in the fuel, the syngas conditioning train, including cleaning and cooling, is determined.

In next sections we discuss the general conditions required to each one of the final syngas applications; different gas cleaning methods and the description of the cleaning gas train considered in this book.

2 Syngas Uses and Treating Unit's Selection Criteria

Syngas requirements mainly include syngas temperature, pressure and pollutants level conditioning before its final application. The pressure of the syngas has been already decided in the gasification reactor. Cleaning is particularly needed when syngas combustion is of concern, to avoid oxides pollutants. *Wet* and *dry*, *hot* and *cold* cleaning systems have been developed and fairly used.

In general terms, BTG [4] stipulates that syngas cooling is required for combustion in gas engines, for filters that have a maximum acceptable temperature, and when syngas requires of compression. Gas cleaning, independently of the scale, ranges from the least to the strictest use as follows: its use in a kiln, a gas engine, a gas turbine, a Stirling engine, a fuel cell, and other syngas uses for chemical production (where the pollutant levels should be from mg/m^3, passing through ppm, and arriving till ppb). These pollutant restrictions include flue gas emission requirements and conditions for certain devices to work properly.

Independently from the final use, *high temperature* and *high pressure* are the desired conditions in the majority of the final applications. According to the categories stated in Fig. 1, from Wang et al. [2] and Highman and van der Burght [3], the following characteristics can be cited for each application:

- *Electricity production*: This production comes mainly from syngas combustion for its profit in a gas turbine or in a combined cycle, or from syngas profit in a fuel cell.
  - *IGCC power plants*: Integrated gasification combined cycle power plants (IGCC) are the specific case of combination of gasification with a combined cycle. They can process a big variety of materials, according to the gasification capabilities described in "Modelling Syngas Generation". Most common IGCC plants use coal and heavy residual fractions from oil if they are close to refineries, for instance. The transition between the gasification and the combined cycle implies a temperature adaptation of the syngas that means a loss of efficiency since the high temperatures obtained in the usually used reactor, an entrained bed, are not profit for the combined cycle (CC). The latter is the main reason for ongoing research on hot, or at least warm, syngas cleaning processes. Ideally, for power generation, pressure and temperature should be decreased to the least possible. IGCC power plants are not in this ideal situation, where usually cryogenic temperatures are needed for the air separation unit (ASU) functions, and where the syngas is cooled down before filtering and desulphurisation units, to be heated once again in the combustor of the gas turbine (GT) cycle. Advanced cycles deal with the enhancement of the CC efficiency, for instance with isothermal compression, or increasing the heat integration.
  - *Fuel cells*: Hot and pressurised syngas is needed in the fuel cell if better efficiencies want to be obtained, and if combining with a gas turbine. There exist several types of fuel cells (solid oxide, molten carbonate, phosphoric acid and proton exchange membranes). As discussed in the specific literature [5, 6] the solid oxide fuel cell (SOFC), which operates at high temperatures (600–1,000°C), is the most extended type of fuel cell to be integrated in a gasification plant for power production. Also molten carbonate fuel cells (MCFC) operate at high temperature, thus being suitable for this application. The syngas, after the gasification process and previously

cleaned up from solid pollutants and sulphur, can be directly introduced into the fuel cell. In principle the fuel cell allows the use of different mixtures containing CH_4, CO and H_2. In fact, only H_2 is converted in the anode, but CO and CH_4 can be converted into H_2 inside the fuel cell (reforming process). The previous cited works refer to small scale plants, with downdraft gasifiers. Much more research has to be done in order to allow fuel cells becoming a well (and bigger scale) extended process.

- *Heat*: The production of heat from syngas simply involves the combustion of the syngas in a kiln to heat water. As already mentioned, this application does not need exhaustive cleaning processes, and not specific conditions of pressure and temperature, since the kiln as technology does not have special pollutants and operation restrictions.
- *Chemical products*: C1 chemistry is an often used name to refer to the chemistry involved in the syngas processing (also involving CH_4 and CO_2) that offers the possible conversion routes to industrial chemicals (ammonia, methanol, oxo-alcohols and hydrogen), basically stating the formation of multi-carbon molecules from one single carbon raw materials [7]. Several chemical products can be obtained through syngas chemical transformation:

  - *Ammonia*: It is mainly obtained from the reforming of natural gas. Nevertheless, a 10% of the world production is performed with coal or heavy oil gasification. It is produced from syngas through a catalysed process at high pressure. In this case, the N_2/H_2 ratio is important to optimise the ammonia production.
  - *Small organic compounds*: In this case CO is the raw material to synthesise organic chemicals like acetic acid, phosgene or formic acid. Due to its toxic nature, CO storage is dangerous and CO processing plants have to be close to the gasification plant. The syngas should have a high purity in CO (more than a 98.5%mol) to be used for this application. The final application of these chemicals is the synthesis of plastics, adhesives, preservatives, paints, etc.
  - *Oxo-alcohols*: Oxo-alcohol is the name given to higher alcohols, used as plasticisers and in the manufacture of synthetic detergents. They are obtained by reacting syngas with olefins, producing aldehydes which subsequently are hydrogenated to finally produce an oxo-alcohol. The quality of the syngas to be used as a raw material for oxo-alcohols is a ratio H_2/CO close to 1.

- *Hydrogen production*: Pure H_2 can be produced from syngas from gasification, if the gasification step is followed by water reforming of CH_4 (in the case that the level of CH_4 is high) and by a water gas shift (WGS) reactor that transforms CO into CO_2 and H_2. Then, a CO_2 removal process separates CO_2 from the main stream. For the WGS reactor, high pressure is needed. Applications of this pure H_2 can be in proton exchange membranes (PEM) that are types of fuel cells which need a feed of high purity hydrogen.
- *Synthetic natural gas*: Abbreviated as SNG, is a natural gas that can be produced from any carbon based feedstock through syngas processing. Methane

synthesis takes place after gas cleaning units, and considers a methanation reaction which transforms CO and H_2 into CH_4 and water generally using supported nickel catalysts [8]. Typical operating temperature is 300–400°C and it is preferably to fulfil the reaction at high pressure. After this step, there is a CO_2 removal unit to purify the methane stream [9].

- *Liquid fuels*: Also called synfuels or abbreviated as gas-to-liquid (GTL) fuels.
  - *Fischer–Tropsch (FT) fuels*: Syngas can be used to produce hydrocarbons of variable chain length, alternative to conventional diesel, kerosene and gasoline. Usually the ratio H_2/CO should be about 2 (sometimes a WGS reactor is needed to obtain the desired proportion). The advantage when comparing with traditional fuels is that FT fuels do not contain or contain little contaminants such as sulphur or aromatics.
  - *Methanol and dimethyl ether (DME)*: They are liquid fuels alternative to gasoline and diesel fuels. Methanol can react with triacylglycerols to produce biodiesel. It can be obtained by CO and/or CO_2 hydrogenation. DME is obtained from methanol in a dehydration step. Molar ratios H_2/CO and CO_2/CO are optimised for each step.
  - *Biobased products*: They are obtained by syngas fermentation, thus biologically converted into organic acids, alcohols and polyesters. Biological process does not need high temperature and pressure such as chemically catalysed process. Syngas fermentation has several barriers to be com mercially extended such as its difficulty to maintain anaerobic conditions or product inhibition.

Finally, the criteria to select the gas cleaning units follow the next indicators. Firstly, *gas purity and composition*; the syngas components ratios, principally H_2/CO, should be adapted to its final application requirement. Secondly, the *selectivity*, which refers to the ability of removing one component while other ones remain in the main flowsheet. For instance, in a gas turbine application, it is necessary to remove acid compounds from S (H_2S, COS) but on the contrary, CO_2 is not of interest since it contributes to the gas turbine power by increasing the mass flowrate. Thirdly, syngas components can cause *parallel effects* such as corrosion, un-wanted co-absorption and solvent losses. And last but not least, come the *economic issues* such as the capital cost and the maintenance costs; for instance, in an absorption unit there always exists a trade-off between investment and solvent regeneration. *Separation Processes, Ullmann's Encyclopedia of Industrial Chemistry, Processes and Process Engineering* [10] define the selection of a separation process according to the following criteria: firstly, the *separation factor* that gives a quick idea about the necessary process to separate a mixture, such as component differences in terms of volatility, solubility, molecular size and shape. Next, the *ordinary or extreme conditions*, such as high or low temperature and high or low pressure which are needed to obtain attractive separation factors. It is also indispensable to know the value of the specific substance to separate, and the final desired value, thus the desired *scale of operation*. It will give us the number and types of alternatives to be used. Finally, the *best practices* or know how of specific experiences already developed are good references when facing a specific problem.

In short the most efficient option in a gasification plant is to determine the pressure in the gasifier itself, and try to maintain it until the syngas is used. High temperature can be profited in downstream heat exchanger integrated with the heat requirements of the plant. Final syngas uses require determined H_2/CO ratios; that is why a WGS reactor should be necessary between the gas cleaning unit and the final application unit. Acid and basic pollutants should be removed. Gas purity and composition, selectivity, corrosion and other negative effects, and economic issues are of concern when choosing one method or another. In the next section we deal with the main principles of syngas cleaning, therefore, of some principles about separation processes.

3 Types of Syngas Purification and Concentration Units

Syngas *purification* mainly includes solids, tars, heavy metals, halogens and alkalines, acid and basic species removal processes. At its turn, CO_2 absorption has the purpose of *concentration*, where H_2 is the desired component. Heterogeneous and homogeneous mixtures require different cleaning methods. In the case of heterogeneous mixtures, a solid–gas mixture, mechanical separation methods such as *filtration* or *water scrubbing* are applied to separate the different phases. On the contrary, for homogeneous mixtures (in this case only gas phase), diffusion based separation processes are suitable, whose aim is to convert a feed mixture into two or more products that differ in composition. The most widely used in syngas cleaning are *absorption* and *adsorption. Physical absorption* and *chemical (reactive) absorption* are the type of separation process typically used for syngas purification, where a liquid solvent is used to selectively remove acid and basic species. The absorption process counts with a regeneration step where the solvent is cleaned from pollutants and recycled to be used again in the absorber. In general they are formed by two columns (one for absorption and the other for desorption) and a set of a heat exchangers and pumps that transform the solvent back to the absorber conditions [10].

Mechanical methods are used to mainly separate fly ash and dust, from the main gas stream. Cyclones are often used, but result ineffective for small size particles (in the order of sub-microns). Ceramic candles or sintered metal filters and wet scrubbers, operate better, in the case of the former ones they can operate close to 800°C.

Absorption methods are divided into physical and chemical processes depending on the type of bond connexion between the selected species with the solvent; simple physical absorption or a stronger chemical bond with the solvent itself can be formed. The loading capacity of the solvents or absorbents depends exclusively on the gas specie solubility (generally modelled using *Henry's law*), and the chemical reactivity of the gas specie with the solution media which could require tackling with *electrolytes formation* (that is, the chemical reactions of the absorbed gases with the solvent). Gas species solubility is mainly influenced by the operating pressure and the composition of the syngas, while for the solvent, the quantity of solvent in the solution is also a decision parameter, which mainly depends on the gas species to be removed. Physical solvents are for example, methanol and dimethyl ethers of polyethylene glycol (DMPEG) that work using common processes called Rectisol and Selexol, respectively. Chemical solvents which are water-based solutions are amines, where MDEA is the most widely used due to its higher selectivity compared with others.

Adsorption systems are normally formed by a solid bed that adsorbs the selected species. The bed has to be periodically changed, or some of them allow for in situ regeneration. This adsorption–desorption process involves changes in temperature and pressure: low T and high P for adsorption, while the contrary conditions (i.e., high T and low P), for desorption. For example, the pressure swing adsorption (PSA) cycle operates at a constant T, and at high P for the adsorption, and at low P for desorption. As will be seen in Sect. 4.5 this unit can be used for H_2 concentration and purification.

According to Sharma et al. [11], a gas cleaning process can be operated at three temperature regimes as a consequence of the syngas final application in a gasification plant; cold (less than 25°C), warm (less than 300°C) and hot cleaning (more than 300°C). Comparatively, all the commercially available processes operate using *cold* and *warm* syngas. It means that for gasification plants where the syngas is obtained at high temperature, there exists a considerable loss of efficiency (energetic and exergetic) due to its cooling before its purification and therefore before its use in the GT. In addition to that, *hot* gas cleaning can lower operational costs when final syngas applications need high temperature (for instance, H_2 production by steam reforming and WGS, or combined heat and power generation with a fuel cell). Nevertheless, the work of Pisa et al. [12] centres its attention in IGCC power plants alternative designs and in desulphurisation processes in particular, evaluating the hot desulphurisation process such as the one with ferrite ($ZnFe_2O_4$). This bed needs oxygen to convert H_2S and steam to proportionate the optimal humidity for the optimal operation work. The final result shows that high steam consumption finish by penalising the steam turbine power production, thus penalising the global efficiency of the plant.

Absorption processes can require a temperature around 200°C; on their side, adsorption processes require nearly ambient temperatures. The syngas cooling has several problems inherent to ashes presence, mainly their slagging and non-slagging conditions when the syngas passes through a critical temperature range.

In the case of tars control, Brown et al. [13] point out that they cause environmental and operation trouble, being condensable tars the main problem since they can damage units of the whole flowsheet. Recalling from "Modelling Syngas Generation", Sect. 2.2.1, there are two possibilities for tars removal: primary and secondary methods. *Catalytic conversions* outside the gasifier are representative type of processes intensification, since they allow not only for tars transformation but also for solids removal. It has been proven that particulates recirculation to the gasifier contributes to tar control [14]. For nitrogen, cyanide and halide species removal, a wet scrubbing process is well extended for cold and warm conditions.

Due to the importance of CO_2 emissions in the context of energy production we devote a whole section (3.1) to carbon capture and storage, which is discussed as one of the possibilities to include during syngas cleaning operations. In the next Sect. 4, a typical syngas cleaning train is described. It is worth emphasising that in "Emerging Technologies on Syngas Purification: Process Intensification" and "H_2 Production and CO_2 Separation" new emerging technologies are described namely the use of dolomite and CO_2 removal for H_2 recovery through membranes.

3.1 Carbon Capture and Storage

Climate change and alternative energy sources are nowadays a great need. It is a reality that CO_2 is one of the most important greenhouse gases (GHG). Nevertheless, it is also true that many current industrial processes, not only for energy production, but also to release CO_2. Virtually all industries emit (directly or indirectly) CO_2. Typical examples are power plants via combustion or gasification of different fuels (fossil such as coal, oil or natural gas; or renewables such as biomass which could be considered as a CO_2 neutral source due to its relatively small time of formation), refineries; iron and steel industries; oil and gas extraction; cement production; petrochemicals production and others such as paper mills, or fertilizers industries. On the one hand, even if there are alternatives to power production from fossil fuels, a complete substitution is still a challenge, and the dependence from oil, natural gas and coal is and will be an issue in the developed and developing countries. On the other hand, the industrial processes that have inherent CO_2 emissions may have no innocuous alternatives. For these reasons, nowadays there is an increasing interest on carbon capture and storage (CCS) processes that aim at liquefying the CO_2 stream before its release to the atmosphere, its transport to a reservoir, and a final geological storage. In this context, a CCS approach for gasification plants is an option to diminish global CO_2 emissions. In order to implement such a solution, it is necessary to have an integrated approach considering the whole supply chain. It means that a CCS process in a factory, which can be done with existing and well proved technology in the field of gas purification, should be directly linked with the localisation of the possible geological reservoir for the captured GHG. It also has to consider the different CO_2 transportation network possibilities, by pipelines or by boats (similar to the ones used in natural gas).

Besides the technical and logistic aspects, also the public acceptance is important to be considered together with the requirements on legal and politic developments which altogether have a key role to implement CCS as a climate change solution. For example, to decide the obligatory nature of the CO_2 capture measure and the purity of the CO_2 to be injected, the subject has to be extensively discussed and assessed from legal and technological points of view. Specifically, for power plants, the implementation of a carbon capture technology penalises the global efficiency of the plant in around a 7% based on the LHV [15].

Fig. 2 Post-combustion carbon capture configuration

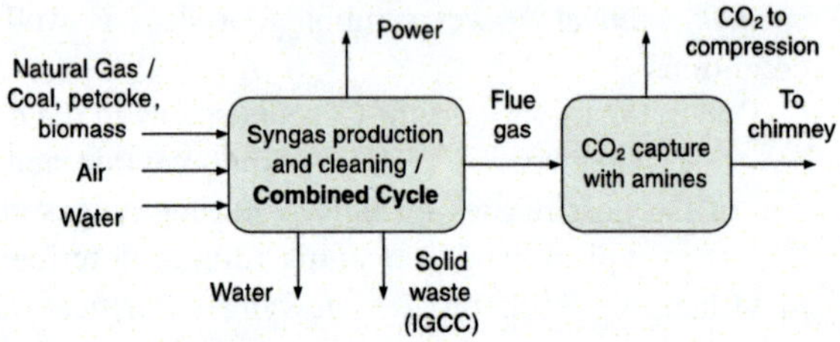

The partial pressure of the CO_2 in the gas mixture is a key parameter that represents the CO_2 concentration and it is crucial to know how difficult CO_2 capture will be. In general, the higher the CO_2 partial pressure is, the easier to separate it from other species will be. In syngas power plants, the difference between syngas CO_2 partial pressure before and after the gas turbine is complex to assess directly, but in general, higher partial pressures are found before the gas turbine. The GT CO_2 partial pressure in the outlet is lower, besides the CO_2 generation due to syngas combustion in the GT the mixture with air dilutes the gas mixture and it is also lowered due to the flue gas expansion in the turbine to atmospheric pressure. Thus, a noticeable difference exists when considering pre or post-combustion carbon capture techniques. Other CO_2 capture principles are based on oxy-combustion, metal oxidation and its capture using membranes. This last one is described in "H_2 Production and CO_2 Separation".

Pre and post-combustion are the typical candidates for syngas power plants [16]. Usually, chemical solvent processes are used for CO_2 partial pressures below around 15 bar. Physical solvent processes are applicable to gas streams which have high CO_2 partial pressure and/or high total pressure. *Post-combustion* techniques are represented mainly by chemical absorption, where amines play an important role. The outlet CO_2 stream is treated, compressed and liquefied to be prepared for transport to its final disposal location. As shown in Fig. 2, after the gas turbine combustion, in our case, after syngas production and use, the flue gas is treated.

Pre-combustion, as Fig. 3 shows, requires a reactor, which produces CO_2 and downstream a process for CO_2 capture that separates CO_2 and H_2. This hydrogen is then sent to the CC to produce power. And analogously to the post-combustion scenario, CO_2 is sent to a compression system to be liquefied for its transport. The pure hydrogen obtained could be sold to the market, or sent to a fuel cell application which will require further purification, for instance, by using a PSA system. The pre-combustion technique counts with a physical solvent that absorbs acid compounds. Consequently and by means of process intensification, the work of Huang et al. [17] evaluates the same absorption process for both, CO_2 and H_2S abatement, finding that the main drawback is the effect of sulphur compounds in the WGS reactor.

In general, for oxygen blown gasifiers at high operating pressures and relatively high CO_2 concentrations, the physical solvent absorption system is the predominantly used technology. According to Metz et al. [16], the most extended technology to capture CO_2 before the gas turbine combustion is the Selexol process

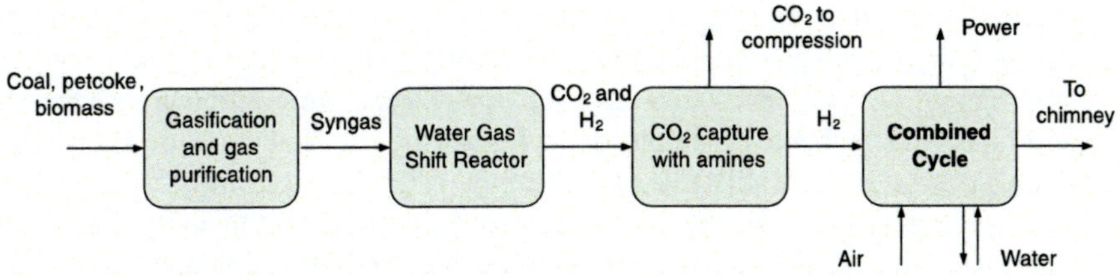

Fig. 3 Pre-combustion carbon capture configuration

that uses dimethyl ether of polyethylene glycol (dimethyl ether of PEG, the key ingredient of Selexol) as solvent, achieving a CO_2 capture efficiency of more than 90%. The optimum pressure for hydrogen purification is in the interval of 15–30 bar. The hydrogen concentration in the outlet stream of a modern PSA unit usually lies between 80 and 92%. PSA is mainly based on the adsorption behaviour difference between molecules. There exists a gap between the knowledge and the suitability of the already known process, with practical projects in the field in an integrated way. Nevertheless, some research has been already developed around this subject, specifically around carbon capture in gasification plants:

Several works can be found in the field of CCS applied to power plants. Desideri and Paolucci [18] is one of the first works developed in the field and in the framework of modelling. They model a carbon capture technology for flue gas from conventional power plants cleaning. The post-combustion configuration is consequently modelled, using Aspen Plus© software. Their approach contemplates the exhaustive description of the system, model validation with literature data, the whole plant performance evaluation and a cost analysis. It allows for optimisation when input characteristics change. It is concluded that a 90% of CO_2 emissions can be reduced, but capital costs are significant and penalise the final COE (cost of electricity). The work of Hamelinck and Faaij [19] is based on biomass gasification and on production of methanol, hydrogen and electricity. This last is produced taking profit of the remaining gases after methanol or hydrogen production units. They have a relatively low LHV if compared with fossil fuels, but they offer the possibility of being self-sustained in electricity consumption. The considered process steps are pre-treatment, gasification, gas cleaning, reforming of higher hydrocarbons, a shift step to obtain proper H_2/CO ratios and the final gas separation for hydrogen production or methanol synthesis and purification. The used software is again Aspen Plus©. The main purpose of the work is to identify biomass to methanol and hydrogen conversion concepts that may drive to higher efficiencies with lower costs. The work of Kanniche and Bouallou [20] takes into account an IGCC power plant with CCS technology in pre-combustion configuration, fuelled with coal. They perform a scenarios evaluation considering different physical and chemical solvents and comparing them in technical and economic terms. Aspen Plus© is again the chosen simulation tool. The premise they follow is to be as realistic as possible, avoiding big modifications of an already existing IGCC power plant, consequently conserving as much as possible the existing operating conditions. The conclusion of the article states that physical processes, concretely Selexol and Rectisol ones, and activated amines have lower thermal consumption (mainly in the desorption column). Capturing CO_2 leads to a 24% of power generation efficiency reduction. Therefore, CCS technology should be included carefully integrated in the already existing power plant. The work of Descamps et al. [21] carefully describes the Rectisol process, with methanol as solvent, for CO_2 abatement in a pre-combustion configuration in an IGCC power plant. Previously to the absorption process, the CO_2 removal train counts with a series of WGS reactors: three reactors to obtain a high CO conversion rate. The necessary steam is produced considering integration with the CC. The performed sensitivity analyses demonstrate that CO conversion is related with the amount of water, concretely in a way that a H_2O/CO mole relation of 1 in the first reactor optimises the conversion. The final conversion achieved is around 92% in mole basis. The CO_2 absorption rate is varied between 77 and 98% in mole basis. Higher rates imply a slight increase of gas turbine power, and a slight decrease in the steam turbine. The validation with former studies is difficult, since there exist a lack of details concerning specific operating conditions and integration heat and mass streams. As it is derived from the already cited references, the work is mainly focused on optimisation and design of power plants counting with CO_2 capture systems. It seems that few real experiences can be cited in this domain, which are mainly blocked by the power efficiency loss. Following the main tendency, Chen and Rubin [22] develop an integrated platform to evaluate CCS costs and performance for IGCC power plants. Their base model counts with a Selexol system and in general with modelled units based on commercial units, considering no integration between the ASU and the CC. Their WGS reactors system counts with two stages (one for syngas steam consumption, and the other one for external steam supply), and the Selexol unit uses the process intensification philosophy by also including two stages for sulphur and carbon removal separately. They observe that a redesign of the heat integration system of the plant should be done when adding new units. As a consequence, even if one of the premises is to maintain as much as possible the already existing conditions of a plant, the most optimal way to proceed is recalculating and redesigning the possibilities of streams integration. A probabilistic uncertainty analysis demonstrates that most of the uncertainty in costs estimation comes from the plant itself rather than from the carbon capture system. Design optimisation process is also seen in Biagini et al. [23] work. It considers different biomass conversion processes to produce hydrogen: gasification and combustion, with pre and post-combustion configuration, at small scale. Sensitivity analyses are performed taken into account the most influencing parameters: the equivalence ratio in the case of gasification, steam addition and moisture content in the biomass.

4 Modelling of Syngas Cleaning Units

This chapter then deals with the synthesis gas cleaning, necessary to produce electricity and hydrogen in an IGCC plant. The pursued objective in syngas application to power production is to clean the gas before its combustion in a GT, mainly to avoid as much as possible nitrogen and sulphur emissions. A gas cleaning train typically involve steps that are discussed in Sect. 3. In a first step, the gas is cleaned from solids in a *ceramic filter*. Then, in the *venturi scrubber*, syngas is placed in contact with a water stream that absorbs and removes cyanide, halide, acid (mainly H_2S) and basic (mainly NH_3) pollutants. Polluted water is treated in *sour water stripper* and recycled back to the scrubber which closes a water loop and decreases the overall plant-wide water consumption. The sour water stripping unit needs to be purged due to the build up of pollutants. The purged water is then treated and disposed off. Syngas is further purified from the acid species through the *COS hydrolysis reactor*. This unit converts COS into H_2S, which is removed in a *MDEA absorber*. Polluted gas streams from the stripping unit, COS hydrolysis reactor and from the MDEA absorber containing high amounts of H_2S are sent to a Claus plant, where sulphur is recovered in liquid form. The clean gas obtained, after the MDEA absorber, is sent to the GT for power production.

In the case of hydrogen generation application, the pursued objective is to separate CO from H_2, the deprived hydrogen stream could be sent to the GT, or the high content H_2 stream could be further purified and sold as pure hydrogen. Considering CCS possibilities, Cormos et al. [24], point out that pre-combustion CO_2 capture method is more suitable for gasification process than post-combustion capture (lower energy penalty, possibility to co-generate power and hydrogen, higher degree of plant flexibility, etc.). In this sense an IGCC power plant is suitable for a CO_2 pre-combustion capture method, given that the high pressure present along the cleaning flowsheet could be profited to separate the CO_2 with a physical absorption method (Kanniche et al. [25]).

Due to the appearance of non-ideal behaviour of the liquid phase due to the occurrence of pH changes related to the speciation of dissolved gasses in water or other solvents, the physical property method chosen to calculate thermodynamic and transport properties of the streams should be based on an activity coefficient model. Several different possibilities are available such as Wilson, UNIQUAC or non-random two liquid (NRTL) (see [26]). However due to the consideration of gas species solvation and subsequent electrolytes formation, an extension of the NRTL model called ELECNRTL is suitable. This thermodynamic model selection allows modelling unit operations where electrolyte presence is notorious with ELECNRTL, while the remaining with NRTL. Moreover, for other cleaning units that do not require or consider the appearance of a liquid phase the Peng-Robinson EOS is used, as recommended in Aspen Plus© for hydrocarbon processing applications such as gas processing, between others.

Separation of vapour and liquid phases during equilibrium (VLE) is considered here as the main separation method for *absorption processes*, therefore considering the *Henry's law* together with the ELECTRNL and NRTL property methods for this gas–liquid interaction. Henry's law states that at a constant temperature the partial pressure of species i in a volume of gas, in equilibrium with a liquid, is directly proportional to the species mole fraction in the liquid phase (see Eq. 1, for an ideal behaviour and for a specific solvent).

$$p_i = y_i \times P = x_i \times H_i(T) \tag{1}$$

In Eq. 1, p_i is the partial pressure of i in the gas phase, y_i is the mole fraction of i in the gas phase, P is the total pressure, x_i is the mole fraction of i in the liquid phase and $H_i(T)$ is the Henry's law constant for i. The Henry's law constant expressed like in Eq. 1 has units of pressure. If we consider a non-ideal behaviour, the *fugacity* (f_i) describes better the behaviour of gas species partial pressure in the gas phase. It implies the substitution of the P term in Eq. 1, for the fugacity itself, usually determined experimentally. ϕ_i is the fugacity coefficient, which has a dimensionless value (see Eq. 2). At its turn, for the liquid phase, this effective ("real") behaviour for the species concentration is given by the *activity* (a_i). In an analogous way as in the gas phase, $\gamma_i$is the activity coefficient, and relates the activity with the mole fraction (see Eq. 3).

$$p_i = y_i \times \frac{f_i}{\phi_i} = x_i \times H_i(T) \tag{2}$$

$$p_i = y_i \times \frac{f_i}{\phi_i} = \frac{a_i}{\gamma_i} \times H_i(T) \tag{3}$$

Thus, an expression like (3) is the one that uses Aspen Plus©, the pair of binary interaction between species and solvent is proposed, and the Henry's constants are available from the Aspen Plus physical property system databanks. $H_i(T)$ is calculated as shown in Eq. 4. The Henry's constants for a specific solvent $(a_i, b_i, c_i, d_i, e_i)$, are explicit for the binary interactions of the species with water and methanol in their respective cases, and are summarised in Table 1. The expressions are suitable for a specific range of temperatures, T_{lower} (T_{L}), and T_{upper} (T_{U}), AspenTech [27].

$$\ln H_i(T) = a_i + \frac{b_i}{T} + c_i \times \ln T + d_i \times T + \frac{e_i}{T^2} \tag{4}$$

ELECTRNL and NRTL property methods are different due to the characteristic of handling with electrolytes. The ions formed can be estimated by Aspen Plus© using the Electrolytes Wizard command in the selection of components sections, when all the components have been already introduced. This command also generates the appropriate chemical equilibrium reactions for electrolyte appearance. NRTL model can describe VLE and LLE of non-ideal solutions.

The *RadFrac model* from Aspen Plus© is used to model all the absorption units except the COS hydrolyser, the Claus plant, the WGS reactor, the PSA and the liquefactor. This unit is based on the principle that different phases have different compositions at *equilibrium*, therefore using the Henry's law together with ELECNRTL each stage phase separation is estimated. The main stream is added to the RadFrac column, while a separating agent is also added to achieve separation; it can be as the form of energy or matter. Generally, this energy is the reboiler heat used to recirculate the bottoms of the column, or/and an absorbent as a matter. The *separation factor* (α_{ij}) evaluates the degree of separation of species between

Table 1 Henry's constants ($H_i(T)$) for each binary mixture, with water and methanol as solvents, in N m^{-2}, T in K [27]

	a_i	b_i	c_i	d_i	e_i	T_L	T_U
Solvent: H_2O							
H_2	152.2	−5,312.5	−20.3	1.3×10^{-2}	0	273.15	344.85
CO	568.8	−17,742	−91.8	1.2×10^{-1}	0	273.15	333.15
CO_2	192.7	−8,982	−25.8	1.2×10^{-1}	0	273.15	347.85
H_2S	143.9	−8,281.7	−16.5	-1.4×10^{-1}	0	273.15	333.15
COS	232.7	−12,025	−30.4	0	0	273	303
NH_3	93.27	−1,096.8	−16.56	0.6×10^{-1}	0	273	373
HCl	−36	1,215	8.4	-9.5×10^{-3}	0	253.15	293.15
HCN	53.80	−8,136.80	0	4.5×10^{-2}	0	283	383
Solvent: CH_3OH							
H_2	−49.9	1,867.4	12.6	-2.7×10^{-2}	0	213.15	343
CO	15.7	1,144.4	0	0	0	293.15	298.15
CO_2	27	−3,426.7	1.5	-2.5×10^{-2}	0	273.15	293.15
H_2S	22	−2,050.8	0	0	0	263.15	298.15
COS	323.7	−12,025	−30.4	0	0	273	303
NH_3	18.5	−1,669.4	0	0	0	273.15	301.55
HCl	−36	1,643.8	7.5	0	0	275.25	307.35

the two matters in equilibrium, gas and liquid in our case in each stage. In our approach we are taking into account the whole column separation effect, and we have defined the *split fraction* as the quotient between the flow of one specific species in the gas (head stream) and the inlet amount of the same component.

This equilibrium approach, whose purpose is to absorb species, has to be seen as a source (the syngas) or feed that transfer components into a receiving phase. One common equilibrium attribute is the *saturation* of a phase; it implies that the reactant in a chemical absorption, as the receiving phase, has a limited capacity [10, 27]. See in Eq. 5 the expression of the equilibrium constant (K_{eq}), analogous to the expression of H_i. Therefore, as the column has a specific temperature profile, the equilibrium constant has different values for each one of the reactions along the column.

$$\ln K_{eq} = A + \frac{B}{T} + C \times \ln T + D \times T \tag{5}$$

In the following sections we are dealing with the detailed model of each syngas cleaning unit, describing the used unit, the equations involved, as well as the specific input conditions, following the ELCOGAS IGCC power plant operation conditions. Following the *hypotheses* already mentioned in "Modelling Syngas Generation", in this modelling approach we are assuming that no char either tar are produced. In the ceramic filter all the solids are assumed to be removed. We are also considering that the syngas obtained in the PRENFLO gasifier has similar characteristics between different mixtures of coal, petcoke and biomass thus, not changing the operating conditions of cleaning units.

4.1 Cleaning Unit Remarks

In general to grasp the results of a plant section different metrics that represent their overall behaviour are required. These metrics can range from simple mass flows to complex relationships between species concentrations in different streams. In chemical plants where treatment of a given stream is required, usually these metrics are related to concentration of inlet and outlet streams or to ratios of moles of specific species. In general for a given block that has one inlet A and two outlets B and C the following ratios and split fractions can be defined for given specie i, as follows.

$$\text{split}_i = \frac{m_i^B}{m_i^A} \tag{6}$$

$$\text{recovery}_i = \frac{m_i^B - m_i^C}{m_i^A} \tag{7}$$

where m_i^X, represents the mole or mass flow of component i in stream X. These recovery fraction and split fraction are used to calibrate and validate a model.

Table 2 Equilibrium constants for the ionisation equations present in the venturi scrubber and the MDEA absorber, in mole fraction basis. K_{eq} units depend on the stoichiometry of the reaction, T in K [27]

Equation	A	B	C	D
8	132.9	−13,445.9	−22.5	0
9	−1.26	−3,335.7	1.5	-3.7×10^{-3}
11	231.5	−12,092.1	−36.8	0
12	216.1	−12,431.7	−35.5	0
13	214.6	−12,995.4	−33.6	0
14	−9.8	−8,585.5	0	0
15	22.9	−9,945.5	0	-5×10^{-2}
20	−9.4	−4,235	0	0

4.2 Venturi Scrubber

The main objective of the venturi scrubber is the *abatement of* NH_3, *cyanide and halide species*, although some H_2S is also absorbed. This syngas cleaning unit is placed after the gasifier and the candle filter. The species transfer occurs taking into account the VLE, following the Henry's law, and considering the electrolyte reactions that take place in the liquid phase, therefore the property package used is ELECTRNL, using a true species approach that reports ionic species composition. A solution of NaOH 15 wt% in is used as acid capturer. The main Henry components are NH_3, CO_2, and H_2S (see in Table 1 the Henry's constants for each species and water). Indeed, the ionisation reactions are represented in Eqs. from 8 to 16. See in Table 2 the K_{eq} values for the ionisation equations.

$$2H_2O \longleftrightarrow H_3O^+ + OH^- \tag{8}$$

$$NH_3 + H_2O \longleftrightarrow NH_4^+ + OH^- \tag{9}$$

$$HCl + H_2O \longrightarrow H_3O^+ + Cl^- \tag{10}$$

$$CO_2 + 2H_2O \longleftrightarrow H_3O^+ + HCO_3^- \tag{11}$$

$$HCO_3^- + H_2O \longleftrightarrow H_3O^+ + CO_3^{2-} \tag{12}$$

$$H_2S + H_2O \longleftrightarrow H_3O^+ + HS^- \tag{13}$$

$$HS^- + H_2O \longleftrightarrow H_3O^+ + S^{2-} \tag{14}$$

$$HCN + H_2O \longleftrightarrow H_3O^+ + CN^- \tag{15}$$

$$NaOH \longrightarrow Na^+ + OH^- \tag{16}$$

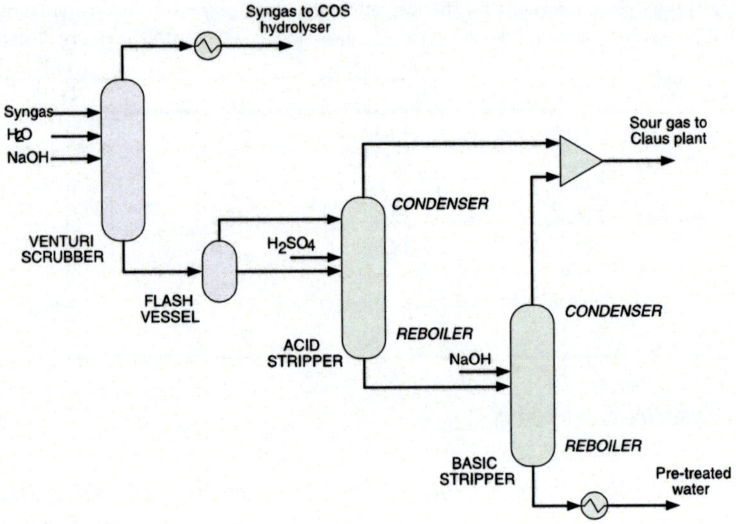

Fig. 4 Venturi scrubber and sour water stripper

The RadFrac model, which represents the venturi scrubber behaviour, works at a constant pressure of 23.6 bar. As the syngas arrives at 24.8 bar, it is assumed a pressure drop of 1.2 bar. The syngas temperature is around 235°C, as a consequence of its usage in the WHB, see "Modelling Syngas Generation". Water and NaOH flowrates are related to the syngas flowrate that enters the venturi scrubber, thus with the gasifier load. The feeds relations have been included into the model by means of two FORTRAN blocks (two calculator blocks) that establish the amount of NaOH in a 0.12% the flowrate of syngas in mass basis, and the flowrate of water in a 9% the same flowrate and the same basis, according to ELCOGAS experience. The contaminated water goes to the sour water stripping system that works at approximately 1.5 bar, to be pre-treated before its final disposal. For this reason, water is depressurised in a flash vessel before entering the first stripping system column. This flash is considered to work at 1.5 bar and 53°C, and it is modelled as a 2 phase flash model. The cleaned gas after the venturi scrubber process has to be cooled down till around 140°C, before the COS hydrolyser. See the layout of the venturi scrubber—water stripping system with the units included in Aspen Plus© and the main material streams in Fig. 4.

4.3 Sour Water Stripper

Its main function is to *pre-treat the used water* in the venturi scrubber by desorbing the absorbed species: NH_3, HCl, CO_2, H_2S and HCN. It is formed by two main units: an acid stripper that abates acid species, and a basic stripper that abates basic species. The species transfer occurs between water and steam that is injected to the column through the column's bottoms. H_2SO_4 and NaOH solutions, of 96 and 15% of purity in weight, regulate the pH in the two columns, thus achieving the selective acid and basic absorption. Analogously as in the venturi scrubber, the species transfer takes place through VLE following the Henry's law, to model such behaviour the ELECTRNL property package is used, with true species approach. The electrolyte reactions are the same than the previous ones (see in Table 1 the Henry's constants for each binary mixture), Eqs. from 8 to 16, plus Eqs. 17 and 18, related with the sulphuric acid dissociation.

$$H_2SO_4 + H_2O \longrightarrow H_3O^+ + HSO_4^- \tag{17}$$

$$HSO_4^- + H_2O \longrightarrow H_3O^+ + SO_4^{2-} \tag{18}$$

The sour water stripper has been modelled using two RadFrac units with condenser and reboiler, working at 1.5 bar and assuming no pressure drop along the column. Solvents mass flowrate are proportional to the water mass flowrate to be treated based on ELCOGAS experience. The H_2SO_4 solution is fixed as a 0.3% of the water in flow; while the NaOH has a proportional factor value of 1.2%. As in the previous unit, these proportionality conditions have been introduced in Aspen Plus© using calculator blocks. After the stripping process two main streams are obtained: the treated water that goes to the final treatment unit and a sour gas that goes to the Claus plant to recover sulphur. The final water temperature is around 40°C; and the sour gas goes to the next cleaning unit at around 105°C. See in Fig. 4 the layout the process as modelled in Aspen Plus©.

4.4 COS Hydrolysis and Amines Absorption

The COS hydrolyser and the MDEA (methyldiethanol amine) units represent the syngas' *sulphur removal* step. Firstly, it is necessary to transform the COS, which is the other sulphur compound formed during gasification apart from H_2S, into H_2S, since the amines processes are effective for this second compound.

The COS reaction with water, see Eq. 19, takes place in a catalysed bed, which in this case is modelled as a stoichiometric reactor from Aspen Plus©, with a COS fractional conversion of 0.99, adjusted to a previous kinetic model, described elsewhere [28]. A pressure drop of 1.6 bar is assumed in this reactor. Peng-Robinson property package is used here.

$$COS + H_2O \longrightarrow H_2S + CO_2 \tag{19}$$

Amines absorption, in this case an MDEA-water solution, acts as the chemical absorber for H_2S at high pressure and low temperature (22 bar and 33°C). Absorption and desorption columns are simulated, with lean stream recirculation. The MDEA solvent is a 50% water solution in mass basis. The polluted MDEA solution stream is decompressed before entering the desorption column that works at 1.5 bar and relies on a condenser and a reboiler. As it is recycled, it must be re-compressed. Also temperature is conditioned before expansion and compression. As in the venturi scrubber and the sour water stripper, ELECTRNL property package is used, with true species approach, and with the Henry's law for the species absorption calculation. The Henry components considered here are the same than in the previous units (see in Table 1 the Henry's constants for each binary mixture). Moreover Eq. 20 is added to the list of Eqs. 8–15. See in Table 2, its K_{eq} expression value.

$$MDEA^{+} + H_2O \longleftrightarrow MDEA + H_3O^{+} \quad (20)$$

The absorption and desorption columns are modelled with RadFrac units. No pressure drop is taken into account. The amount of MDEA solution that is recycled has been calculated through sensitivity analysis for the base case based on ELCOGAS results. Due to MDEA losses through the gasses streams and liquid entrainment, the net consumption of MDEA is around 3 kg h^{-1}. Two streams result from the process: the clean gas from the absorber that goes to the gas turbine, and the Claus gas from the desorber that goes to the Claus plant. The clean gas has about 60°C, and is heated till around 150°C before the CC with the heat excess from the COS outlet stream cooling, assuming a 0.7 bar of pressure loss. A recycle stream comes from the Claus plant process, and is added to the clean gas that goes out of the MDEA absorber. Between the COS hydrolyser and the absorber there is a small fraction of clean gas that goes to a hydrogenation reactor in the Claus plant. See the process layout as modelled in Aspen Plus© in Fig. 5.

4.5 Claus Plant

The Claus plant is the plant section that is responsible for *sulphur recovery*. Its main purpose is to convert H_2S into elemental sulphur. Second, it aims at NH_3 and HCN conversion into nitrogen. The process relies on two parallel kilns and two series reactors catalyzed by alumina, called Claus reactors. Sulphur is recovered after each step with flash separators. Finally, the gas with a small amount of acid and basic compounds goes to a hydrogenation reactor, also catalysed, to increase the overall sulphur recovery by means of the residual gas recycling, thus giving another chance to the remaining sulphur in the gas to be recovered. The property package used is Peng-Robinson, and both thermal and catalytic stages are modelled by taking into account the degree of advance of the involved reactions, deduced from a previous kinetic model (see [28]). The hydrogenation reactor considers the conversion of any remaining S or SO_2 in the gas, into H_2S. Therefore, hydrogen is needed for this transformation. It has been modelled considering the equilibrium chemical compounds composition. See main equations of the kilns and the catalysed reactors in Eqs. 21–24. The hydrogenation step can be described by Eq. 25 and 26.

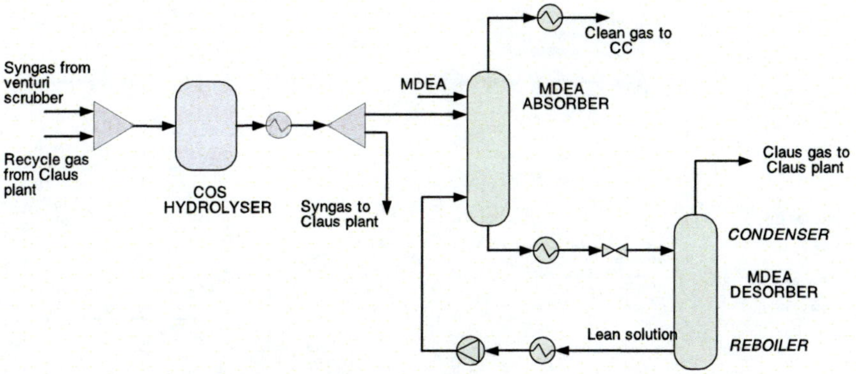

Fig. 5 COS hydrolyser and MDEA plant

$$H_2S + 1.5O_2 \longrightarrow SO_3 + H_2O \tag{21}$$

$$NH_3 \longrightarrow 0.5N_2 + 1.5H_2 \tag{22}$$

$$HCN + H_2O \longrightarrow 1.5H_2 + 0.5N_2 + CO \tag{23}$$

$$2H_2S + SO_2 \longrightarrow 3S + 2H_2O \tag{24}$$

$$S_2 + 2H_2 \longleftrightarrow 2H_2S \tag{25}$$

$$SO_2 + 3H_2 \longleftrightarrow H_2S + 2H_2O \tag{26}$$

Thermal and catalytic stages have been modelled in Aspen Plus© using three Stoichiometric reactors. The thermal stage has been considered as one unique step that considers Eqs. 21–24. It receives two residual gas streams from previous cleaning units: the sour gas from the sour water stripper and the stripped gas from the desorption column in the MDEA absorption process. Both polluted streams go to the kiln at 1.5 bar. Another required input to the kiln is a stream of air, which is necessary for the conversion of H_2S into SO_2. The necessary amount of air has been established as a fraction of the necessary stoichiometric air to burn the amount of H_2S present in the gas mixture. Thus, the property set COMB-O2 from Aspen Plus© has been used to calculate such value and is set in the simulation by means of a calculator block. For modelling purposes, the gas stream has been divided into H_2S on the one hand and the rest of species on the other (see Fig. 6), considering combustion at 1,000°C, modelling this last stream combustion by means of a Gibbs reactor. Then, the mixture of the two streams resulting from combustion is sent to the next step. The catalytic stage, with two stoichiometric reactors, contemplates Eq. 24 at two different temperatures, 270 and 210°C. Before it, a cooling step with its subsequent sulphur recovery step is needed. In an analogous way, after each catalyzed reactor there exists a flash reactor to recover

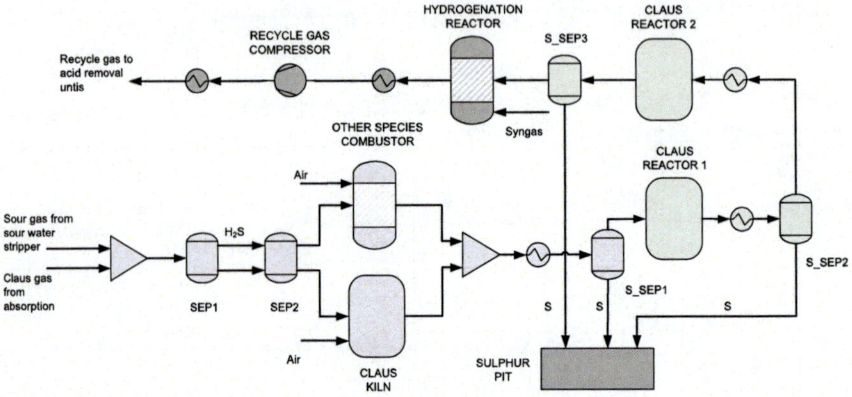

Fig. 6 Claus plant

the obtained liquid sulphur. See in Table 3 a compilation of the conversion factors considered for each reaction and its reference component, in all cases the reactors have been considered to operate under adiabatic conditions.

Finally, the remaining gas is fed to the hydrogenation reactor, where a small portion of clean gas, a 0.5% in mass basis from the outlet stream of the COS reactor, is used to transform the not recovered sulphur, and recycle the resultant to the gas cleaning process before the COS hydrolysis reactor, for H_2S removal. This step is simulated using a Gibbs reactor. Before recycling, it is necessary a temperature and pressure conditioning step: the resultant gas stream must be compressed till 23.6 bar and heated till 141°C, which are the conditions of the inlet stream to the sulphur removal step. See in Fig. 6 a schema of the Claus plant as modelled in Aspen Plus©, only including the material streams.

4.6 WGS and CO_2 Capture

As generally explained in Sect. 3.1, the chosen actuation way for the IGCC power plant to remove CO_2 thus, to obtain H_2 is the *pre-combustion configuration*. The first step before CO_2 capture is the conversion of CO into H_2 and CO_2. It is achieved with the *WGS reaction,* Eq. 27, in this case, in two reactors in series.

$$CO + H_2O \longleftrightarrow CO_2 + H_2 \tag{27}$$

This shifting step follows the configuration found in Chiesa et al. [29]. The transformation takes place in two adiabatic reactors, considering Eq. 27 in equilibrium at two different temperatures. The chosen temperatures have been 400°C for the first reactor and 210°C for the second one. Downstream the first reactor, the gas stream is cooled down to be adapted to the conditions of the second reactor,

Table 3 Calibration parameters in the different syngas cleaning units

Unit	Variable	Basis	Reference	Parameter
Venturi	H_2O consumption	Mass	Syngas flowrate	9%
	NaOH consumption	Mass	Syngas flowrate	0.12%
	Stages	Number	–	10
Stripper	H_2SO_4 consumption	Mass	H_2O flowrate	0.3%
	Acid stripper stages	Number	–	5
	NaOH consumption	Mass	H_2O flowrate	1.2%
	Basic stripper stages	Number	–	8
Hydrolyser	COS	Conversion	Eq. 19	0.99
Absorber	MDEA solution flow	Mass	–	118,000 kg h^{-1}
	Absorber stages	Number	–	5
	Desorber stages	Number	–	12
	Desorber reflux ratio	Mass	–	1.9
	Desorber boilup ratio	Mass	–	0.3
Claus plant	O_2	Conversion	Eq. 21	1
Kiln	NH_3	Conversion	Eq. 22	0.96
	HCN	Conversion	Eq. 23	0.96
	H_2S	Conversion	Eq. 24	0.50
	Air consumption	Mole	H_2S flowrate	0.33 stoich.
Claus 1	SO_2	Conversion	Eq. 24	0.80
Claus 2	SO_2	Conversion	Eq. 24	0.60
WGS reactor	β	Mole	–	1.25
Rectisol process	Methanol flow	Mass	CO_2 flowrate	4.5
	Absorber stages	Number	–	8
	Flash vessel 1	Pressure	–	10 bar
	Flash vessel 2	Pressure	–	2.7 bar
	Flash vessel 3	Pressure	–	1.4 bar
	Desorber stages	Number	–	15
	Desorber reflux ratio	Mass	–	5
	Desorber boilup ratio	Mass	–	0.3
PSA Bed 1	CO_2 and H_2O flows	Separation	H_2 flow	0
Bed 2	H_2 flow	Separation	H_2 flow	0.84

while downstream the second reactor, the product stream is again cooled down to be adapted to the acid gas removal conditions (see Fig. 7). The assumed pressure drop is about 0.4 bar into each reactor. The high temperature proposed for the first reactor is obtained using intermediate pressure (IP) steam. The amount of steam to be introduced has been fixed according to the data found in Descamps et al. [21], and corresponds to a value of $\beta = 1.25$, being β defined as the steam to CO ratio in mole basis, in the obtained mixture. Two reactors are used in series due to the highly exothermic conditions, which require of intermediate cooling. Due to this characteristic, the desired products are favoured when lower temperatures are considered, but reaction rates are slow then, consequently intermediate cooling allows to take advantage of the heat released during the WGS reaction and to enhance the production of H_2. The step is modelled in Aspen Plus© with two

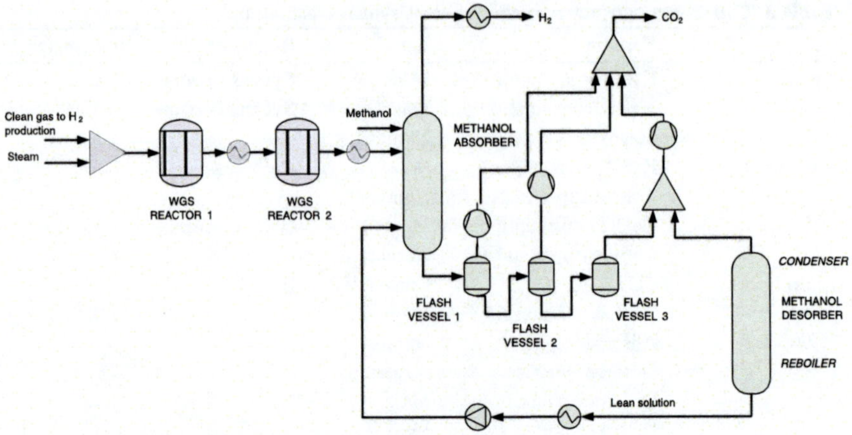

Fig. 7 WGS reactors and carbon capture with Rectisol process

Equilibrium Reactors that take into account the abovementioned equation and determine products by specifying temperature approach to equilibrium, in this case assumed as 0 degrees. The Peng-Robinson property package is used here.

Acid species removal is done using physical absorption through the *Rectisol process* that uses pure methanol as absorption solvent. The physical absorber process follows the same principles as the described for the MDEA, however no ionisation occurs. Consequently, the property package used is NRTL. VLE follows again the Henry's law, and as in previous units, the value of Henry's constants for each pair of involved species is reported in Table 1 (considering the solubility of the gas specie in methanol). In contrast to the MDEA process, the solvent expansion required before the desorption column is done here by means of three pressure drums (which are modelled as three flash vessels in Aspen Plus©). These drums recover the CO_2 on the one hand and also allow for recycling the solvent. These intermediate flash drums pressures have been settled in order to obtain the desired methanol purity. The high concentration H_2 stream is obtained from the top of the absorption column. According to Hamelinck and Faaij [19], a pressure drop of 0.5 bar and of 0.2 bar are customarily considered in the absorption and regeneration columns, respectively. The final pressure of the H_2 stream is 21.7 bar, and that of CO_2 is 10 bar. Please note in Fig. 7 that after the flash vessels, the CO_2 stream is compressed. The absorption column works at low temperatures around −30°C, that is why after the second WGS reactor the shifted gas is cooled down from approximately 320°C till 200°C. The desorption column works at 1.2 bar, and counts with a condenser and a reboiler. The amount of methanol required is proportional to the mass flowrate of CO_2 contained in the WGS product shifted stream; this value is introduced through a calculator block; to obtain the desired purity. In this sense the methanol mass flowrate has been fixed as a 4.5 times the mass flowrate of the CO_2. The methanol flowrate has been determined according to the final purity conditions. The Rectisol plant layout follows the configuration and information of Descamps et al. [21]. The two main streams leaving this flowsheet part have other species that are important to mention: the H_2 stream has a small fraction of N_2, while the CO_2 rich streams count with a small fraction of syngas compounds. Any improvement in concentration of both streams implies more power use and a plant-wide efficiency penalisation. At this moment, we have considered this approach as representative enough at this conceptual modelling stage. The H_2 purity has been fixed in an 80 mol%, and the CO_2 purity in a 70 mol%.

Hydrogen stream, if it is not feed to the CC, should be treated to increase its purity. It is performed by means of a PSA system, as in ELCOGAS power plant (further explained in "Examples of Industrial Applications"). Independently to the application, after the Rectisol absorption column, the stream is heated till 40°C. If the stream goes to the CC it is heated till 150°C, to be around the same temperature as the syngas after the MDEA absorber. The topology simulated and the working conditions that are commonly used in a PSA system can be obtained from the work of Hamelinck and Faaij [19]. Their system is formed by two reactors that use two different solid beds: the first one uses activated carbon and the second one a zeolites molecular sieve. In PSA systems there exists a trade-off between the number of reactors and the flue gas recycling for enhancing the H_2 recovery. Here the flue gas is not recycled; it can be released, or it can be compressed and sent to the CC. We assume 0.35 bar of pressure loss in each unit. Each PSA column is simply simulated in Aspen Plus© using a component separator. The first one operates separating all CO_2 and all H_2O, while the second one adsorbs the remaining gas species, except for H_2, whose final purity is of 99.99 mol%. The final H_2 pressure is around 0.95 bar.

For the case of CO_2, this stream requires liquefaction to ease its transportation needs. CO_2 should have supercritical conditions to meet storage conditions. The *liquefaction* system considers the work of Desideri and Paolucci [18] and Chiesa et al. [29]. The assumed conditions in the proposed model are as follows: the operating pressure around 150 bar, an isentropic efficiency of 82% for the compressor and of 75% for the pump and a temperature before pipeline transportation around 35°C. Before its compression, water is separated from CO_2 in order to avoid pipeline problems. Thus, before the compressor unit, the stream from the desorption column in the Rectisol system is cooled down from 183 to 25°C, being at 10 bar of pressure. Then, a flash vessel is used to separate the liquid water from the CO_2. At this stage, the following step is to compress and cool down the CO_2 to obtain it in liquefied state, at 150 bar and 30°C. The final CO_2 stream has a purity of 95.5 mol%.

4.7 Calibration and Validation

For cleaning gas calibration, as well as for the gasifier model, real ELCOGAS power plant data, the works of Chiesa et al. [29] and Descamps et al. [21], and

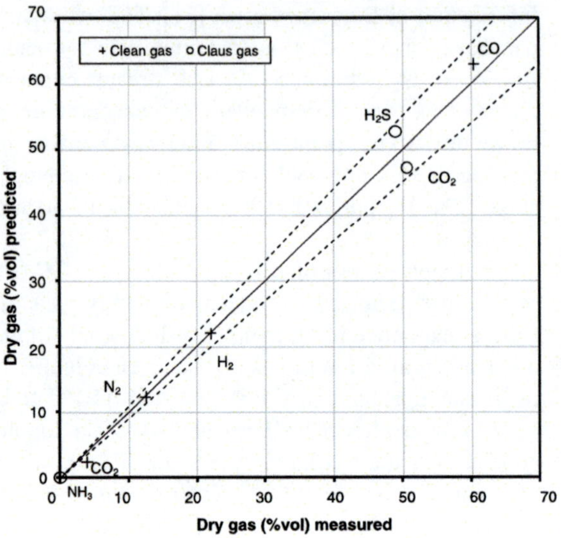

Fig. 8 Model results comparison with industrial data. *Error bar* corresponds to the 10% of difference

exhaustive kinetics models [28] (in the case of the stoichiometric reactors in COS hydrolyser and Claus plant) have been used to adjust the variables of the different models. It means that even if the modelled units are general enough, in order to adjust them to other real installations the parameters shown in Table 3 are specific values suitable for changes, and also for optimisation.

Results validation contemplates the *base case*, which corresponds to gasifier fed with a fuel feedstock formed by a 50–50% of coal and petcoke in mass basis. The syngas treated in these cleaning gas units comes from the base case feedstock, at around 1,400–1,500°C and at 25 bar, as described in "Modelling Syngas Generation". For the gas cleaning train including venturi scrubber, sour water stripper, COS hydrolyser and MDEA absorber, validation can be carried out by comparing ELCOGAS information with model outputs. Concretely, clean gas composition after the absorption process and Claus gas composition after desorption process are of concern. Both streams are from the last cleaning unit applied to the main stream before CC or before further purification to obtain H_2. It can be seen from Fig. 8 that considering the whole flowsheet from feedstock preparation till MDEA absorption, the final gases composition error is within a margin of 10%. Regarding "big number" from outlet flows, Claus plant validation is done by regarding the final sulphur flowrate: in the case of the simulation a value of 3.286 tons h^{-1}. The industrial value is 3.110 tons h^{-1}. In the case of the sour water gas stripping, pre-treated water in the model has a flowrate of 7.277 tons h^{-1}; in ELCOGAS power plant this value corresponds to 7,617 tons h^{-1}. In these two last numbers, the error is 5.7 and 4.5%. All in all, at this stage, the results show a maximum error of 10%, which is better than the range of a preliminary design stage, typically between 15 and 20% according to Wells and Rose [30]. Regarding the carbon capture system, as stated before, the works of Chiesa et al. [29] and Descamps et al. [21] are of concern. The PSA unit is validated in its simplified way of modelling, which already takes into consideration the results of those references. Concerning the

WGS reactors, the work of Descamps et al. [21] is considered, for a value of $\beta = 1.25$, the CO conversion rate obtained is around 92%. In this simulation, a conversion of 91% is obtained. Concerning the Rectisol system, a CO_2 absorption rate of 95% is obtained, while a value of 99% is obtained in the simulation. The CO_2 product has a purity of 99 mol%, the rest being traces of syngas. This purity should be adapted to the reservoir convenience and the legislation parameters. Nevertheless, ELCOGAS states that a purity of the 99% ("Examples of Industrial Applications") should be convenient for the final storage. At this stage of modelling, we assume that a purity of 97% is enough for final disposal.

In chapter "Modelling Superstructure for Conceptual Design of Syngas Generation and Treatment", Sect. 2 a brief introduction to the use of metamodels in process simulation is done, one of the most common approaches to ease simulation requirements is the use of component splitters, where different species splits are entered to mimic the behaviour of more complex models. By looking at each of the former purification unit separately and their complex models involving ion chemistry, the following split fractions, as defined in Sect. 4.1, are found for the specific components of interest. Please note that in the case of the Claus plant, the split fraction is referred to the final H_2S composition in the recycle gas after the hydrogenation step. See Table 4 for a summary of these values. In the first unit, the species NH_3 and HCl are fairly depleted from the gas stream. The stripper cleans water stream from all acid species; nevertheless its efficiency with the basic ones shows worse values. The MDEA absorption process scrubs the syngas from all the species of concern (being the reported stream the Claus gas, see Fig. 5), while half of the CO_2 composition remains in the main stream. In the Claus plant, almost all the sulphur recovered. The remaining NH_3 is recycled; it is transformed into N_2 and H_2 which is also recycled. CO_2 is all recycled, with a split fraction greater than one mainly due to the combustion in the Claus kiln. Finally, Rectisol process recovers all the H_2, and nearly all the present CO_2 in the shifted stream.

Summing up, the proposed syngas cleaning units models have been deeply explained and validated with ELCOGAS power plant data or with reliable data from the literature. The main conclusion here is that the conjunction of the different explained blocks till now; feedstock preparation, gasification and gas cleaning with carbon capture, is well represented through the proposed modelling approaches for each individual unit, and thus, the modelling approach gives an accuracy enough for conceptual modelling.

4.8 Results

Representative data of this IGCC power plant approach with CO_2 capture is the total amount of energy contained in the syngas that goes to the gas turbine, and how the production of H_2 affects this value. In Fig. 9 we have represented this amount of calorific value in MW distributed into each stream: syngas to CC, pure H_2 (after the PSA unit), CO_2 after the compression system and a "residual" gas

Table 4 Split fractions for the species of interest in the in the different syngas cleaning units

Unit	Component	Split fraction
Venturi	NH_3	0.207
	HCl	0
	CO_2	0.999
	H_2S	0.998
	HCN	0.977
Stripper	NH_3	0.188
	HCl	0
	CO_2	1
	H_2S	1
	HCN	1
Absorber	NH_3	0.999
	CO_2	0.503
	H_2S	0.997
	HCN	0.924
Claus plant	NH_3	0.318
	CO_2	1.120
	H_2S	0.029
	HCN	0
Rectisol process	H_2	0.996
	CO_2	0.999

that can be compressed and sent into the CC or emitted to the atmosphere (called here PSA gas), result of the PSA process, thus containing traces of syngas, CO_2 and H_2O. The obtained LHV for each one the streams are, in order, 9.29, 119.96, 0.34, 8.15 MJ/kg.

It is observed that all the streams variation follow a linear behaviour. The slopes have values of 5 for the syngas, −4 for the H_2 stream, −0.2 for the CO_2 stream and −1 for the PSA gas. It is seen that the same variation does not affect the different streams with the same proportion: it is due to the fact that the slope is dependant of the different LHV and of the proportion of mass flowrate that goes into each process, as well as the efficiency of the H_2 purification unit of concern. This last value can be defined as in Eq. 28 for the whole train of units, where *cg* is the clean gas, and H_2 is the final H_2 stream after the PSA process. The inlet calorific value in MW (for the feedstock) is a constant value of 678.42 MW. After the gasification process, with a CGE of 75.6%, the syngas calorific value is of 512.64 MW. That is the value before the gas cleaning processes.

$$\mathrm{Eff}_{\mathrm{CCapture}} = \frac{m_{\mathrm{H_2}} \times \mathrm{LHV}_{\mathrm{H_2}}}{m_{\mathrm{cg}} \times \mathrm{LHV}_{\mathrm{cg}}} \tag{28}$$

If plotting the total calorific value that is finally contained in the outlet streams after the H_2 obtaining process, with the calorific value contained in the syngas, a linear behaviour is again observed, as in Fig. 10. Note that the value correspondent to the 0% is the final energy value contained in all the outlet streams of the carbon removal train; and the value correspondent to the 100% is the value contained in

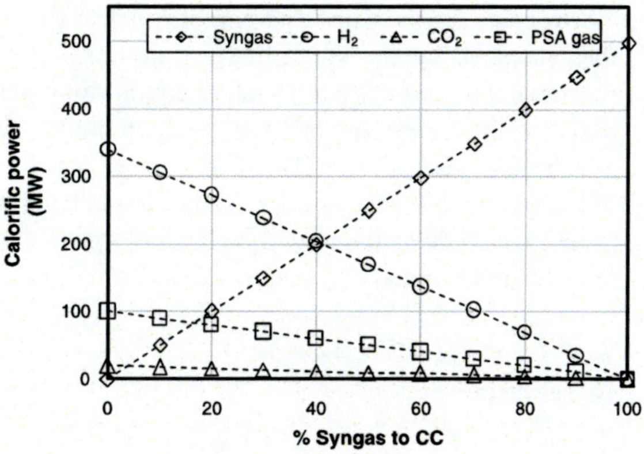

Fig. 9 Calorific value variation of the purification gas unit's outlet streams with the percentage of syngas that goes to the CC

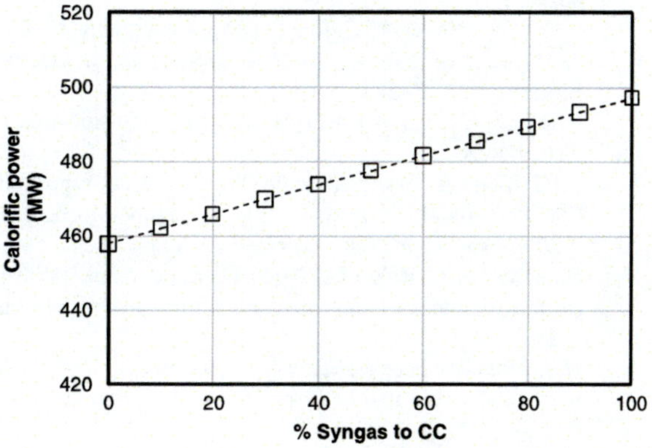

Fig. 10 Calorific value variation of all the outlets streams after the H_2 purification units: WGS reactors, Rectisol process and PSA and CO_2 compression, plus the calorific value of the syngas

the syngas after the cleaning train. It can be directly deduced that if sending the final H_2 stream to the CC (which represents a 75% of the final carbon removal train calorific value), it would lead to less power produced if using a CC with the same efficiency than the one used for syngas. By means of Eq. 28, this last carbon removal step has an efficiency of about 69%. If the calorific value contained in the PSA gas is also considered, the efficiency increases 20%, thus being a total of 89% (The downside is instead a combustion gas that is not CO_2 free).

5 Conclusion

This chapter starts with a brief summary of the final possible applications of the syngas and the subsequent conditions required. A presentation of a possible gas cleaning train with carbon removal is then presented and modelled. The final modelling approach is accurate enough since validation with real gasification plant shows an error around 10%, which is accurate enough in a conceptual modelling stage. This approach considers a extensive modelling of cleaning unit by use of electrolytes, Henry's law and complex activity coefficient formulations which are then replaced by simpler split fractions for separation processes. In the case of reactions it also relies on chemical equilibrium or on reaction degrees of advance based on previously modelled exhaustive models. Due to a longer units train, and even though the LHV of H_2 is higher than that of syngas, the final energy content of the stream that goes to the CC has a lower calorific value. After the PSA process there exists a flue stream with high energy potential, thus suitable to be introduced into the gas turbine.

References

1. Demirbas MF (2009) Biorefineries for biofuel upgrading: a critical review. Appl Energ 86:S151–S161
2. Wang L, Weller CL, Jones DD, Hanna MA (2008) Contemporary issues in thermal gasification of biomass and its application to electricity and fuel production. Biomass Bioenerg 32:573–581
3. Highman C, van der Burght M (2003) Gasification. Elsevier Science, Amsterdam, the Netherlands
4. BTG Biomass Gasification (2008). Technical report, Biomass Technology Group (BTG), The Netherlands. http://www.btgworld.com/uploads/documents/Gasification%20Attachment%20Website%20v2.pdf. Accessed 31 Aug 2010
5. Bang-Moller C, Rokni M (2010) Thermodynamic performance study of biomass gasification, solid oxide fuel cell and micro gas turbine hybrid systems. Energ Convers Manage 51:2330–2339
6. Hernández-Pacheco E, Mann MD, Hutton PN, Singh D, Martin KE (2005) A cell-level model for a solid oxide fuel cell operated with syngas from a gasification process. Int J Hydrogen Energy 30:1221–1233
7. Wender I (1996) Reactions of synthesis gas. Fuel Process Technol 48:189–297
8. Watson GH (1980) Methanation catalysts, report number ICTIS/TR09, Feb 1980. IEA Coal Research, London
9. Gassner M, Maréchal F (2009) Thermo-economic process model for thermochemical production of synthetic natural gas (SNG) from lignocellulosic biomass. Biomass Bioenerg 33:1587–1604
10. Separation processes, Ullmann's encyclopedia of industrial chemistry, processes and process engineering (2004). Volume 1, Wiley-VCH Ed
11. Sharma SD, Dolan M, Park D, Morpeth L, Ilyushechkin A, McLennan K, Harris DJ, Thambimuthu KV (2008) A critical review of syngas cleaning technologies: fundamental limitations and practical problems. Powder Technol 180:115–121
12. Pisa J, Serra de Renobales L, Moreno A, Valero A (1997) Evaluación de alternativas en un diseño de planta GICC con gasificador PRENFLO. XII Congreso Nacional de Ingeniería Mecánica.
13. Brown D, Gassner M, Fuchino T, Maréchal F (2009) Thermo-economic analysis for the optimal conceptual design of biomass gasification energy conversion systems. Appl Therm Eng 29:2137–2152
14. Arena U, Zaccariello L, Matellone ML (2010) Fluidized bed gasification of waste-derived fuels. Waste Manage 30:1212–1219
15. IEA-GHG (2008) Co-production of hydrogen and electricity by coal gasification with CO_2 capture updated economic analysis. Technical report, UK.
16. Metz B, Davidson O, Coninck H, Loos M, Meyer L (2005) Special report on carbon dioxide capture and storage, Chapter 3. IPCC, Switzerland

17. Huang Y, Rezvani S, McIlveen-Wright D, Minchener A, Hewitt N (2008) Techno-economic study of CO_2 capture and storage in coal fired oxygen fed entrained flow IGCC power plants. Fuel Process Technol 89:916–925
18. Desideri U, Paolucci A (1999) Performance modelling of a carbon dioxide removal system for power plants. Energy Convers Manage 40:1899–1915
19. Hamelinck CN, Faaij APC (2002) Future propects for production of methanol and hydrogen from biomass. J Power Sources 111:1–22
20. Kanniche M, Bouallou C (2007) CO_2 capture study in advanced integrated gasification combined cycle. Appl Therm Eng 27:2693–2702
21. Descamps C, Bouallou C, Kanniche M (2008) Efficiency of an integrated gasification combined cycle (IGCC) power plant including CO_2 removal. Energy 33:874–881
22. Chen C, Rubin E (2009) CO_2 control technology effects on IGCC plant performance and cost. Energy Policy 37:915–924
23. Biagini E, Masoni L, Tognotti L (2010) Comparative study of thermochemical processes for hydrogen production from biomass fuels. Bioresour Technol 101:6381–6388
24. Cormos CC, Cormos A-M, Agachi S (2009) Power generation from coal and biomass based on integrated gasification combined cycle concept with pre- and post- combustion carbon capture methods. Asia-Pac J Chem Eng 4:870–877
25. Kanniche M, Gros-Bonnivard R, Jaud P, Valle-Marcos J, Amann J-M, Bouallou C (2010) Pre-combustion, post-combustion and oxy-combustion in thermal power plant for CO2 capture. Appl Therm Eng 30:53–62
26. Prausnitz JM, Lichtenthaler RN, Gomez de Acevedo E (1986) Molecular thermodynamics of fluid-phase equilibria, 2nd edn. PTR Prentice Hall, Englewood Cliffs, p 269–278
27. AspenTech Incorporation (2010) Aspen Plus user's guide. http://support.aspentech.com Accessed 20 Sept 2010
28. Pérez-Fortes M, Bojarski AD, Velo E, Puigjaner L (2010) IGCC power plants: conceptual design and techno-economic optimization. In: Clean energy: resources, production and developments. Energy science, engineering and technology. NOVA, NY
29. Chiesa P, Consonni S, Kreutz T, Williams R (2005) Co-production of hydrogen, electricity and CO_2 from coal with commercially ready technology. Part A: performance and emissions. Int J Hydrogen Energy 30:747–767
30. Wells GL, Rose LM (1986) The art of chemical process design. Elsevier, Amsterdam
31. Mastellone ML, Zaccariello L, Arena U (2010) Co-gasification of coal, plastic waste and wood in a bubbling fluidized bed reactor. Fuel 89:2991–3000

Emerging Technologies on Syngas Purification: Process Intensification

Ramón Álvarez-Rodríguez and Carmen Clemente-Jul

Abstract Syngas normally contains a series of contaminating gases, depending on the raw materials used, the most abundant one usually being H_2S, accompanied by COS and, also, HCl, HF, etc. Normally, purification should be performed before its combustion in the gas turbine (in the case of combined cycle plants) and the classic procedure, as performed at present in some installations, uses the wet process, which demands a reduction in the temperature of the gas to be purified and, therefore, gives rise to a series of thermodynamic losses. The trend is to research high-temperature purification processes that avoid or reduce this loss in performance. In particular, there are two research lines for sulphur compounds: (i) The use of low-value metal adsorbents that may be discarded once they have been stabilised and without contaminating properties, such as calcium compounds that may produce CaO that captures the hydrogen sulphide and (ii) the 'important value' adsorbents that therefore require the ability to be regenerated and reused, and whose base are metals with high affinity with sulphur, such as Zn, Fe and Cu but whose cost is much higher than the previously mentioned calcium compounds.

Notation

IGCC	Intergrated Gasifuration Combined Cycle
MDEA	Methyldiethanolamine

R. Álvarez-Rodríguez (✉) · C. Clemente-Jul
ETSI Minas, Alenza 4, Madrid, 28003, Spain
e-mail: ramo.alvarezr@upm.es

C. Clemente-Jul
e-mail: Carmen.clemente@upm.es

L. Puigjaner (ed.), *Syngas from Waste*, Green Energy and Technology,
DOI: 10.1007/978-0-85729-540-8_6, © Springer-Verlag London Limited 2011

1 Adsorption of Sulphur Compounds Over Solid Adsorbents

1.1 General Considerations

In reactions with the intervention of solid sorbents, it is very important for possible practical applications to consider the influence of the different state of aggregation of the reactants and the reaction products. In the case of syngas, the reactant to be eliminated is in a gaseous state and the eliminating substance is in solid state. The reaction products will normally be solid or gaseous (water is in vapour form at high temperatures).

For the sorbent, in the form of solid particles, it is sought for it to react as fast and completely as possible. Initially, the action of gases on particles occurs without any problem on their surfaces but, afterwards, the reaction must progress towards the grain core. This will be according to its porosity and the size distribution of its pores, it is also important to consider its size when considering the reaction at the grain core because, if it is very large, the core will be far away from the surface, hindering the passage of reactant gases and those produced. This justifies the sorbent preparation methods seeking an improvement in these porosity characteristics.

For the process to occur normally, the solid product formed in the reaction must have a molar volume such that the pores are not obstructed, which would create a gas-tight layer preventing the reaction from progressing towards the core.

On the other hand, the gases produced on exiting the grain to become part of the general gas current must pass through the layer that has already reacted and, at this stage, secondary reactions may occur in some cases because there is good contact between the new gas produced (if it is the case) and the new solid in its pores. This is outlined in Fig. 1.

In addition, the grains of the sorbent can be found more or less packed depending on the type of reactor in which the reaction takes place. In this way, there will be strong packing in a fixed or moving bed, whereas this will not be the same in a fluidised bed. The grain size distribution of the sorbent and its bed height will have an influence in the case of packing, because the porosity of the column will be low with very low grain size distributions and preferential gas passage paths may be created, leaving areas with bad gas supply that will partly cancel the advantages of small grain sizes having a more complete reaction, while also increasing a drop in the pressure necessary for a specific gas flow.

There may also be an influence between the different grains of the sorbent, being greater the more packed they are, as it can be seen in Fig. 2. If a product from the reaction of the first grains is a gas, it flows in the direction of the gas flow and will go on into contact with the next ones and, if there is any possibility, there may be secondary reactions.

Another important issue, particularly in the cases in which the sorbent is to be regenerated and used during several cycles, is to maintain its mechanical resistance, together with its porosity. Because of the alternative changes in composition of the solid, advancing cracks or splits could appear, mechanically degrading the sorbent and

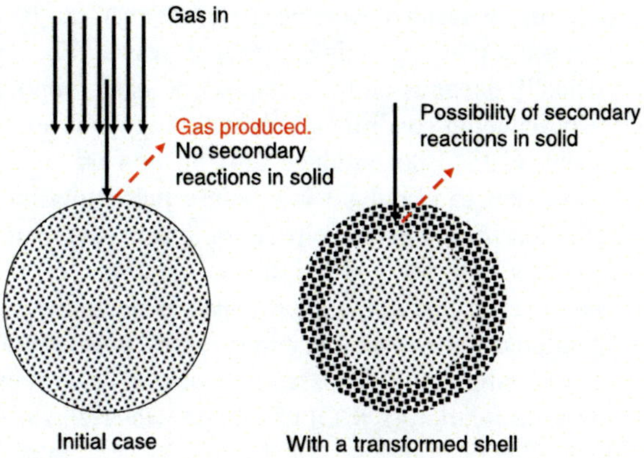

Fig. 1 Possibility of secondary reactions in a solid-transformed outer shell

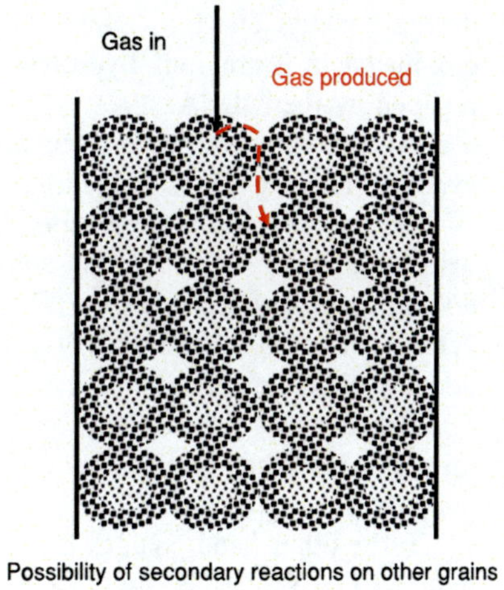

Fig. 2 Possibility of secondary reactions over the solid-transformed outer shell of other grains

creating fines that hinder the passage of gases, dragged along, and are also, continuously subjected to high temperatures favouring sorbent sintering, leading to a gradual reduction in its porosity and, as a result, in its efficiency. This is the reason for using sorbents that can be regenerated in these cases, using product mixtures which, with an adequate preparation process, on one hand improves the porosity and on the other hand contributes towards an improvement in mechanical resistance and resistance to sintering, which explains the large variety of proposals in existence regarding this issue.

1.2 Adsorption of Sulphur Compounds Over Cheap Solid Adsorbents

In the production of syngas, there is an entire range of contaminants that pollute gas among which sulphur compounds should be stated as the main ones, as this element is present both in coal and in petroleum coke, heavy fuel and in the majority of residual organic materials that can be used as raw material for gasification, including animal organic waste. The content of these contaminants basically depends on the composition of the mixture to be gasified.

In reducing conditions, sulphur compounds are mainly H_2S but also, to a lesser extent, COS. Both produce SO_2 in the case of combustion and, therefore, it is convenient to eliminate this before the combustion process to reduce problems regarding corrosion in pipes, equipment and particularly in the turbines as the tendency is to use gas turbines in combined cycles to obtain greater energy efficiency. For other applications, such as obtaining H_2 for fuel-cells, the elimination of sulphur must reach very low values, normally under 1–10 ppm.

The elimination of these contaminants is currently carried out in the combined cycle installation (IGCC) of Elcogas (Puertollano, Spain) [1] with water cleaning in a Venturi type set of equipment and later H_2S through extraction with Methyldiethanolamine (MDEA), but requiring a reduction in the temperature of the gas, which leads to inherent losses in performance. In this case, COS is eliminated in a previous hydrolysis stage, transforming it into H_2S for it to be retained by the MDEA.

Conventional wet process technologies are proven ones that are almost fully available although, as stated earlier, they are thermally inefficient.

The use of cheap adsorbents is essentially based on the use of calcium oxide found in two mineral substances widely present in nature, such as dolomite and calcite minerals. These compounds, such as calcite or CaO, have been used to eliminate the SO_2 caused by burning the sulphur or its compounds contained in the fuels of thermal power plants. There are many researches trying to increase the reactivity of these compounds by means of mixing with other products and special preparation methods to increase their porosity and reactivity [2].

On the other hand, syngas has a variable composition, to a large degree containing CO, CO_2, H_2, N_2 and to a lesser extent H_2S, COS, together with some other compounds, such as HCl and HF.

As a whole, there are a series of reactions and their corresponding balances controlling the composition of the gas, the main reactions in which the solids participate are the following:

$$CaCO_3 \leftrightarrow CaO + CO_2 \quad \Delta H^0 = +178.3\ \text{kJ mol}^{-1} \tag{1}$$

$$CaMg(CO_3)_2 \leftrightarrow CaO + MgO + 2CO_2 \quad \Delta H^0 = +302.7 \tag{2}$$

$$CaO + H_2S \leftrightarrow CaS + H_2O \quad \Delta H^0 = -59.2\ \text{kJ mol}^{-1} \tag{3}$$

$$CaCO_3 + H_2S \leftrightarrow CaS + CO_2 \quad \Delta H^0 = +119.1\ \text{kJ mol}^{-1} \tag{4}$$

$$COS + CaO \leftrightarrow CaS + CO_2 \quad \Delta H^0 = -93\ \text{kJ mol}^{-1} \tag{5}$$

The main reactions between the main gaseous compounds and those implying the reaction with sulphur compounds are the following:

$$CO_2 + H_2 \leftrightarrow CO + H_2O \quad \Delta H^0 = +41.1\ \text{kJ mol}^{-1} \tag{6}$$

$$CO_2 + H_2S \leftrightarrow COS + H_2O(g) \quad \Delta H^0 = +33.6\ \text{kJ mol}^{-1} \tag{7}$$

$$CO_2 + 2H_2S \leftrightarrow CS_2 + 2H_2O(g) \quad \Delta H^0 = +68\ \text{kJ mol}^{-1} \tag{8}$$

$$CO + H_2S \leftrightarrow COS + H_2 \ \Delta H^0 = -7.4\ \text{kJ mol}^{-1} \tag{9}$$

Reactions (1) and (2) imply that the minerals we consider may lose their CO_2 at high temperatures to provide the corresponding oxides. The temperature at which this occurs is normally high and depends on the specific mineral and also on the partial pressure of CO_2, these reactions being reversible. This decomposition temperature is lower in dolomites than in calcites, with the possibility of reaching a state of semi-calcining in the dolomite.

Once these minerals have been calcined, and as the reaction is reversible, CaO may be recarbonated in the presence of CO_2 at low temperatures, such as 400 or 650°C as syngas contains a certain proportion of CO_2. Certain processes for CO_2 sequestration from flue gases in thermal power stations are based on this reversibility.

Reactions (3) and (4) imply that the hydrogen sulphide can react both with the calcium oxide and with the carbonate to turn into calcium sulphide and, therefore, retains this gas. Water or water and carbon dioxide are produced by these reactions, which go on to form part of the gaseous mass, modifying their proportions in terms of the balance constants controlling the composition of such gaseous state. MgO does not react with H_2S under the sulphidisation conditions.

In fluidised bed gasifiers (normally 800–1,050°C), the standard practice to retain H_2S is to enter these minerals, mainly calcite, in the reactor and under the corresponding temperature conditions and depending on these, therefore being in non-calcining or calcining conditions, the reactions (3) and (4) take place.

Normally, it is necessary to add an excess amount of calcite in order to reach good H_2S elimination conditions and to get close to the balance values, which translates into the appearance of solid waste residue with a notable proportion of calcite [3, 4]. Yrjas [5] studying the capture of H_2S by calcite (limestone) and dolomite in a pressurised thermogravimetric apparatus found that 0.125–0.180 mm particles at 950°C and 2 MPa in non-calcined calcite and half-calcined dolomite, dolomite was significantly more efficient than calcite. With calcined dolomite and calcite, capture was fast and highly efficient.

Pinto et al. [6] gasifying mixtures of Puertollano (Spain) coal (1.2% S), petcoke (6% S), pine waste (0.2% S) and polyethylene waste (0% S), using a pilot gasification plant with a bubbling fluidised bed at 850°C, found that the proportion of sulphur going into the gaseous state strongly depends on the sulphur contents of the mixture. Mixtures of sorbents or catalysts with sand from the bed (calcined dolomite, dolomite enriched with Ni [7], one called Ni–Mg [8], natural and calcined olivine and other commercial ones) are used to study the sulphur retained and the reduction in tars and hydrocarbons. The greatest reduction in sulphur in the

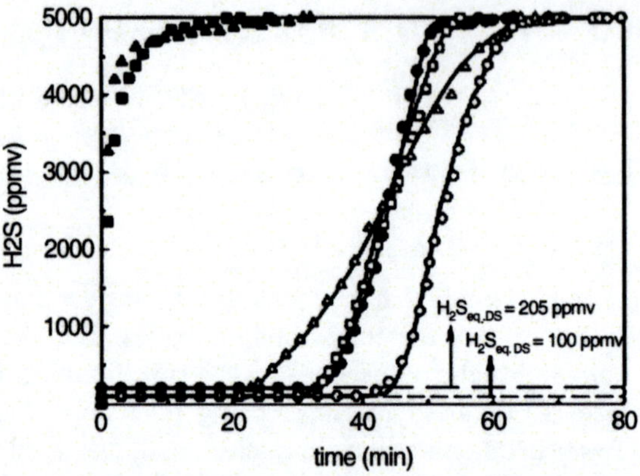

Fig. 3 Breakthrough curves obtained at calcining conditions, 900°C, (*open circles*) dolomite, (*open squares*) 'sástago' limestone and (*open triangles*) 'omyacarb' limestone, and non-calcining conditions, 850°C, (*filled circles*) dolomite, (*filled squares*) sástago limestone and (*filled triangles*) omyacarb limestone. Grain size 0.8–1.0 mm, 1.0 MPa, 18.5 cm bed length [14]

gaseous state is with dolomite (90%), being lower with Ni-dolomite, very low with Ni–Mg and calcined olivine and not at all with natural olivine. At 900°C, the proportion of sulphur going into the gaseous state increases and dolomite is the one retaining the most. These adsorbent compounds have also been used for the production of hydrogen from coal [9].

In entrained-flow type gasifiers, the temperatures reached are much higher and, normally, calcite additions to the mixture to be gasified are basically used to control the fluidity conditions of the slag extracted in molten state. In these gasifiers, the greater part of sulphur from feeding goes into gaseous state and only a small part is eliminated in the condensed state. In the IGCC of Puertollano, it has been determined that the passage of sulphur to the gaseous state is of around 95% of the contents during feeding [10] or 68–96% depending on the conditions (Font et al. 2010) and that the greatest part of that is retained in the slag, essentially as troilite (FeS) and pyrrhotite ($Fe_{1-x}S$) and, to a lesser extent, in the fly ash as pyrrhotite and also galena and sphalerite, if there are Pb and Zn in the mixture to be gasified, which usually occurs with coal and pet-coke mixtures, and there seems to be no correlation between the proportion of sulphur going into the gaseous state and calcite addition (Font et al. 2010).

There are many studies on the adsorption of H_2S in beds containing CaO, some of which have been carried out using very small amounts of oxides (from 1 mg to 1 g) to obtain the initial influence of the main parameters or to obtain mathematical models [11–13]. In other cases, greater amounts of adsorbent were used but mixed with silica to obtain a better distribution of the gas flow within the mass of the adsorber and to obtain a better profile conversion, also with the purpose of obtaining results for the preparation of mathematical models [14].

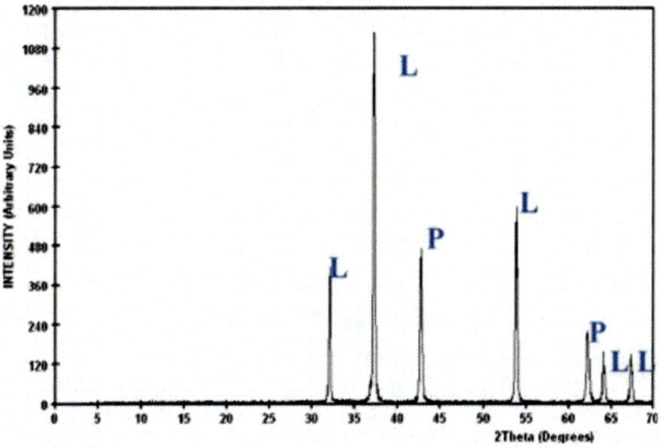

Fig. 4 Calcined granada dolomite, *L* lime, *P* periclase

Greater amounts, from 100 to 150 g, were used other times without mixing with inerts so as to obtain a better approximation to their real use [15].

Adánez et al. [14] have already shown, Fig. 3, that in experiments at 1 MPa and 0.5 vol% of H_2S, 18.5 cm of bed length and 37 cm s^{-1} gas velocity and using dolomite and calcite from Aragón (Spain) at 900°C, there is a greater conversion of dolomite (almost a full conversion) than that of calcite (85%), making the breakthrough curve of dolomite appear to the right to that of calcite. At 850°C (non-calcining conditions), the behaviour of dolomite can be stated to be similar to that of calcite at 1,173°K but that the behaviour of calcites at this temperature of 1,123°K is much worse and only very low conversions are achieved.

In general, the better behaviour of dolomite than that of calcite can be explained, apart from the lower calcining temperature, by the fact that the mixture remains after emitting the CO_2 is one of MgO and CaO and, as MgO does not react with H_2S, it leaves a greater porosity available for the access of H_2S towards the core of the grains. Figure 4 shows the diffractogram of a calcined dolomite from Granada (Spain) in which it can be seen that the final composition is of CaO (Lime) and MgO (periclase).

The use of dolomite instead of calcite, therefore, seems advantageous and, with regard to the adsorption of H_2S in a fixed bed, there are several considerations that should be made [15]: An important issue is the grain size distribution of the dolomite because fine grains (e.g., less than 1 mm) cause a non-uniform gas flow distribution in the bed, therefore, creating preferential paths for gas passage and giving rise to lower efficiency.

With regard to the H_2S adsorption, the breakthrough curve presents an initial section corresponding to a very low concentration of H_2S in the output gas that later increases progressively until the adsorbent becomes saturated and the input and output become equal. This can be seen in Fig. 3. Considering a possible industrial use, this initial section of the curve will condition the minimum H_2S contents that may be achieved in the output gas, and once a certain level is set, the adsorber, still with capacity to adsorb H_2S, cannot continue to be used as an output stage for the gas to be purified and, by means of a valve system, the output gas will

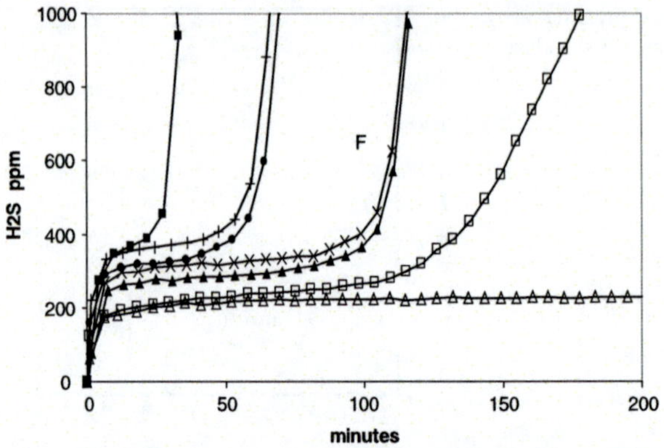

Fig. 5 Influence of gas velocity, sample weight, temperature and H_2S concentration: 2% H_2S, 2–2.5 mm: 29.11 cm s^{-1} gas velocity, *filled square* 100 g, 850°C; 14.55 cm s^{-1} gas velocity, *filled circle* 100 g, 850°C, '+' 100 g, 900°C, *filled triangle* 150 g 850°C, '×' 150 g, 900°C; 0,4% H_2S, 2–2.5 mm: 29.11 cm s^{-1} gas velocity, *open square* 100 g, 850°C, *open triangle* 150 g, 900°C [15]

go through a new tank with a new adsorbent, while the 'partially used up' goes on to receive the initial gas to become saturated and to be used in the best possible way. When the adsorbent is practically depleted, it must then go on to another treatment stage to provide a non-contaminant stable product.

The main factors conditioning the H_2S concentration emitted with the output gas (initial section of the curve) are the following:

- The length of the bed the gas passes through
- The gas circulation speed
- The H_2S concentration of the input gas
- The temperature

All these can be seen in Fig. 5 [15]. The greater the length of the bed, the lower the circulation speed of the gas, the lower the concentration at entry and lower the H_2S concentration in the output gas. Temperature has little influence on the interval between 850 and 950°C, although it is better for it to be at 850°C.

As can it be seen, it is feasible to achieve residual H_2S levels of around 250 ppm in the syngas, which would lead to a much lower concentration in the output of gas already burned in the chimney, because of the dilution caused by the combustion agent. The height of the bed, between specific limits, would be a parameter easy to control and optimise in a possible industrial use of this system.

An important fact is that one of the factors influencing a better or worse use of the adsorbent is its grain size distribution and its position in the column. In this way, very fine-grain size distributions, such as those of around 0.5 mm and particularly if they are found in the lower part of the column because of the effect of the mechanical pressure, give rise to preferential gas passages leading the breakthrough curves to become distorted in their inclination, tending to give lower

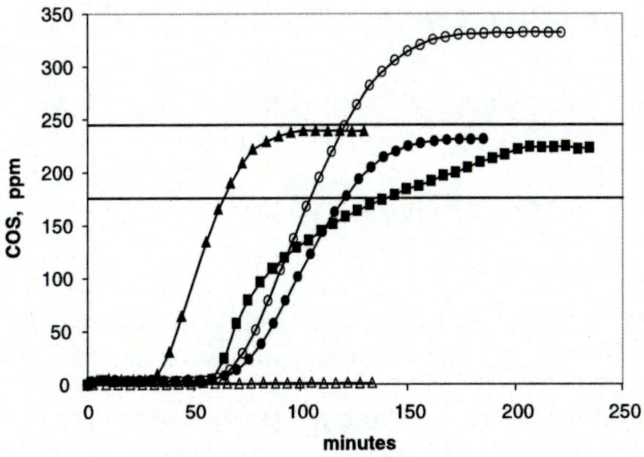

Fig. 6 COS in the outlet gas (11.4 cm bed length). COS breakthrough curves with the usual inlet gas composition (2% H_2S, 6% CO_2, 10% H_2O, 18% H_2, 176 ppm COS) for (*filled square*) 0.4–0.5 mm grain size, 11.4 cm bed lengh, 14.5 cm^{-1} gas velocity, (*filled triangle*) 2–2.5 mm grain size, 11.4 cm bed length, 29.1 cm^{-1} gas velocity and (*filled circle*) 2–2.5 mm grain size, 11.4 cm bed length and 14.5 cm s^{-1} gas velocity. Upper horizontal *solid line* when silica is used instead dolomite. Lower horizontal *solid line* when COS is introduced. *Open circle* as *filled circle* but without CO_2, (*open triangle*) without CO_2 and H_2O [15]

adsorption performance up to saturation [14, 15]. This problem no longer exists in coarser grain size distributions, for example, 2–2.5 mm, and practically all the CaO present in CaS can be converted, because of the presence of non-reacting MgO and the fact that the molar volume of CaS is lower than that of the initial $CaCO_3$ at the start, meaning that the pores do not become obstructed or prevent the passage of H_2S to react inside the grain.

Likewise, COS control is performed in the same way as for H_2S. COS reacts with CaO giving CaS according to reaction (5), and the shape of the breakthrough curves is identical (at a different scale) to those corresponding to H_2S. Figure 6 [15] shows the curves corresponding to an initial gas with 2% H_2S, 176 ppm COS, 6% CO_2, and 10% H_2O (g). The highest level of COS with saturation of the adsorbent is a function of the gas composition, as the component gases react with each other to reach a balance. It is because of this that the COS level at the output when the adsorbent becomes saturated is higher than that entered, all this depending on the composition. In this way, if H_2O is not entered, the hydrolysis reaction (inverse to reaction (7)) is made difficult and the final COS level increases. If neither CO_2 nor H_2O is entered, the COS content of the output gas will be very low, around 1.5 ppm, as the COS reaction with H_2 (inverse to reaction (9)) will practically eliminate this.

The COS content of the emerging gas is of approximately 1.5 ppm in the initial horizontal section of the curves, that is to say, a very low level that will have practically no influence on the total contents of sulphur compounds, basically controlled by the H_2S.

2 Inertisation of the Cheap Adsorbents Used

In the case of using CaO at high temperatures, the process will not end with the adsorption of H_2S or COS, but the product generated, CaS, will not be a stable one in atmospheric conditions because it reacts with water vapour or with the surface water to regenerate H_2S and emits this into the atmosphere [16], according to the inverse reaction of adsorption reaction 6, preventing its deposition as landfill material. Other studies used the reaction of CaS with water and Methyldiethanolamine (MDEA) to recover the H_2S at room temperature [17].

The CaS oxidation reactions by means of oxygen are the following:

$$CaS + 2O_2 \leftrightarrow CaSO_4 \quad \Delta H^0 = -952.2\ \text{kJ mol}^{-1} \tag{10}$$

$$CaS + 1.5O_2 \leftrightarrow CaO + SO_2 \quad \Delta H^0 = -458.9\ \text{kJ mol}^{-1} \tag{11}$$

The use of combustion flue gases with excess O_2, normally between 4 and 6%, from syngas combustion as oxidising gas at high temperatures has been considered, but they also contain important amounts of CO_2 and $H_2O_{(g)}$ with the increase, therefore, in the number of possible reactions. H_2S (inverse to (3)) and COS (inverse to (5)) may appear in the gas currents, apart from CaO in the solid. These gases can react with oxygen, if this is present in their formation, in accordance with the following reactions:

$$COS + 2O_2 \leftrightarrow CO_2 + SO_2\ \Delta H^0 = -552\ \text{kJ mol}^{-1} \tag{12}$$

$$H_2S + 1.5O_2 \leftrightarrow H_2O + SO_2\ \Delta H^0 = -515.2\ \text{kJ mol}^{-1} \tag{13}$$

The presence of CaO generated on the surface of the solid and later inside this because of oxidation, plus the H_2S and COS are also generated, makes reactions (3) and (5) commented on the sulphidisation case possible.

Other possible reactions are:

$$CaS + 3CO_2 \leftrightarrow CaO + 3CO + SO_2\ \Delta H^0 = +390\ \text{kJ mol}^{-1} \tag{14}$$

$$CaS + 3H_2O \leftrightarrow CaO + 3H_2 + SO_2\ \Delta H^0 = +266\ \text{kJ mol}^{-1} \tag{15}$$

Also, the following may occur at high temperatures:

$$CaO + 3CaSO_4 \leftrightarrow 4CaO + 4SO_2\ \Delta H^0 = +1{,}021\ \text{kJ mol}^{-1} \tag{16}$$

$$CaSO_4 \leftrightarrow CaO + SO_3 \quad \Delta H^0 = +394.3\ \text{kJ mol}^{-1} \tag{17}$$

Reaction (16) is a reaction between solids and, therefore, it is difficult for this to occur at temperatures that are not very high as it is highly conditioned by diffusion phenomena between solids. In this way, no signs of carbonates in the X-ray diffraction have been found in the products oxidised at 850°C [15]. Similarly, CaS oxidation reactions (10) and (11) are highly exothermic, causing a relatively

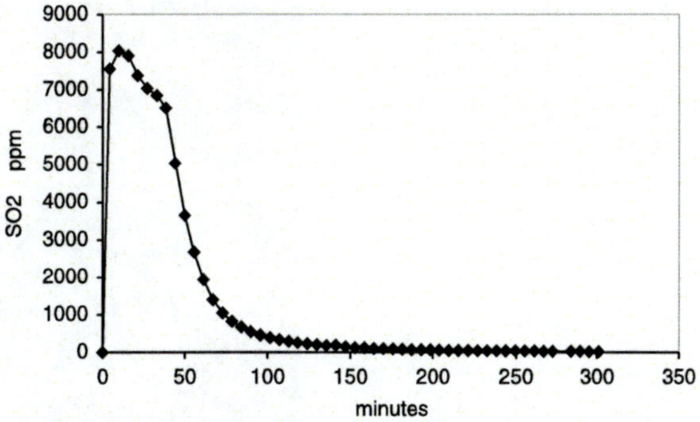

Fig. 7 Gas emission during oxidation of sulphurised dolomite 2–2.5 mm grain size, 14.55 cm s^{-1} gas velocity and 11.4 cm bed length, with 4% O_2 and 96% N_2, 850°C

important rise in temperature even when mixing oxidisers with only 4% O_2, which must be taken into account in a possible industrial scenario.

CaO regeneration and its reuse could be considered but this requires oxidation with oxygen at temperatures greater than 1,400°C in accordance with reaction (17) for decomposition of the possible calcium sulphate that may have been formed [18], but the H_2S retention capacity of the regenerated sorbent drops after a few cycles, probably because of a sintering phenomenon resulting from heating to a high temperature for a long time, which reduces its porosity.

There are studies measuring the kinetic parameters of the CaS reaction with oxygen or other oxidisers, usually performed with very low amounts of substances in thermobalance type or chemical reactor systems [19–22].

Other studies get closer to a possible industrial reality using greater amounts of CaS produced previously in a sulphidisation reaction [23].

The main difference of oxidation with regard to the sulphuration stage (where, in principle, sulphuration took place down to the core of the grain even in relatively thick grains of perhaps 2–2.5 mm) is that now the molar volume of the calcium sulphate is higher than that of calcium sulphide and even higher than that of the original calcium carbonate from which calcium oxide derives (these molar volumes with 46.0, 28.9 and 36.9 cm^3/mol, respectively). Therefore, there is a tendency to clog the pores of the sulphur dolomite or calcite grains hindering the access of the oxidiser towards the core of the grain, leaving residual calcium sulphide [16, 19–21, 23].

It was found that greater degrees of sulphation up to pressures of 2 MPa are achieved with dolomite than with calcite [24], which could be attributed to the fact that, as MgO has not reacted, the set presents better porosity.

In studies on a fixed bed and oxidising with a 4% O_2 and 96% N_2 mixture (Álvarez-Rodríguez and Clemente-Jul Figure not published), sulphured dolomite grains with a size of 2–2.5 mm at 850°C and 11.4 cm long with a fluid passage speed of 14.55 cm s^{-1} obtains the breakthrough curve as shown in Fig. 7, showing how an emission of SO_2 occurs, increasing until it reaches to a maximum then slowly decreases down to unappreciable values.

This means that an oxidation of CaS to CaO occurs with SO_2 elimination, which also occurs with other finer-grain size distributions. Afterwards, there will

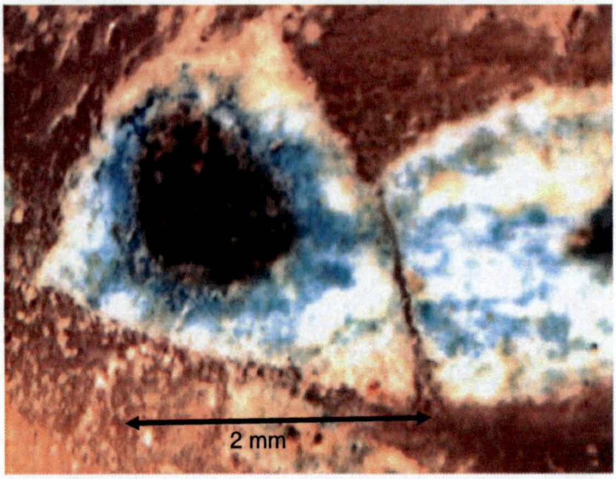

Fig. 8 Grain sections of a previously sulphurised dolomite, oxidised with 8% O_2 and 92% N_2, with 2.5 mm grain size, 11.4 cm bed length 29.11 cm s^{-1} gas velocity, and 850°C

Table 1 Oxidation with a mix of O_2 and N_2: Influence of grain size, temperature, bed length and oxygen partial pressure (oxygen content) [23]

Test	Size (mm)	Bed length (cm)	Gas velocity (cm s^{-1})	O_2 (%)	Temperature (°C)	CaO (%)	$CaSO_4$ (%)	CaS (%)	CaO/ $CaSO_4$
1	0.4–0.5	11.4	29.11	4	850	14.76	57.09	0.45	0.258
6	0.71–1	11.4	29.11	4	850	15.14	56.51	0.51	0.268
11	2–2.5	11.4	29.11	4	850	10.18	58.5	4.32	0.174
14	2–2.5	17	29.11	4	850	8.68	60.22	4.55	0.144
15	2–2.5	11.4	29.11	4	700	4.14	47.58	20.07	0.087
5	0.4–0.5	5.7	29.11	4	700	12.12	54.43	5.57	0.223
19	2–2.5	11.4	29.11	4	950	17.61	27.46	21.95	0.641
12	2–2.5	11.4	14.55	4	850	7.94	57.61	7.53	0.138
16	2–2.5	11.4	29.11	8	850	16.08	53.2	2.23	0.302
17	2–2.5	11.4	29.11	2	850	7.48	56.55	8.92	0.132

be SO_2 in the output gas that must be removed by either transforming it into sulphuric acid or by adsorbing it with milk of lime or calcite (limestone) as it is done sometimes with flue gas of power stations. On the other hand, part of the calcium sulphide will reduce to sulphate and, because of the clogging of the pores, part of the calcium sulphide will remain without reacting in spite of the long period of exposure to the oxidiser. Figure 8 shows a cross-section of a grain with grain size distribution of 2–2.5 mm, oxidised with 8% O_2 and 92% N_2 mixture at 850°C and it can be seen how the grain has three different areas, a central black one corresponding to unoxidised calcium sulphide, some blue areas corresponding to CaO produced by the oxidation and other white ones corresponding to the calcium sulphate, these being the ones obstructing the passage of the gases towards the core of the grains.

The Table 1 shows an analysis of tests on a fixed oxidation bed with O_2 and N_2 mixtures at several temperatures, bed lengths and grain sizes.

Fig. 9 COS and SO_2 trend in oxidation with 21% CO_2 (79% N_2), grain size 2–2.5 mm, 850°C, COS (*filled circle*) and SO_2 (*filled square*)

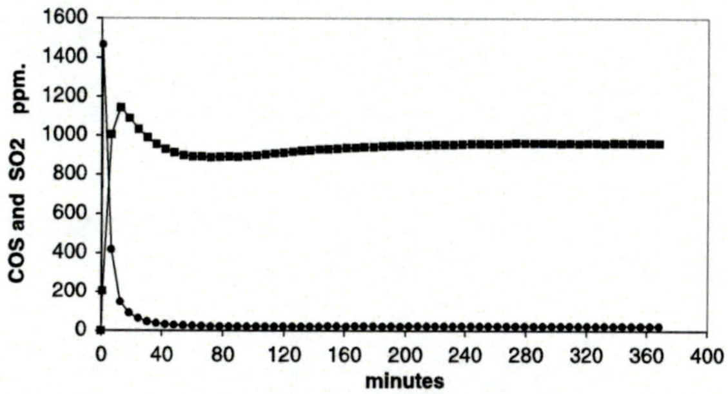

It can be seen that the main parameter controlling the residual CaS is the size of the grain, the smaller the grain lesser the residual CaS with all other conditions remaining the same. The greater oxygen concentration favours a lower residual CaS but increases the $CaO/CaSO_4$, indicating a greater SO_2 emission. With regard to temperature, the best one seems to be the one around 850°C, as the amount of residual CaS increases noticeably at 700 or 950°C.

In the cases already commented, oxidation was performed just with the O_2 and N_2 mixture, but the reactions change if an attempt is made to use a combustion flue gas as the oxidising gas. In a fixed bed, oxygen, particularly in low concentrations (4–8%), is depleted because of its fast reaction with CaS to give SO_2 and, afterwards, this gas (SO_2) plus CO_2 and H_2O, which react much more slowly, are the ones reaching the next CaS layers, then causing the reactions to inverse to (3), generating H_2S, and to (5), generating COS.

Álvarez-Rodríguez and Clemente-Jul [23], oxidising CaS previously obtained in a calcined dolomite sulphidisation stage, with a 21% CO_2 and 79% N_2 mixture at 850°C, found, Fig. 9, that a strong emission of COS occurs initially, meaning that the inverse reaction of (5) takes place, although it drops quickly, becoming substituted by a SO_2 emission that rises initially and then drops, stabilising at a relatively high value of 100 ppmv for hours. This change in behaviour could be explained by the fact that, initially, the entire surface of the grains is CaS, and the inverse reaction of (5) takes place in competition with reaction (14), the main one being the former.

The check for the existence of this reaction is that the similar reaction (18)

$$CaS + COS \leftrightarrow CS_2 + CaO \quad \Delta H^0 = 93.4\ \text{kJ mol}^{-1} \tag{18}$$

also occurs at first (when the grains surface are only CaS and there is a great quantity of COS generated in previous grains), detecting a small CS_2 peak, Fig. 10. As the reactions are scarcely intense, the CO_2 consumption is very low and, therefore, reaches the entire column of grains from the beginning, with the entire external surface of the grains becoming loaded with CaO. During the following moments, the COS produced finds CaO on the surface of other grains and reacts according to reaction (5), regenerating CaS. The same occurs when the COS produced inside the grains must go through areas with CaO. In this way, the COS

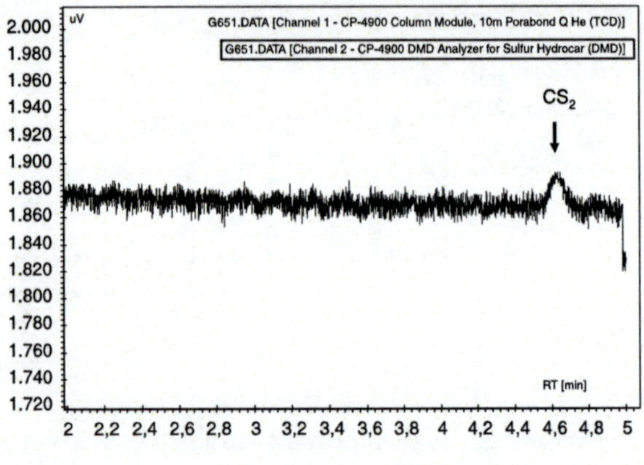

Fig. 10 CS_2 detection at the beginning of oxidation with 21% CO_2 (grain size 2–2.5 mm, 850°C). Chromatogram showing CS_2 peak (DMD detector)

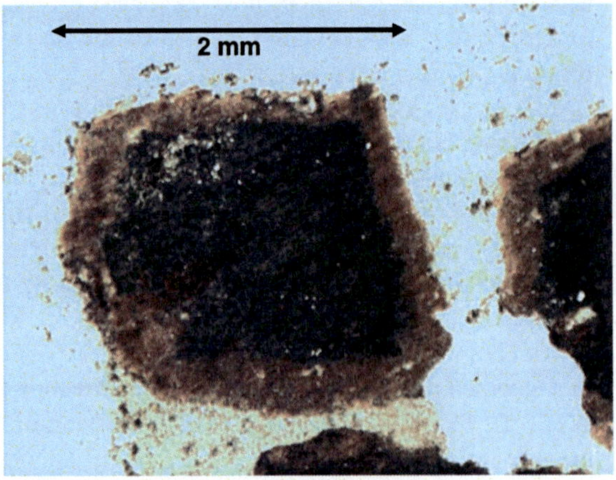

Fig. 11 Grain sections of a previously sulphurised dolomite, oxidised with 21% CO_2 79% N_2, 2.5 mm grain size, 11.4 cm bed length 14.55 cm^{-1} gas velocity, 850°C

is progressively trapped and only SO_2 can exit in a predominant manner with which this reaction returns to the main one (Figs. 1 and 2).

Figure 11 (Álvarez-Rodríguez and Clemente-Jul not published) presents the photo of a cross-section of the grains in this test showing that almost the entire core of the grain is black, indicating it is CaS and only an external layer is lighter, indicating there is less CaS. Naturally, the rest is CaO, but there is no clear delimited blue area of CaO, indicating that part of this is becoming sulphurised (this layer seems to be a mix of CaO and CaS).

With the oxidation of H_2O and N_2 vapour mixtures, something similar occurs in such a way that a high emission of H_2S is generated initially because of the inverse reaction of the adsorption over the CaO (reaction (3)) and low SO_2 because of reaction (15), whereas, later, H_2S falls rapidly because of the adsorption over the new surface of CaO of other grains or to the contents in the more external layers of the grain itself, in such a way that both stabilise their emissions to levels of around 700 ppmv. The reaction of oxidation speeds up only when O_2 is added.

When a mixture similar to a flue gas is used, it should be expected and in fact does initially occur that, when the O_2 is used up in the first layers, the CO_2 and H_2O react causing a H_2S and COS emission, but the presence of SO_2 and later the residual oxygen immediately makes these gases disappear in the output gas current [15].

The presence of water vapour in the oxidising gas mixture, all other conditions remaining the same, leads to a greater residual contents of CaS. According to Álvarez-Rodríguez and Clemente-Jul that is more noticeable when the size of the grain is greater.

Given all these conditions in the CaS sulphuration and oxidation operations, of which a part are inverse, a fixed or moving bed system could be used for sulphuration in the case of their possible industrial use, with grains of a specific size, for example, 2–3 mm, as they are going to react almost completely down to their core, whereas for oxidation and to achieve a low presence of residual CaS, grains should be ground to much smaller sizes and oxidised in a flash type reactor, injecting them by means of an air current to basically produce CaO and some $CaSO_4$. The SO_2 produced could be used to produce sulphuric acid or an attempt could be made to adsorb it using milk from the actual CaO produced.

3 Use of Other Adsorbent Metals (Zn, Cu, Mn, Fe, etc.)

There are many researches regarding this, particularly resulting from the attractive idea of reusing these adsorbents during many cycles, although they are initially much more costly than those made up of Ca.

Among the work dealing with this issue, the following could be mentioned for Zn [25–28], for iron [29] and for copper and manganese and their mixtures [30–32].

Metal oxides are normally not used pure but are mixed with other substances to improve their physical and chemical properties, in particular to improve their mechanical stability, resistance to sintering and to increase their porosity.

For Zn compounds, the main sulphuration and oxidation reactions to regenerate the adsorbent are the following:

$$ZnO + H_2S \leftrightarrow ZnS + H_2O \quad \Delta H^0 = -79\ kJ\,mol^{-1} \tag{19}$$

$$ZnS + O_2 \leftrightarrow ZnO + SO_2 \quad \Delta H^0 = -439.1\ kJ\,mol^{-1} \tag{20}$$

Also, it must be taken into account that zinc oxide is a substance that can be reduced easily at high temperatures because of the reducing gases present in syngas:

$$ZnO + CO \leftrightarrow Zn + CO_2 \quad \Delta H^0 = +65.3\ kJ\,mol^{-1} \tag{21}$$

$$ZnO + H_2 \leftrightarrow Zn + H_2O \quad \Delta H^0 = +106.44\ kJ\,mol^{-1} \tag{22}$$

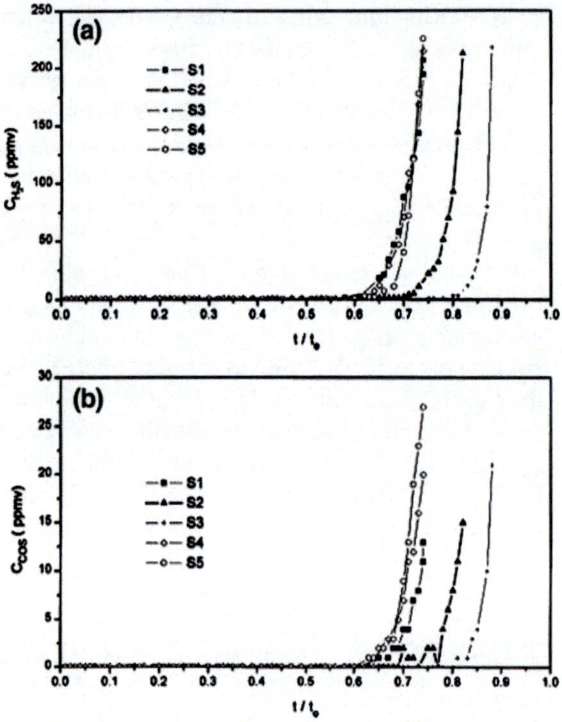

Fig. 12 Breakthrough curves in sulphidation of sorbent ZFT(0.8:1:0.2). Evolution of **a** H_2S; **b** COS [27]

Metal Zn presents high volatility at high temperatures. Another important factor is that these compounds tend to sinter reducing the porosity of the system.

To avoid or reduce adverse effects, many different adsorbent compositions have been formulated with a zinc base because the addition of other metal oxides, such as TiO_2, F_2O_3 or CuO, seems to provide better stability and improve the behaviour in the sulphidisation and regeneration cycles.

In the case of Ti [33, 34], they proved this improvement in stability is because of the formation of a spinel structure of the Zn_2TiO_4 type. For Fe additions to the formation of $ZnFe_2O_4$ [34], Pineda [27] proved the reactivity of these compounds is very good, performing an almost full H_2S reduction using adsorbents ZnO/TiO_2 0.8/1, 1/0.8/0.2 and $ZnO/Fe_2O_3/CuO$ 0.86/1/0.14 at a sulphidisation temperature of 600°C and a gas composed of 1% H_2S, 8% H_2, 15% CO, 15% H_2O (v) and the rest N_2, and a regeneration temperature of 710°C with a gas made up by 3% O_2, 30% H_2O (v). In the case of the ZnO/TiO_2 mixture, an output gas concentration of around 20 ppmv (thermodynamic balance concentration) is achieved. In the case of the $ZnO/Fe_2O_3/CuO$ mixture, the balance concentration is of 1.2 ppmv of H_2S and the results get close to this, only with a small concentration of COS now appearing.

In the case of the $ZnO/Fe_2O_3/TiO_2$ mixture, the results are similar to the previous ones only with a bit less COS. The presence of a noticeable percentage of H_2S mixed with the SO_2 in the output gas of the regeneration stage can be detected

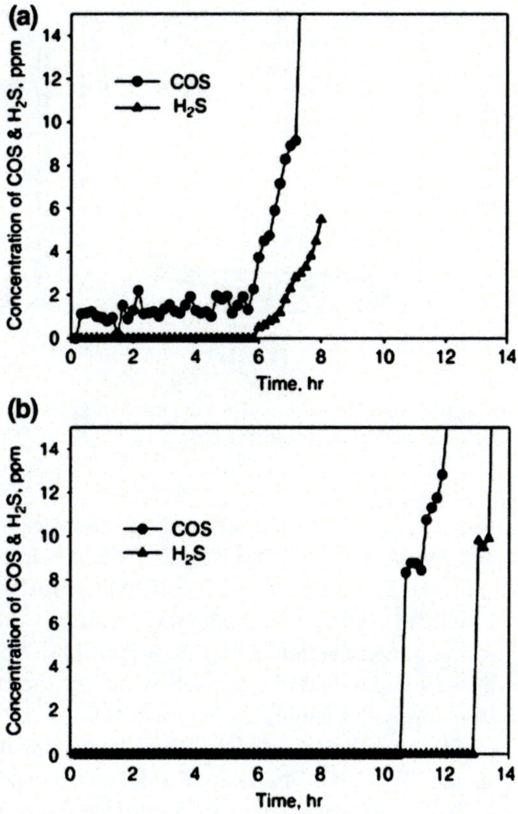

Fig. 13 The breakthrough curves if H_2S and COS in the outlet of **a** fluidised bed (ZZF sorbents) and **b** fixed-bed (ZCA-2 sorbents) desulsurisation system [35]

in compounds containing iron in the regeneration stage. In general, a decrease can be detected in the sorbent performance according to the number of cycles, which seems to be the lowest in the last compound. Illustrating this is Fig. 12, showing a reproduction of Fig. 8 from the aforementioned piece of work.

Alonso et al. [26] proved that the efficiency of spinel-type adsorbents with Ti, Fe or Cu already studied by Pineda [34] increases very substantially when mixed with 5% graphite during the preparation of the adsorbents. This addition has several effects, the first being that of increasing the porosity of the material and the second the reduction of the appearance of a network of cracks in the material, which degrade it mechanically during the different sulphidisation/regeneration cycles.

Based on the low H_2S concentration that can be achieved in gases purified with substances having the participation of Zn, Park et al. [35] have proposed a two-stage proposal to obtain ultraclean gases with the idea of using syngas to produce, for example, extremely pure H_2 for fuel-cells or for the chemical industry.

The first stage uses an adsorbent specially prepared with $ZnO/Fe_2O_3/CaO$ and natural zeolite (ZZF) in a fluidised bed, reducing the sulphur contents from 10,000 ppm to just 3 ppmv, of which the majority are COS with temperatures of

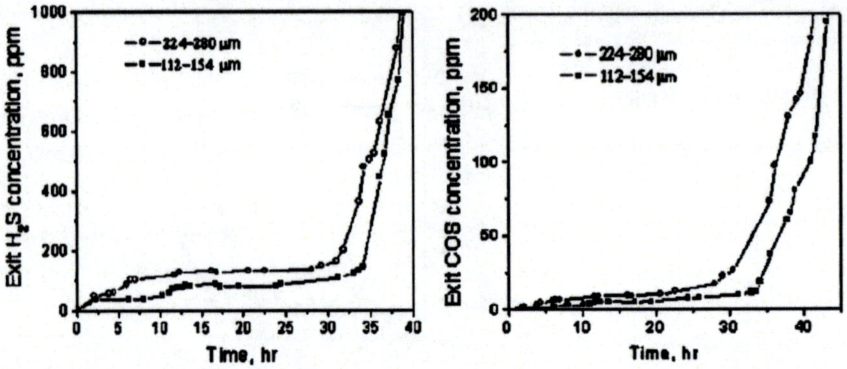

Fig. 14 Concentrations of H_2S and COS in the exit gas streams during sulfidation at 620°C using F_8C_2AS and F_8C_2AB sorbents [36]

480°C. For the second stage of ultrapurification in a fixed bed reactor, several adsorbents are used, finding that the adsorption speed increases with the Cu contents. In the case of ZCA-2 ($ZnO/CuO/Al_2O_3$ 45/45/10) at 400°C, the content of sulphur compounds at output is practically nil. To illustrate this process, Fig. 13 shows a reproduction of Fig. 9 of Park [35]. For iron, already used to adsorb H_2S in the old gas factories for cities, one of the most studied compounds is Fe_3O_4. Iron forms a sulphide in nature that is pyrite FeS_2, not produced by industrial processes, it being normal to find iron sulphides such as troilite (FeS) and pyrrhotite (Fe_{1-x} S). The lack of a defined composition of these sulphides makes it difficult to formulate precise reaction equations.

Pan et al. [29] studied the kinetic behaviour of the Fe_3O_4 oxide stabilised with other additions in its desulphurisation and compared it with a mixture of zinc titanates ($Zn_2TiO_4 + ZnTiO_3$ mixture). They also found that using a mixture of 2% H_2S and the rest N_2, increased the sulphidisation capacity notably at temperatures from 500 to 650°C and that this capacity is 1.9 times greater than that of Zn-titanate. Regeneration is performed at 650°C with 2% O_2 and the rest N_2, reaching a conversion of around 75%.

Xie et al. [36] studied the removal of sulphur compounds using a mixture of iron oxide, cerium oxide and coal combustion ashes, with the Fe/Ce molar relationship being 8/2 and of this mixture with the ashes being variable but also of around 8/2, with which the majority of the adsorbent is made up of iron oxide. They found that from 420 and 620°C, the best was 620°C. As an example of these results, Fig. 14 shows a reproduction of Fig. 7 in this piece of work, where it can be seen that the H_2S concentration at the output is of around 150 ppmv and that of COS of 15 ppm, depending on the size of the grains, making them reach the conclusion that these adsorbents are also good at eliminating COS. Regeneration is performed at 720°C with 5% O_2 and the rest N_2.

Regarding this capacity to reduce COS, there is an exception using a gas mixture without an H_2O addition (H_2S 0.47%, CO 32.69%, H_2 39.58%, CO_2

18.27% and N_2 8.92%), which may be produced by the water gas reaction readjustment, essentially (6) and (7). The similarity of the H_2S and COS breakthrough curves is a constant in all the work on adsorption using any adsorbent, what varies is the relative H_2S and COS concentration and the temperature used.

With regard to the copper and manganese oxides, copper oxides show a greater capacity for H_2S reduction, both as CuO and as Cu_2O, and the thermodynamically based balanced contents can reach values under 1 ppmv [37] according to the following equations:

$$CuO + H_2S \leftrightarrow CuS + H_2O \quad \Delta H^0 = -113.8\ \mathrm{kJ\,mol^{-1}} \tag{23}$$

$$Cu_2O + 2H_2S \leftrightarrow 2CuS + H_2O \quad \Delta H^0 = -127.1\ \mathrm{kJ\,mol^{-1}} \tag{24}$$

The following reaction would be used for regeneration:

$$CuS + 3/2O_2 \leftrightarrow CuO + SO_2 \quad \Delta H^0 = -404.3\ \mathrm{kJ\,mol^{-1}} \tag{25}$$

This reaction is very exothermic and the oxides have a low melting point that could really complicate the regeneration stages.

But in conditions of high temperatures and with the presence of reducing gases, copper tends to be reduced easily because of the copper metal:

$$CuO + CO \leftrightarrow Cu + CO_2 \quad \Delta H^0 = -127\ \mathrm{kJ\,mol^{-1}} \tag{26}$$

$$Cu_2O + CO \leftrightarrow 2Cu + CO_2 \quad \Delta H^0 = -112.3\ \mathrm{kJ\,mol^{-1}} \tag{27}$$

$$CuO + H_2 \leftrightarrow Cu + H_2O \quad \Delta H^0 = -85.8\ \mathrm{kJ\,mol^{-1}} \tag{28}$$

$$Cu_2O + H_2 \leftrightarrow 2Cu + H_2O \quad \Delta H^0 = -71.1\ \mathrm{kJ\,mol^{-1}} \tag{29}$$

Cu may also adsorb H_2S:

$$Cu + H_2S \leftrightarrow CuS + H_2 \quad \Delta H^0 = -28\ \mathrm{kJ\,mol^{-1}} \tag{30}$$

This adsorption by the Cu is thermodynamically less efficient and leads to residual values of H_2S in the output gas of around 50 ppmv.

However, it has been seen [34] that copper increases, for example, the stability of zinc ferrites but there does not seem to be a method for its use as the main component.

Manganese oxides also adsorb H_2S and, in the reducing conditions of the gasification gases, it will have valence II. But if it is not balanced, it leads to a content of around 50 ppmv greater than Cu and Zn for example.

Alonso et al. [30] studied the mixture of manganese oxides with copper to lower the residual H_2S concentration of manganese with the presence of copper and that of stabilising copper, but they found that the presence of copper does not alter the behaviour of manganese oxides in a substantial manner, apparently because of the reduction to metal copper and, therefore, not achieving the benefit of the decrease in residual H_2S that copper oxides would have when balanced even though the metal is maintained. Another problem of manganese oxides is the tendency to become Manganese Sulphate during regeneration that is stable at less than 780°C, for this reason regeneration should take place at higher temperatures. With a sorbent that is a mixture of CuO/MnO 10/1, García et al. [32] achieved good results in multi-cycle sulphidisation/regeneration operations without any apparent decay.

4 HCl Adsorption

It is clear that both calcium oxide and other calcium compounds may react with HCl to produce $CaCl_2$

$$CaO + 2HCl \leftrightarrow CaCl_2 + H_2O \quad \Delta H^0 = -217.9\ kJ\,mol^{-1} \tag{31}$$

$$Ca(OH)_2 + 2HCl \leftrightarrow CaCl_2 + 2H_2O \quad \Delta H^0 = -108.9\ kJ\,mol^{-1} \tag{32}$$

$$CaCO_3 + 2HCl \leftrightarrow CaCl_2 + CO_2 + H_2O \quad \Delta H^0 = -39.6\ kJ\,mol^{-1} \tag{33}$$

Regarding HCl control, there is ample research on the influence of the different parameters, particularly for its control in flue gases, but the process is not as simple as it may seem and, in this way, [38] in research simulating a fluidised bed combustor, it became patent that dolomite obtains a better conversion than calcite and that, at 850°C, the presence of H_2O exercises a noticeable influence on the reversible reaction (31).

The conversion of the material is greater at lower temperature, i.e., from 650 to 850°C, in spite of the recarbonation effect at lower temperatures. There is a $CaCl_2$-melting effect (already melting at 772°C when in pure state) which makes the particles sticky, obstructing the pores and, also in this case, with the possibility of reacting with the sand of the bed. The conversion of the material is much faster with small-sized grains than with large-sized particles.

For gasification in a pilot plant bubbling fluidised bed gasifier at 850°C using diverse mixtures including vegetable waste, Pinto et al. [6] concluded that the greater the chlorine contents of the mixture, the greater the contents in the gas for this reason a good selection of the mixtures to be gasified would be adequate. The greatest chlorine retention is obtained when using dolomite as adsorbent additive in the bed and only 30–35% is emitted in the gaseous state. They attribute the relatively high emission to the fact that $CaCl_2$ starts to decompose at 740°C and, as work is carried out at 850°C, this may cause part of the calcium chloride to decompose, emitting more chlorine and, also, as vapour is used in the gasification, the inverse reaction may take place and new HCl may be released. On the other hand, the melting temperature of $CaCl_2$ is 772°C and may form eutectic with CaO and the $CaCl_2$–CaO system with 6% molar volume of CaO has a lower melting temperature, 750°C [39], which may reduce the adsorption capacity of CaO in this liquid state.

5 Conclusions

Among the cheap adsorbents that are disposable after their use and subsequent inertisation, dolomite seems to be preferable as opposed to calcite because of its lower use temperature and greater conversion. Not only does it retain the H_2S but also the COS.

With the sulphidisation of dolomite, it is possible to easily reach the centre of the grains, achieving good conversions even with relatively thick sizes (2–2.5 mm).

The aforementioned implies that dolomite could be used in fixed or moving bed reactors, or in a fluidised bed or a combination of these, in one or several stages, with one of the main factors for control of sulphur compound content in the gas treated being the length of the bed to be run across and the speed of the gas.

In the case of very fine grains, of around 0.5 mm, a fixed bed is not convenient because of the high packing of grains and the creation of preferential channels for the gas to circulate, which would reduce the conversion of the adsorbent.

In inertisation through oxidation of sulphured dolomite and because of the formation of calcium sulphate, which tends to obstruct the pores, it is difficult to reach the centre of the grain and to achieve scarce residual calcium sulphide. Therefore, it would be convenient to use very fine-grain size distributions to enable a practical full conversion.

Given the conditions regarding the grain size distribution between the sulphidisation process and the oxidation process and, in a way, opposing these conditions, if thick dolomite is used for adsorption in a fixed or moving bed, it would be convenient to grind the resulting product before the oxidation stage.

The best gas is the mixture of oxygen and nitrogen for the oxidation of sulphured dolomite. The presence of CO_2 and/or water may lead to small amounts of H_2S or COS to appear at the start of the process, for which reason, the use of combustion flue gases is less convenient than that of the oxygen and nitrogen mixture.

SO_2 is always released during oxidation, to a greater or lesser extent depending on the oxidation conditions, for which reason, a system must be foreseen to extract it from the gas used for oxidation (conversion to sulphuric acid or adsorption in milk of calcium carbonate or lime).

The residue of oxidation would be a mixture of magnesium oxide, calcium oxide and calcium sulphate (and possibly some CaS residue) that could be used in certain processes in which CaO is required, for example, in acid effluent neutralisation.

The more costly adsorbents based on other metals, Zn in particular, have the advantage of achieving incredibly low H_2S or COS levels in the syngas treated. This is useful for certain applications that are more sensitive to the presence of sulphur compounds such as fuel-cells.

These adsorbents that, in principle, can be regenerated or reused are still under research to lengthen the number of cycles during which they can be used, and reduced, by means of better formulations, the mechanical degradation processes and the obstruction of pores normally because of the thermal and chemical stress arising from their continuous change in composition and to the sintering process resulting from the continuously high operating temperatures.

The adsorbents based on calcium may also retain other acid gases such as HCl, but the use conditions must be particularly adjusted because of problems regarding calcium chloride fusibility.

References

1. Méndez-Vigo I (2002) Operational experience of the Puertollano 330 MW IGCC power plant. In: International conference on clean coal technologies for our future. Chia Laguna, Italy, October 21–23
2. Ogenga DO, Mbarawa MM, Lee KT, Mohamed AR, Dahlan I (2010) Sulphur dioxide removal using South African limestone/siliceous materials. Fuel 89:2549–2555
3. Fenouil LA, Lynn S (1995) Study of calcium-based sorbents for high-temperature H2S removal. 1. Kinetics of H2S sorption by uncalcined limestone. Ind Engng Chem Res 34:2324–2333
4. Thambimuthu KV (1993) Gas cleaning for advanced coal-based power generation. IEA Coal Research, London
5. Yrjas P, Iisa K, Hupa M (1996) Limestone and dolomite as sulphur adsorbents under pressurized gasification conditions. Fuel 75:89–95
6. Pinto F, Lopes H, André RN, Gulyurtlu I, Cabrita I (2008) Effect of catalyst in the quality of syngas and by-products obtained by co-gasification of coal and wastes. 2: Heavy metals, sulphur and halogen compounds abatement. Fuel 87:1050–1062
7. Wang S, Lu GQ (1998) Catalytic activities and coking characteristic of oxides-supported Ni catalysts for CH_4 reforming with carbon dioxide. Energy Fuels 12:248–256
8. García L, Benedicto A, Romeo E, Salvador ML, Arauzo J, Bilbao R (2002) Hydrogen production by steam gasification of biomass using Ni–Al coprecipitated catalysts promoted with magnesium. Energy Fuels 16:1222–1230
9. Lin S, Harada M, Suzuki Y, Hatano H (2004) Continuous experiment regarding hydrogen production by coal/CaO reaction with steam (I) gas products. Fuel 83:869–874
10. Álvarez-Rodríguez R, Clemente-Jul C, Martín-Rubí JA (2007) Behaviour of the elements introduced with the fuels in their distribution and immobilization between the coal–petroleum coke IGCC solid products. Fuel 86:2081–2089
11. Abad A, Adánez J, García-Labiano F, de Diego LF, Gayán P (2004) Hot coal–gas desulphurization with calcium-based sorbents in a pressurized moving-bed reactor. Energy Fuels 18:1543–1554
12. Adánez J, Abad A, de Diego LF, García-Labiano F, Gayán P (2004) Direct sulfidation of half-calcined dolomite under pressurized conditions. Ind Engng Chem Res 43:4132–4139
13. García-Labiano F, Adánez J, Abad A, de Diego LF, Gayán P (2004) Effect of pressure on the sulfidation of calcined calcium-based sorbents. Energy Fuels 18:761–769
14. Adánez J, Abad A, García-Labiano F, de Diego LF, Gayán P (2005) H2S retention with Ca-based sorbents in a pressurized fixed-bed reactor: application to moving-bed design. Fuel 84:533–542
15. Álvarez-Rodríguez R, Clemente-Jul C (2008) Hot gas desulphurisation with dolomite sorbent in coal gasification. Fuel 87:3513–3521
16. García-Calzada M, Marbán G, Fuertes AB (2000) Decomposition of CaS particles at ambient conditions. Chem Eng Sci 55:1661–1674
17. Brooks MW, Lynn S (1997) Recovery of calcium carbonate and hydrogen sulphide from waste calcium sulphide. Ind Eng Chem Res 36:4236–4242

18. Swerdtfeger K, Brin I (1993) Problems in hot desulfurization of coal gas with lime. Erdöl Khole-Erdgas Petrochem 46:103–110
19. Anthony EJ, Jia L, Qiu K (2003) CaS oxidation by reaction with CO_2 and H_2O. Energy Fuels 17:363–368
20. Qiu K, Lindqvist O, Mattisson T (1999) Oxidation behaviour of desulphurization residues from gasification and fuel-rich combustion. Fuel 78:225–2331
21. Qiu K, Anthony EJ, Jia L (2001) Oxidation of sulfided limestone under the conditions of pressurized fluidized bed combustion. Fuel 80:549–558
22. Wu S, Uddin MA, Nagamine S, Sasaoka E (2004) Role of water vapour in oxidative decomposition of calcium sulphide. Fuel 83:671–677
23. Álvarez-Rodríguez R, Clemente-Jul C (2009) Oxidation of the sulfurised dolomite produced in the desulphurisation of the gasification gases. Fuel 88:2507–2519
24. Yrjas P, Hupa M, Iisa K (1996) Pressurized stabilization of desulfurization residues from gasification processes. Energy Fuels 10:1189–1195
25. Ahmed M, Alonso L, Palacios JM, Cilleruelo C, Abanades JC (2000) Structural changes in Zn ferrites as regenerable sorbents for hot coal gas desulphurization. Solid State Ion 138:51–62
26. Alonso L, Palacios JM, Moliner R (2001) The performance of some Zn based regenerable sorbents in hot coal gas desulphurization long-term using graphite as a pore-modifier additive. Energy Fuels 15:1398–1402
27. Pineda M, Palacios JM, Alonso L, García E, Moliner R (2000) Performance of zinc oxide based sorbents for hot coal gas desulfurization in multicycle test in a fixed-bed reactor. Fuel 79:885–895
28. Tomás-Alonso F, Palacios JM (2004) Synthesis and surface properties of zinc ferrite spices in supported sorbents for hot-coal gas desulfurization. Fuel Process Technol 80:191–203
29. Pan YG, Perales JF, Velo E, Puigjaner L (2005) Kinetic behaviour of iron oxide sorbent in hot gas desulfurization. Fuel 84:1105–1109
30. Alonso L, Palacios JM, García E, Moliner R (2000) Characterization of Mn and Cu oxides as regenerable sorbents for hot coal gas desulfurization. Fuel Process Technol 62:31–44
31. Alonso L, Palacios JM (2002) Performance and recovering of a Zn-doped manganese oxide as a regenerable sorbent for hot coal gas desulfurization. Energy Fuels 16:1550–1556
32. García E, Palacios JM, Alonso L, Moliner R (2000) Performance of Mn and Cu mixed oxides as regenerable sorbents for hot coal gas desulfurization. Energy Fuels 14:1296–1303
33. Ibarra JV, Cilleruelo C, García E, Pineda M, Palacios JM (1998) Vibrational spectroscopy study of zinc-containing mixed oxides as regenerable sulphur sorbents at high temperature. Vib Spectrosc 16:1–10
34. Pineda M, Fierro JLG, Palacios JM, Cilleruelo C, García E, Ibarra JV (1997) Characterization of zinc oxide and zinc ferrite doped with Ti or Cu as sorbents for hot gas desulphurization. Appl Surf Sci 119:1–10
35. Park NK, Lee DH, Jun JH, Lee JD, Kim JC, Chang CH (2006) Two-stage desulfurization process for hot gas ultra cleanup in IGCC. Fuel 85:227–234
36. Xie W, Chang L, Wang D, Xie K, Wall T, Yu J (2010) Removal of sulphur at high temperatures using iron-based sorbents supported on fine coal ash. Fuel 89:868–873
37. Swisher JH, Schwerdtfeger K (1992) Thermodynamics analysis of sorption reactions for the removal of sulphur from hot gases. J Mater Eng Perform 1:565–571
38. Partanen J, Backman P, Backman R, Hupa M (2005) Adsorption of HCl by limestone in hot flue gases. Part 1: the effects of temperature, gas atmosphere and adsorbent quality. Fuel 84:1664–1673
39. Weinell CE, Jensen PI, Dam-Johansen K, Livbjerg H (1992) Hydrogen chloride reaction with lime and limestone. Kinetics and sorption capacity. Ind Eng Chem Res 31:164–171
40. Font O, Querol X, Izquierdo M, Alvarez E, Moreno N, Diez S, Álvarez-Rodríguez R, Clemente-Jul C, Coca P, García-Peña F (2010) Partitioning of elements in a entrained flor IGCC plant: influence of selected operational conditions. Fuel 89:3250–3261

H_2 Production and CO_2 Separation

Antonello Di Donato

Abstract A promising technology for H_2 production and CO_2 separation is based on water gas shift reaction operated in water gas shift membrane reactor (WGSMR). In such a reactor the synthetic gas reacts with steam in a catalytic bed to produce additional hydrogen and CO_2. A H_2 selective membrane allows the simultaneous production of hydrogen at a high purity level and a stream of concentrated CO_2. The performance of such a reactor is defined in terms of CO conversion fraction, H_2 recovered fraction and produced H_2 flow rate. The chapter deals with the modelling of a WGSMR. A model developed to assist the design of a pilot scale, tube-in-tube reactor, is described. Simulations with the model are presented and discussed. The simulations were performed to analyse the effect of operating conditions (H_2O/CO ratio, temperature, pressure and syngas flow rate), catalyst characteristics (catalytic bed efficiency, void fraction) and membrane length, on the reactor performance. The results provide quantitative information to define the set of conditions to obtain the target value of the H_2 flow rate, with high values of CO conversion fraction and H_2 recovered fraction, minimising the length of the H_2 selective membrane. A last paragraph is dedicated to a short analysis of the main issues and foreseen solutions for the industrial application of the technology.

Notation

CSS	CO_2 capture and storage
IGCC	Integrated gasification combined cycle
PSA	Pressure swing adsorption
TSA	Temperature swing adsorption
WGS	Water gas shift reaction
WGSMR	Water gas shift membrane reactor

A. Di Donato (✉)
Centro Sviluppo Materiali, S.p.A, Via di Castel Romano,
100, 00128 Rome, Italy
e-mail: a.didonato@c-s-m.it

L. Puigjaner (ed.), *Syngas from Waste*, Green Energy and Technology,
DOI: 10.1007/978-0-85729-540-8_7, © Springer-Verlag London Limited 2011

1 Introduction: H_2 Production and CO_2 Separation in Gasification Processes

Because of the growing interest for hydrogen, to be used as a clean fuel source for electricity generation and transportation or for chemicals production [1], gasification based processes could be designed to produce hydrogen for partial or complete substitution of electricity as the end product. This possibility would improve the flexibility of the gasification plant, which could balance energy and hydrogen production according to the variations of raw materials costs and end product prices [2, 3].

The process scheme for hydrogen production consists of three steps:

(1) Gasification of fuel to produce a synthetic gas (syngas) whose main components are CO and H_2.
(2) Reaction of the syngas with purposely added steam to increase the hydrogen content by means of the water gas shift (WGS) reaction: $CO + H_2O \rightarrow CO_2 + H_2$.
(3) Separation of H_2 and CO_2 obtaining a stream of H_2 rich gas and a stream of concentrated CO_2.

To be attractive such a process should produce hydrogen at a high purity level, to allow its use in fuel cells, and should be combined with the implementation of CO_2 capture and storage (CSS) technologies, to favour the hydrogen production from not renewable fuels, like coal, refinery by-products, municipal wastes [4].

Hydrogen for fuel cells requires a very low concentration of impurities. In particular CO must be below 100 ppm [5]. To make economically suitable CO_2 disposal, the CO_2 stream must be produced at the appropriate pressure and purity, as required by transport, storing in geological formations or for enhanced oil recovering techniques [6, 7].

The existing commercial separation technologies, such as chemical and physical absorption, adsorption, low temperature membranes and cryogenic processes are extremely "energivorous", and the later re-heating and re-pressurisation stages heavily penalise (8–12%) the global efficiency of the plant. To overcome these limitations, Pressure swing adsorption (PSA) and temperature swing adsorption (TSA) processes based on liquid or porous media (e.g., activated carbons), and membrane reactor based processes are considered promising emerging technologies for H_2 and CO_2 separation from the syngas [8, 9].

A membrane reactor for WGS reaction, equipped with a H_2 selective membrane, or a CO_2 selective membrane, or both, operating at adequate temperature and pressure, would allow their separation without an energy cost for heating or compressing the gases [10]. A reactor for carrying out the process is simple, without moving parts, and can be replicated in a sequence of single units to increase the performance. This option would result in a tremendous process intensification. The membrane reactor technology is in an advanced state of development. Both catalysts and membranes have been widely studied and a vast literature about small scale reactors is available. Although not yet mature for industrial application, the findings on catalysts and membranes make possible to move the investigation to pilot scale. In the conclusions of the European HYDROSEP [11] project a pilot reactor, to be integrated into an IGCC plant, has been proposed.

This chapter deals with the modelling of a reactor for catalytic WGS reaction integrated with a H_2 selective membrane. The presented model was developed to assess the feasibility of a pilot reactor and to assist its basic design. It constitutes a good example of the role of mathematical modelling in the evolution of the technology from basic studies to final applications.

2 H_2 Production and CO_2 Separation in a Single Catalytic Membrane Reactor

2.1 Process Concept: Thermodynamic Equilibrium of the Water Gas Shift Reaction

The extent of the WGS reaction can be measured by the CO conversion fraction, X_{CO}, defined as the ratio of CO moles transformed to CO_2 to the initial CO moles.

$$X_{CO} = \frac{\text{mol CO in} - \text{mol CO out}}{\text{mol CO in}}. \tag{1}$$

The CO conversion fraction at the thermodynamic equilibrium can be calculated by the initial composition and temperature of the gas mixture (syngas plus steam) and data of Gibbs free energy, ΔG^0. A simple linear correlation between reported ΔG^0 data [12] and temperature is the following:

$$\Delta G^0 = 37.43T - 39{,}704.21 \quad \text{in}\left(\text{J mol}^{-1}\right) \tag{2}$$

At room temperature the Gibbs free energy of the WGS reaction is negative and the equilibrium constant calculated from Eq. 2 is 91,393.30 at 300 K. The reaction is almost completely shifted towards the products. But at low temperatures the reaction is slow, with no practical use. Increasing the temperature, increases the velocity. As the reaction is exothermic, the yield decreases. At 900 K the equilibrium constant is 2.24.

Figure 1 shows curves of CO conversion fraction as a function of temperature at different H_2O/CO ratios. The composition used in the calculation is representative of the composition obtained adding steam to a syngas from coal gasification (CO 60%, H_2 22%, CO_2 4% and inert 14%). The simple thermodynamic calculations show the decrease of the thermodynamic CO conversion as the temperature increases and H_2O/CO ratio decreases.

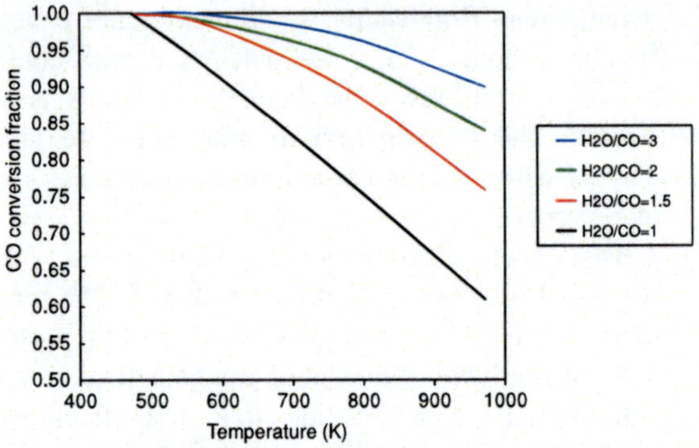

Fig. 1 CO conversion fraction as a function of temperature, calculated from the equilibrium constant of the WGS reaction mixing an initial gas mixture (CO 60%, H_2 22%, CO_2 4% and inert 14%) with H_2O at four different H_2O/CO ratios

Exercise 1 (*Thermodynamic Calculation of the Co-Conversion Fraction*) The objective is the calculation of the CO conversion fraction in a gas mixture obtained mixing a given volume, Q (m^3 STP), of a syngas of prefixed composition with a volume of steam, at a defined H_2O/CO (mol mol^{-1}) ratio and constant temperature, T (K), assuming that the condition of thermodynamic equilibrium is achieved.

The initial moles of the species can be easily calculated from the syngas volume syngas composition, and H_2O/CO ratio (mol H_2O/mol CO).

$$\text{mol CO in} = Q \times \frac{\%\text{CO}}{100} \times \frac{1}{22.414}$$

where 22.414 (mol m^{-3} STP) is the molar volume of the gas at standard temperature and Pressure. Similarly for CO_2 and H_2.

The moles of H_2O are given by the imposed H_2O/CO ratio. N is the sum of the moles of the species in the gas mixture.

According to the WGS reaction stoichiometry:

$$CO + H_2O \rightarrow CO_2 + H_2$$

x moles of CO and x moles of H_2O are transformed in x moles of CO_2 and x moles of H_2. The total number of moles, N, remains the same.

At equilibrium the concentrations of the species (molar fractions) are related to the equilibrium constant.

$$\frac{\frac{(\text{mol CO}_2\text{in}+x)}{N} \times \frac{(\text{mol H}_2\text{in}+x)}{N}}{\frac{(\text{mol CO in}-x)}{N} \times \frac{(\text{mol H}_2\text{O in}-x)}{N}} = K_{eq} = e^{-\frac{\Delta G^0}{R \cdot T}}$$

where ΔG^0 at temperature T can be calculated from Eq. 2 and $R =$ 8.31 J mol^{-1} K^{-1}.

The transformed moles x can be easily found with a numerical algorithm to find the zero of an algebraic equation. x is the difference mol CO in – mol CO out in the Eq. 1. Figure 1 shows the conversion values calculated for different temperatures and initial gas compositions.

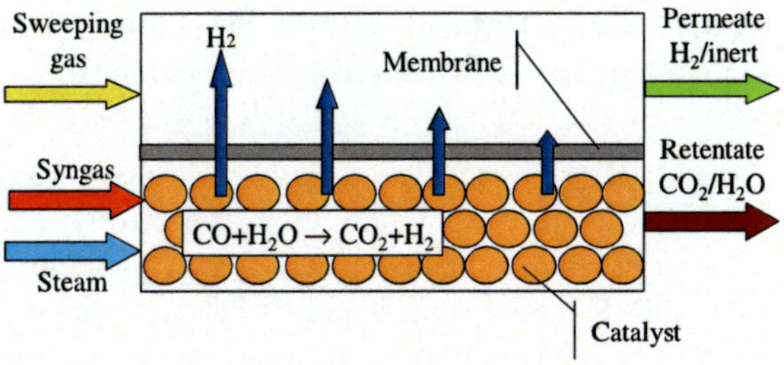

Fig. 2 WGSMR process concept. The WGS reaction occurs in the reaction zone. Hydrogen passes from the reaction zone to the permeation zone through the selective membrane

2.2 Process Concept: Removal of H_2 *to Promote the* CO *Conversion*

If H_2 is continuously removed from the reaction zone, the reaction shifts towards the products and the CO conversion fraction increases. The contemporary H_2 production by WGS reaction, and H_2 removal from the reaction zone, can be achieved in a single unit: water gas shift membrane reactor (WGSMR). In such a reactor a H_2 selective membrane physically separates the reactor in two zones: the reaction zone and the permeation zone. Syngas and steam are fed under pressure into the reaction zone of the reactor, filled with the catalyst, where the WGS reaction occurs. Hydrogen selectively passes through the membrane into the permeation zone. From the permeation zone H_2 is continuously extracted by aspiration. A sweeping gas can assist the extraction. The hydrogen removal maintains a H_2 concentration gradient between the two sides of the membrane, which is the driving force of H_2 transport. Figure 2 schematically shows the process concept. Following a common convention, the two gas mixtures exiting from the permeation and reaction zones are called permeate and retentate, respectively.

The production rate of hydrogen depends on the catalyst efficiency and on membrane permeability. The hydrogen purity depends on the membrane selectivity. The retentate can be re-circulated or can be post-combusted for complete oxidation of residual H_2 and CO. After post-combustion and water elimination, high pressure CO_2 remains for storing. The process becomes more advantageous if the syngas is previously cleaned, especially from dust and sulphur, in hot cleaning operations.

The efficiency of the reactor to separate hydrogen is measured by the H_2 recovered fraction, which is the ratio between the separated hydrogen to the total produced hydrogen (the total hydrogen is the sum of the pre-existing and produced hydrogen).

$$R_{H_2} = \frac{\mathrm{H_2\ separated}}{\mathrm{H_2\ in} + \mathrm{H_2\ produced}} \tag{3}$$

The H_2 flow rate produced by the reactor is directly linked to the CO conversion fraction Eq. 1 and H_2 recovered fraction (Eq. 3).

$$\mathrm{mol\,H_2\,Prod} = (\mathrm{mol\,H_2\,in} + \mathrm{mol\,CO\,in} \times X_{\mathrm{CO}}) \times R_{\mathrm{H_2}} \tag{4}$$

CO conversion fraction (Eq. 1), H_2 recovered fraction (Eq. 3) and produced H_2 flow rate (Eq. 4) are the three main parameters defining the reactor performances.

3 A Mathematical Model for Catalytic Membrane Reactor Design and Performance Evaluation

Many WGSMR models are reported in the literature, dedicated to the design of the reactor or to the optimisation of a process scheme including such a reactor. The most are based and dimensioned on laboratory scale reactors. In what follows a WGSMR model is presented dedicated to the design of a pilot scale reactor for hydrogen production from syngas to be integrated in an IGCC power plant.

Such a unit pilot should supply a flow rate of hydrogen of the order of 180 m^3 STP h^{-1} (i.e., $\sim$2.2 mol s^{-1}) sufficient to operate a 250 kW fuel cell [11].

The model was developed in a thesis [13], integrated with experimental data of membrane permeability purposely measured, and applied in the context of the Agapute [14] and HYDROSEP [11] European projects. An other application of the model can be found in Piemonte et al. [15].

3.1 Basic Reactor Configuration

The basic configuration of the reactor was decided ‘a priori’. The reactor is constituted of two coaxial tubes (the so-called tube-in-tube configuration) with the following dimensions: length 10 m, external radius 0.080 m and internal radius 0.045 m.

This configuration, characterised by a high ratio of length to radius, derived by a preliminary analysis of different configurations (in terms of length and radius of the tubes), carried out in Favetta’s thesis work, which demonstrated that a high length to radius ratio favours high CO conversion fraction. Similar length to radius ratios are currently adopted in the catalytic tubes of methane reforming reactors.

The catalyst can fill either the external tube or the internal tube. The H_2 selective membrane is a dense metallic membrane, perfectly H_2 selective. It can be as long as the whole reactor or shorter. A sweeping gas can be introduced either in co-current or in counter-current to the H_2 extracted flow.

3.2 Model Purposes

The WGSMR model was designed to evaluate the effect of operating conditions, characteristics of catalytic bed and membrane, reactor configuration on CO conversion, hydrogen recovery and H_2 production.

- Operating conditions are temperature, pressure, gas inlet flow rate, ratio of steam flow rate to gas flow rate and sweeping gas flow rate.
- Catalytic bed characteristics are catalyst efficiency, particle size and void fraction.

- Membrane characteristics are hydrogen permeability and thickness
- Reactor configurations refer to the location of the catalytic bed (internal or external tube) and length of the selective membrane.

This set of information is the basis to evaluate the feasibility of the WGSMR and the requirements for its engineering.

3.3 Physical Laws Governing the Reactor Process

The WGSMR model considers heterogeneous chemical kinetics, mass and heat transport phenomena in a porous medium, hydrogen flow rate through the membrane.

3.3.1 WGS Reaction Kinetics

The expression for the reaction rate depends on the catalytic mechanism. A wide variety of reaction mechanisms have been proposed for the various investigated catalysts. Following indications from the literature [16, 17], the present WGSMR model assumes that the catalysed WGS reaction occurs according to the following mechanism.

H_2O and CO moves from the gas bulk to the catalyst surface, where they are adsorbed on the catalyst surface. The adsorbed species react with a reaction rate, forming the products, which are, in turn, desorbed giving the gaseous products, H_2 and CO_2. The limiting step of the process is the reaction of the adsorbed species on the catalyst surface. Consequently, at the catalyst surface an equilibrium condition is established between gaseous and adsorbed species.

According to this mechanism the reaction rate is given by the Langmuir–Hinshelwood expression.

$$r_{CO} = k \times K_{CO} \times K_{H_2O} \times \frac{\left(p_{CO} \times p_{H_2O} - \frac{p_{CO_2} \times p_{H_2}}{K_{eq}}\right)}{(1 + K_{CO} \times p_{CO} + K_{H_2O} \times p_{H_2O} + K_{CO_2} \times p_{CO_2})^2} \times \rho_{cat} \quad (5)$$

where:

r_{CO}	Reaction rate	(mol kg^{-1} s^{-1})
p_X	Partial pressure of component X	(Pa Pa^{-1})
ρ_{cat}	Catalyst density	(kg m^{-3})
$K_{CO}, K_{H_2O}, K_{CO_2}$	Absorption coefficients for CO, H_2O and CO_2 on the catalyst surface	(Pa^{-1})
K_{eq}	Equilibrium constant of the Water Gas Shift reaction	(Pa Pa^{-1})
k	Kinetic coefficient of the reaction	(mol m^{-3} s^{-1})

The values of the absorption coefficients, equilibrium constant and kinetic coefficient are calculated as a function of temperature with the following formulas [16].

$$K_{CO} = 10^{-5} \times e^{\left(\frac{3064}{1.987 \cdot T} - \frac{6.74}{1.987}\right)} \quad (\mathrm{Pa}^{-1})$$

$$K_{H_2O} = 10^{-5} \times e^{\left(\frac{-6216}{1.987 \cdot T} + \frac{12.77}{1.987}\right)} \quad (\mathrm{Pa}^{-1})$$

$$K_{CO_2} = 10^{-5} \times e^{\left(\frac{12542}{1.987 \cdot T} - \frac{18.45}{1.987}\right)} \quad (\mathrm{Pa}^{-1})$$

$$K_{eq} = e^{\left(\frac{4577.8}{T} - 4.33\right)} \quad \text{(dimensionless)}$$

$$k = \frac{1{,}000}{60} \times e^{\left(\frac{-29364}{1.987 \cdot T} + \frac{40.32}{1.987}\right)} \quad \left(\mathrm{mol\,kg_{catalyst}^{-1}\,s^{-1}}\right)$$

Equation 4 represents the reaction rate at catalyst surface, where an equilibrium condition is established between adsorbed species and gaseous species in the gas phase at the interface. The partial pressure of the various species at the interface depends on the fluid dynamics of the system.

The distributions of temperature, pressure and gas composition depend on the dimensions of the reactor, on the size of the catalyst particles, on the void fraction and on the flow mass. Different approaches with different levels of complexity, approximations and simplifications have been proposed to solve the problem. The most comprehensive models are based on solving the complete Navier–Stokes equations, considering both axial and radial mass, momentum and energy transport. This approach is computationally expensive and require computational fluid dynamic (CFD) codes and a large effort of skilled specialists.

The present WGSMR model is based on the plug flow representation of the system. This is a relatively simpler way to formulate and manage a WGSMR model. The key assumption is that the fluid is perfectly mixed in the radial direction but not in the axial direction (forwards or backwards). Plug flow models are written in terms of ordinary differential equations, the solution for which can be calculated providing that appropriate boundary conditions are known. An evaluation of the two approaches, CFD and plug flow, can be found in Raja et al. [18].

Despite its simplicity the plug flow approach has been largely adopted to model catalytic reactors. A recent application to WGSMR has been presented by Gosiewsky et al. [19], In that paper also selected examples from the literature are shortly discussed.

To take into account in a global way the complex fluid dynamic aspects, a ‘catalyst efficiency’ parameter η, specific of the catalytic bed can be introduced. This parameter accounts for the difficulty of the gas to penetrate, in a homogeneous way, into the catalytic bed. The parameter η represents an average catalyst efficiency. It is specific for a given catalyst and bed configuration. η is an adjustable parameter of the model.

3.3.2 H_2 Flow Rate Through the Membrane

In the present model a classical formulation of the hydrogen flow rate through a metallic dense membrane has been adopted, based on the assumption that the step controlling the separation process is the hydrogen diffusion into the membrane.

$$J_{H_2}^{Perm} = \frac{B_H}{\delta}\left(P_{H_2,React}^{0.5} - P_{H_2,Perm}^{0.5}\right) \quad (6)$$

where

$J_{H_2}^{Perm}$	H_2 flow rate through the membrane	mol m^{-2} s^{-1}
B_H	H_2 permeability	mol m^{-1} s^{-1}
δ	membrane thickness	m
P_{H_2}	Hydrogen partial pressure	Pa Pa^{-1}

Accepting some simplifications the Eq. 6 can be derived as presented in the next Exercise 2.

Exercise 2 (*Obtaining the Relationships between the H_2 Flow Rate Through a Dense Metallic Membrane and the H_2 Partial Pressure in the Gas Phase*)

In a metallic alloy hydrogen dissolves in atomic form:

$$0.5H_2 \leftrightarrow H \quad (7)$$

The hydrogen concentration in solid solution is related to the H_2 partial pressure according to Sievert's law.

$$[H] \times \gamma = K \times P_{H_2}^{0.5} \quad (8)$$

where

[H]	Concentration of atomic H in the solid solution	(mol m^{-3})
P_{H_2}	H_2 partial pressure in gas phase	(Pa Pa^{-1})
K	Equilibrium constant	(mol m^{-3})
γ	Activity coefficient	(dimensionless)

The diffusion through the membrane is driven by the concentration gradient of the dissolved hydrogen inside the membrane thickness. The hydrogen flow rate is:

$$J_H = D \times \frac{d[H]}{dx} \quad (9)$$

where

J_H	H flow rate through the membrane	mol m^{-2} s^{-1}
[H]	H concentration in solid solution	mol m^{-3}
D	Diffusion coefficient of H in solid solution	m^2 s^{-1}
x	Dimension along membrane thickness	m

At steady state, at the two membrane/gas interfaces the equilibrium is established according to the Eq. 9. The flow rates of H_2 in the gas phase (on both reaction and permeation sides) and the flow rate of atomic hydrogen in solid solution are linked by stoichiometry.

$$-J_{H_2}^{\text{React}} = J_{H_2}^{\text{Perm}} = \frac{1}{2} J_{H}^{\text{Membrane}} \tag{10}$$

React and Perm stand for reaction zone and permeation zone.

If D is constant along the membrane thickness, the integration of Eq. 9 gives:

$$J_{H}^{\text{Membrane}} = \frac{D}{\delta}\left([H]_{\text{React}} - [H]_{\text{Perm}}\right) \tag{11}$$

The substitution of [H] with P_{H_2} (Eq. 8) gives the equation of the hydrogen flow rate (Eq. 6) through a membrane of thickness δ, as a function the partial pressures of hydrogen in the two sides of the membrane and its permeability B_H.

$$B_H = \frac{1}{2} \times \frac{K}{\gamma} \times \frac{D}{\delta} \tag{12}$$

$$J_{H_2}^{\text{Perm}} = \frac{B_H}{\delta}\left(P_{H_2,\text{React}}^{0.5} - P_{H_2,\text{Perm}}^{0.5}\right) \tag{6}$$

The expression of hydrogen permeability contains the product of a diffusion coefficient and an equilibrium constant. Both vary with the temperature according to an Arrenhius type function. For this reason its dependence on the temperature is commonly expressed by an Arrhenius type relationships:

$$B_H = A \times e^{\frac{E}{R \cdot T}} \tag{13}$$

The simplifying hypotheses in this exercise are that D and γ are constant into the membrane. The literature on hydrogen in metals is immense. Values of hydrogen solubility, equilibrium constant and diffusivity coefficient in palladium alloys can be found in many books and papers. See, for example, Alefeld and Volkl [20]. Both measurement and prediction of equilibrium constant, activity coefficient and diffusion coefficient are complex and strongly dependent on metal composition and metal crystalline structure [21]. Consequently, although in principle B_H could be calculated from measurements of the quantities in Eq. 12, the usual practice is the direct experimental determination of the A and E parameters in Eq. 13. As a matter of fact in many cases permeability measurements are used to derive the values of hydrogen solubility or diffusivity [22].

3.3.3 Conservation Equations

The mathematical model is a set of ordinary differential equations at steady state. The equations are the mass balances for the considered chemical species (CO, CO_2, H_2, H_2O and inert) in the reaction and permeation zones, the energy balance in both zones and the pressure drop along the reactor in the reaction zone. The membrane temperature is assumed constant along the thickness. The initial condition is the set of composition, temperature and pressure of the gas mixture entering the reactor calculated assuming instantaneous mixing of syngas and steam.

The model calculates the evolution of gas composition, temperature and pressure along the reactor.

The mass balance in the reaction zone for the chemical species X = CO, H_2O, CO_2 is:

$$\frac{dF_x}{dz} = \pm\eta \times (1-\varepsilon) \times (r_{CO}) \times A_{React} \tag{14}$$

The positive sign is for the formed species CO_2, the negative sign is for H_2O and CO. The mass balance for H_2 is:

$$\frac{dF_{H_2}}{dz} = \eta \times (1-\varepsilon) \times (r_{CO}) \times A_{React} - J_{H_2}^{Perm} \times 2\pi R_{Membrane} \tag{15}$$

where

F_X	Flow rate of species X (CO, H_2O, CO_2) along the reactor	mol s^{-1}
F_{H_2}	Flow rate of H_2 along the reactor	mol s^{-1}
z	Position along the membrane	m
ε	Catalyst void fraction	Dimensionless
r_{CO}	Reaction rate in terms of CO Eq. (5)	mol m^{-3} s^{-1}
A_{React}	Area of the section of the catalytic bed $A_{React} = \pi\,(R_e{}^2 - R_i{}^2)$ or $A_{React} = \pi R_i{}^2$	m^2
$J_{H_2}^{Perm}$	H_2 flow rate through the membrane	mol m^{-3} s^{-1}
$R_{Membrane}$	Radius of the membrane	m
η	Catalyst efficiency	Dimensionless

The energy balance was introduced to calculate the temperature profile in the reactor. The equation is based on the enthalpy variation of the WGS reaction, on the enthalpy variation associated to the H_2 flux through the membrane and on the heat transfer between the gas flow and the reactor tubes. The thermal exchange between gas and tube walls mainly depends on gas velocity, gas physical characteristics and catalyst particle size. The heat exchange in the packed bed is treated by Whitaker [23].

The model includes an externally source of energy, positive or negative, to heat or to cool the reactor, in order to control the wall temperature. The detailed mathematical formulation of the present model is described in Favetta's thesis [13].

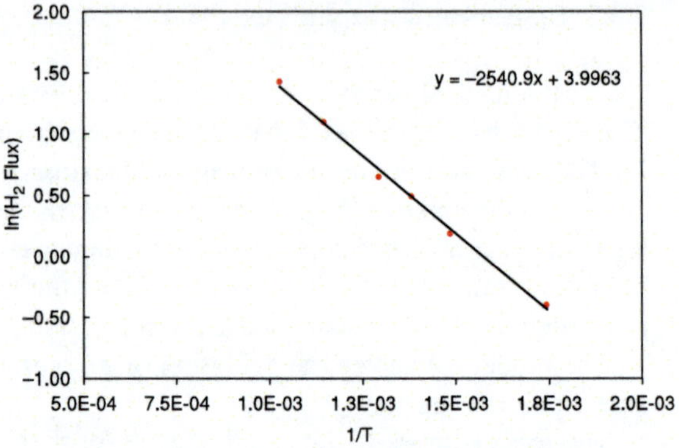

Fig. 3 Plot of the $\ln\left(J_{H_2}^{Perm}\right)$ as a function of $1/T$ and linear fitting of the experimental values. The slope of the curve is the term E/R in Eq. 13

The pressure drop was calculated with the classical Ergun equation [24, 25]. The main factors affecting the pressure drop are void fraction of the catalytic bed and catalyst particle size.

3.4 Experimental Measurement of Membrane Permeability

The WGSMR presented here was envisioned equipped with a H_2 selective dense metallic (Pd/Ag alloy) membrane, whose A and E parameters were experimentally determined on tubular membrane of dimensions 10 cm length and 2 cm diameter. The determination was carried out by means of an apparatus consisting of a gas feeding system, a membrane reactor maintained at constant temperature by means of an external controlled heating system, and a gas analysis section to measure the composition of the gas before and after the membrane reactor and the H_2 flow rate from the permeation zone. Experiments were performed at different pressure, composition and temperature of the feed gas.

The experimental details and results can be found in Agapute [14] and Hydrosep [11] reports.

Figure 3 shows the fitting of the H_2 flow rate as a function of temperature.

4 Model-Based Water Gas Shift Membrane Reactor Design and Performance Evaluation

4.1 Process Objectives

The process objective is to obtain a target value of H_2 production flow rate (see Sect. 3) with maximum efficiency of H_2 and CO_2 separation. The parameters CO conversion fraction (Eq. 1) and H_2 recovered fraction (Eq. 3) measure the reactor performance.

$J_{H_2}^{Perm}$	H_2 flow rate through the membrane	mol m^{-2} s^{-1}
B_H	H_2 permeability	mol m^{-1} s^{-1}
δ	Membrane thickness	m
P_{H_2}	Hydrogen partial pressure	Pa Pa^{-1}

Table 1 Reference conditions in the model simulations

Reactor	Unit	Values used when not differently specified
Length	m	10
Internal radius of the external tube	m	0.08
External radius of the internal tube	m	0.045
Catalyst		
Particle average diameter	m	0.005
Density	kg m^{-3}	2400
Void fraction	–	0.6
Efficiency	–	0.28
Membrane		
Activation energy[a]	kJ mol^{-1}	22.35
Pre-exponential factor[a]	mol m^{-1} s^{-1} $Pa^{-0.5}$	1.27×10^{-9}
Thickness	m	5×10^{-6}
Permeability (B_H/δ)	mol m^{-2} s^{-1} $Pa^{-0.5}$	2.55×10^{-4}
Reaction zone		
Gas inlet temperature	K	628
Gas inlet pressure	Pa	2,376.53
Syngas flow rate	mol s^{-1}	3.6
Syngas composition		
CO	mol %	60
CO_2	mol %	4
H_2	mol %	22
Inert	mol %	18
H_2O/CO	–	3
Permeation zone		
Sweeping gas flow rate	mol s^{-1}	3.9
Sweeping gas inlet temperature	K	628
Pressure permeation side	Pa	101,325

[a] From fitting of experimental data. See Fig. 3

4.2 Reference Conditions

The reactor is a tubular reactor described in Sect. 3.1. The catalytic bed is made of spherical particles of constant diameter and density. The H_2 flow rate through the membrane is calculated by means of Eqs. 6 and 13. The inlet syngas has fixed composition and temperature.

Table 1 lists the main variables of the model. The reported values are used in the calculations when not differently specified.

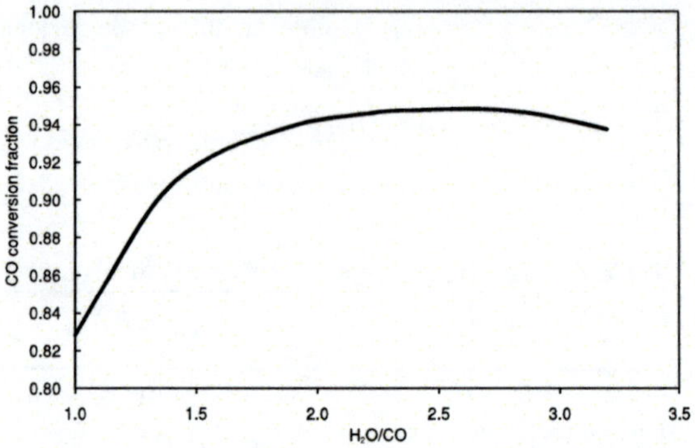

Fig. 4 CO conversion fraction as a function of H_2O/CO ratio

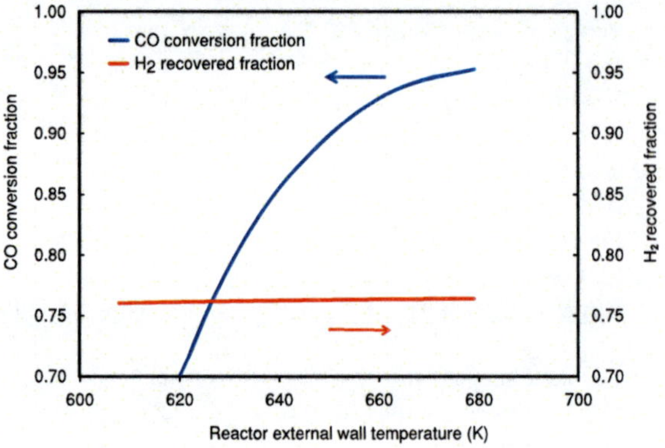

Fig. 5 CO conversion fraction and H_2 recovered fraction as a function of temperature of the external wall of the reactor

4.3 Influence of Operating Conditions on Reactor Performance

According to thermodynamics, increasing the ratio H_2O/CO, the CO conversion fraction increases (see Fig. 1). The model was used to explore the effect of increasing the H_2O/CO ratio on the reactor performance. Figure 4 shows the result. The curve is not monotone, but a maximum in CO conversion is obtained at $H_2O/CO = 2.8$.

The model indicates that the kinetic slows down when the H_2O/CO ratio exceeds a maximum value. This is an effect of the expression rate of the WGS reaction written according to the Langmuir–Hinshelwood mechanism. The experimental verification of the existence of a maximum of CO conversion fraction as a function of H_2O/CO ratio is a validation of the mechanism.

The reactor performance can be improved by setting different conditions, like higher pressure and higher temperature. Figures 5 and 6 show the effect on these parameters of the reaction zone pressure and reactor external wall temperature, respectively. These results indicate that a combination of high temperature and high pressure improves the reactor performance.

The WGS reaction is exothermic, and the equilibrium is shifted to lower CO conversion as temperature increases. But the hydrogen removal sustains the

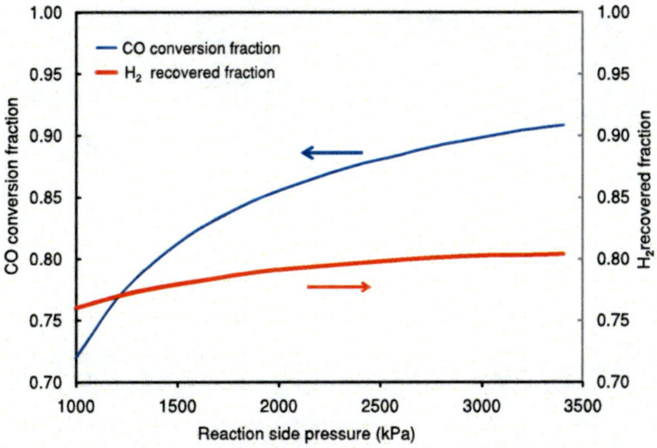

Fig. 6 CO conversion fraction and H_2 recovered fraction as a function of pressure in the reaction zone

reaction overcoming the thermodynamic limit. Figure 5 shows the effect of temperature on CO conversion fraction and H_2 recovered fraction. The presence of the membrane allows greater hydrogen production in the reaction zone. But the effect on the H_2 separation efficiency is negligible. A better exploitation of the higher CO conversion is obtained increasing the pressure in the reaction zone that means to increase the driving force for the hydrogen separation and consequent higher H_2 flow rate through the membrane. Figure 6 shows the effect of pressure on CO conversion fraction and H_2 recovered fraction. This result means that increasing reaction rate is a necessary but not sufficient condition. The process effectiveness strongly depends on the operating pressure.

4.4 Influence of Catalytic Bed Characteristics on Reactor Performance

The evaluation of the effect of parameters such as catalyst particle size, void fraction, catalyst efficiency, membrane thickness on CO conversion fraction and H_2 recovered fraction is necessary to define the requirements for catalyst and membrane.

Figures 7 and 8 show the CO conversion fraction as a function of catalyst efficiency and void fraction in the catalytic bed, respectively. To obtain a CO conversion fraction higher than 0.8 the catalyst efficiency must be higher than 0.3 and the void fraction lower than 0.6. This requirements must be obtained designing the catalytic bed, in terms of particle size and shapes.

4.5 Performance Evaluation of Different Reactor Configurations

A typical application of this type of model is the simulation of reactor configuration. The next two examples concern the geometry of the system reactor plus

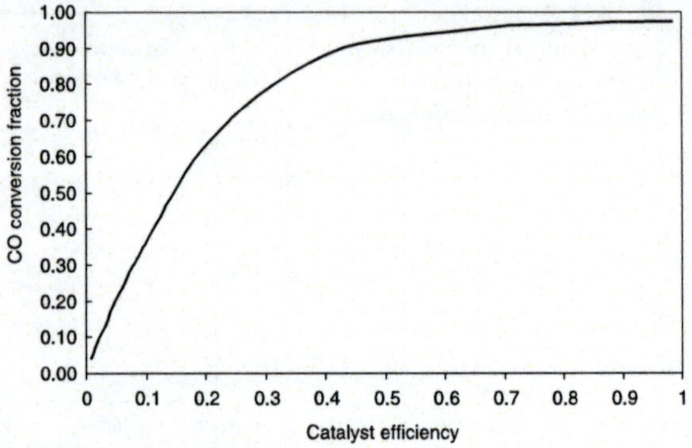

Fig. 7 CO conversion fraction as a function of catalyst efficiency

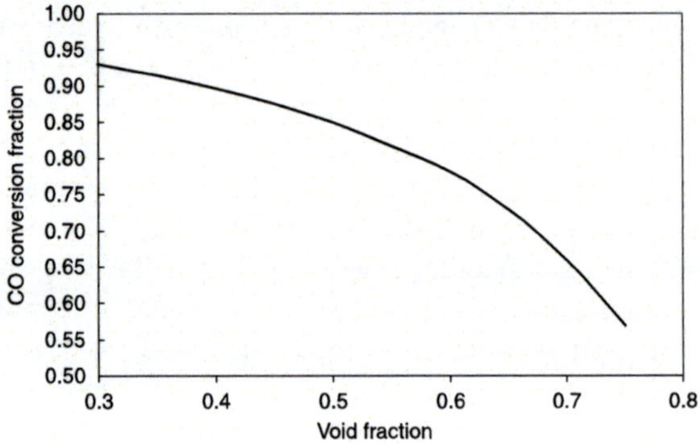

Fig. 8 CO conversion fraction as a function of void fraction in the catalytic bed

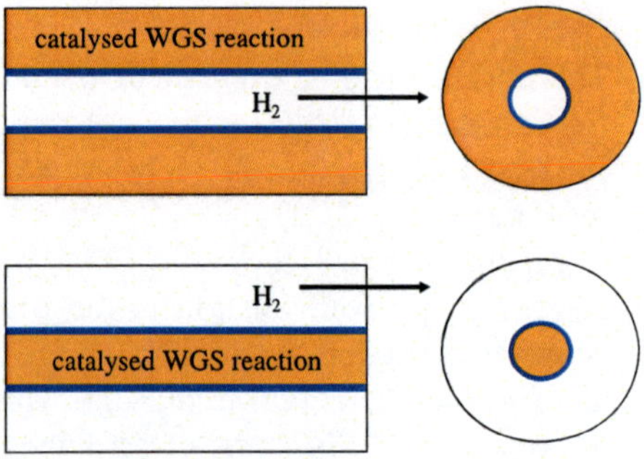

Fig. 9 Different configurations of the WGSMR

membrane. In the adopted tube-in-tube configuration, the catalytic bed can be in the external zone or in the internal zone, as schematically depicted in Fig. 9.

Both configurations are reported in the literature. Here the model has been used to compare the performances of the two configurations in similar conditions. Figure 10 shows the difference in terms of CO conversion fraction.

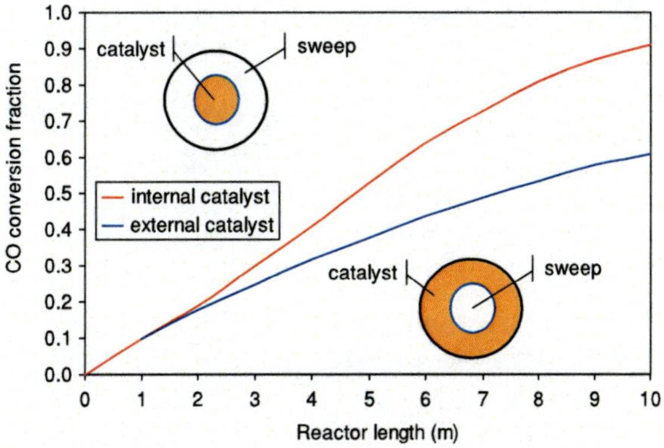

Fig. 10 CO conversion fraction for the two WGSMR configurations along the reactor length

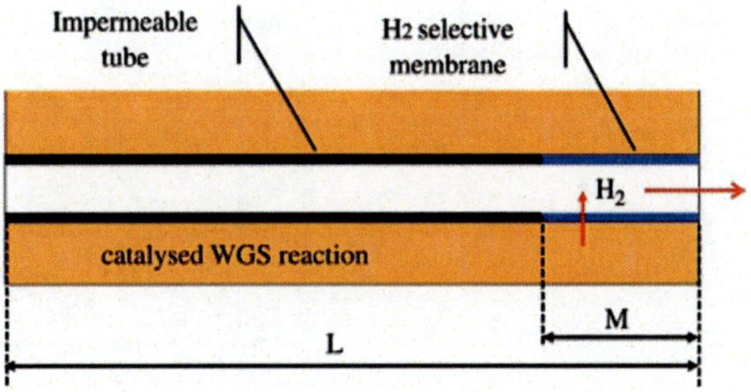

Fig. 11 Schematic representation of the tube-in-tube reactor with the membrane shorter than the reactor

The configuration with the catalyst in the internal tube is clearly more efficient. This is a consequence of the lower energy dissipation when the reaction zone is the internal tube, which causes a higher temperature in the reaction zone. As seen above, both CO conversion factor and H_2 recovered factor increase as the temperature increases. The corresponding H_2 flow rate increase is of the order of 10%.

The second example shows the use of the model to optimise the length of the membrane. The high cost of the membrane requires an analysis of the most suitable dimensions and placement, without reducing its performance. The question is if the membrane must have the same length as the reactor or it can be shorter. Figure 11 schematically shows the configuration. Simulations with the WGSMR model using different membrane lengths (*M*) inside the reactor (length $L = 10$ m) were performed. The CO conversion fractions for the four calculated *M*/*L* ratios are plotted in Fig. 12. The red curve is the X_{CO} as a function of the reactor length when the membrane inside the reactor has the same length of the reactor itself. The blue curves have been calculated for membranes of different lengths inside the reactor. The black curve is the X_{CO} of the WGS reactor without the membrane.

A membrane with a length of 40% of the total length of the reactor gives almost the same X_{CO} of the membrane as long as the reactor. A longer membrane does not produce a significant X_{CO} improvement.

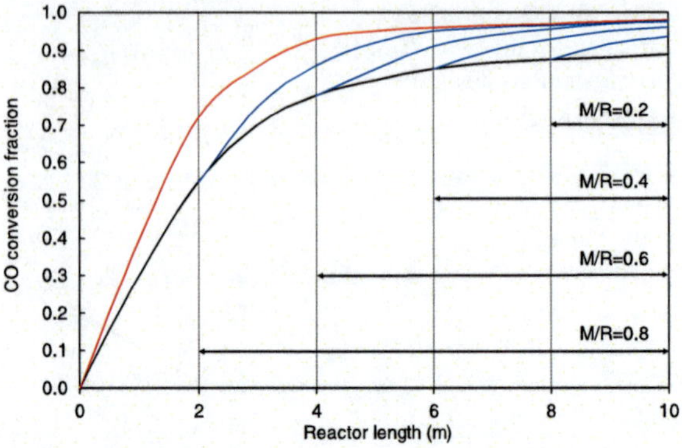

Fig. 12 CO conversion fraction calculated along the reactor length in the following cases: reactor without H_2 selective membrane (*black curve*), reactor with the H_2 selective membrane as long as the whole reactor (*red curve*), reactor with H_2 selective membrane shorter than the whole reactor (*blue curves*)

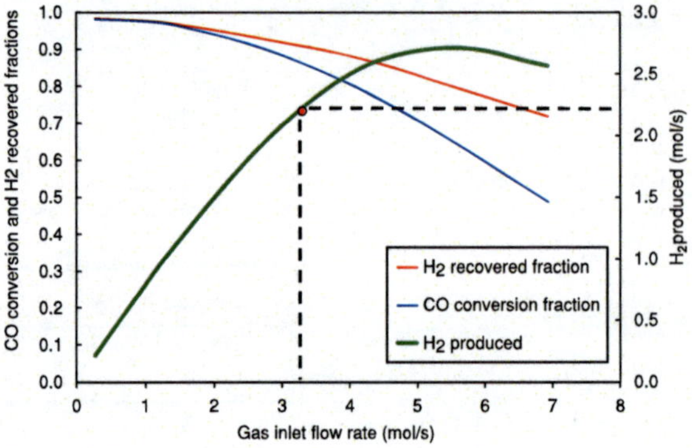

Fig. 13 CO conversion fraction, H_2 recovered fraction and produced H_2 flow rate as a function of gas inlet flow rate

In conclusion the model suggests to build the WGSMR using the H_2 selective membrane for only the final 60% of the internal tube, with a corresponding reduction of the membrane cost. The rest of the tube is an impermeable tube.

4.6 Definition of a Working Point

The reactor working point is a set of operating conditions. Some of these conditions, like temperature and pressure, could be imposed by operational constraints of catalyst and membrane, and could be different from the optimum values. Once defined the applicable temperature and pressure, the model can calculate the syngas flow rate to be fed into the reactor to obtain the target value of hydrogen flow rate.

Figure 13 shows an example. CO conversion fraction, H_2 recovered fraction and produced H_2 flow rate (in moles per second) are calculated as a function of the gas inlet flow rate. The calculated curves allow the identification of a working point to obtain the above defined target of 180 m^3 STP h^{-1} H_2 production (equivalent to 2.2 mol s^{-1}). With the set of operating conditions used for the calculation, a syngas flow rate of 3.3 mol s^{-1} must be fed to the reactor (i.e., $\sim$270 m^3 STP h^{-1}).

5 Feasibility and Foreseen Technological Solutions for Catalyst, Membrane and Reactor Configuration

Catalysts for the WGS reaction have been studied for over a century. A large variety of catalyst can be used, including reducible oxides, oxidable metals and noble metals dispersed in oxides.

The most common catalyst at higher temperature, used since 1914, is a mixture of iron oxide and chromium oxide (Fe_2O_3 and Cr_2O_3). Metallic Ni dispersed in an oxide substrate (e.g., CaO, Al_2O_3 and Ce_2O_3) is also extensively studied.

To improve the performance, especially in terms of long term stability and resistance against polluting species such as sulphur and nitrogen compounds, a large number of catalysts are continuously studied, the most present in the literature being copper, zinc, nickel and noble metals (Au, Pt, Pd) deposited on oxides supports. Experimental and modelling studies dedicated to WGSMR report the use of both commercial and purposely developed catalysts. The main issues in the current research are the experimental determination of kinetics laws and parameters and the improvement of the stability and life of catalysts [26].

A recent review on WGS catalyst can be found in Smith et al. [27]. A systematic analysis of the molecular mechanisms of the WGS reaction has been proposed by Callaghan [28].

On the basis of the transport mechanism of chemical species inside the wall, the membranes for H_2 separation can be basically divided in two classes: porous and dense membranes.

In porous membranes molecular hydrogen diffuses through the pores from the side at higher concentration to the side at lower concentration. Permeability and selectivity depend on pore size and pore structure, according to complex mechanisms. Roughly speaking, permeability increases and selectivity decreases as pore size increases. Membrane materials and pore structure (which means preparation techniques) must be selected taking into account the requirements in terms of hydrogen purity and hydrogen flow rate. An acceptable compromise seems to be microporous membranes with pore diameter of the order of one nanometer. Materials for hydrogen selective membranes working at high temperatures (500–900°C) are carbon, and ceramic (silica, alumina, zirconia, titania and zeolites).

Dense membranes are made of ceramics or metals. In both cases the separation is based on a mechanism of bulk dissolution/diffusion. In ceramic membranes the hydrogen is transported as an ion (proton/electron diffusion); in metallic membranes the hydrogen is dissolved and transported in atomic form in solid solution. Ceramic membranes are made of oxides of Ce, Sr, Ba or mix. Metallic membranes are based on thin layer of alloys of Pd with Ag or Cu, deposited on porous ceramic substrates. Dense membranes are characterised by a very high selectivity. Critical issues are physical and chemical stability of the materials.

At present metallic membranes are the most developed. Pd alloy-based tubular membranes of length of the order of one meter are reported [29]. Pd-based membranes are commercialised for hydrogen purification. Ceramic dense and microporous membranes exist only as laboratory prototypes.

Metallic membranes at high hydrogen selectivity are typically made by depositing the selective materials as a thin layer on a substrate, in flat or tubular configurations. Several flat parallel membranes can be assembled to maximise the packing density (i.e., membrane surface per reactor volume). Barriers can be used to guide the gas flux in order to eliminate stagnant zones and preferential paths. Tubular membranes can be deposited in the internal or external side of a tubular support. Several tubes can be assembled in a single separation unit.

A clear description of the principle and technological application of selective membrane for H_2 and CO_2 separation can be found in Kluiter [30] and in Ocwig and Menoff [31].

WGSMR have been realised with both flat and tubular H_2 selective membranes, although the second option is more popular and is adopted in a larger number of applications. A recent review on patents for hydrogen production using membrane reactors has been made by Gallucci et al. [32].

In the simplest configuration the catalyst is placed in the volume where the feeding gas travels inside the reactor. Other configurations, with the catalyst on the membrane surface or inside the membrane itself, have been experimented. These configurations increase the technological problems of manufacturing and handling of the reactor (for example, replacement of the exhaust catalyst) without clear benefits in terms of efficiency and stability. The separated hydrogen is recovered by suction. The H_2 extraction can be enhanced by using a gas sweeping the permeation zone. Inert gas or steam have been used for this purpose.

Good sealing of the membrane, mechanical stability at the temperature and pressure in service must be guaranteed. Tubular membranes seem more able to satisfy these requirements. The manufacturing technique and the requirement of good sealing at high temperatures imposes a practical lower limit to the diameter of the tubes, preventing the use of capillary systems. Typical tubular membranes have diameters of the order of centimetres.

6 Conclusions

The technological evolution of the WGSMR for H_2 production and CO_2 separation from syngas is mature enough to allow the realisation of a pilot plant. The knowledge about both WGS catalysed reaction and membranes for selective hydrogen separation can be applied in a unit at pilot scale to be integrated in a gasification plant.

The performances of a WGSMR, in terms of CO conversion and H_2 production is the global result of the mutual interaction of chemical kinetic mechanisms and hydrogen transport through the membrane, strongly depending on pressure, temperature and gas distribution inside the reactor. These conditions depend, in turn, on geometry of the reactor, characteristics of the catalytic bed, characteristics of the membrane and operating conditions. To analyse the effect of these variables on the reactor performance, a model based on appropriate physical laws and reliable data of catalytic bed efficiency and membrane permeability is necessary.

In this chapter a mathematical model of a WGSMR has been presented, developed to assist the basic design of a pilot scale reactor. The results of the calculations confirmed the feasibility of the pilot reactor, supplied the operating conditions to obtain the target values of hydrogen flow rate, and gave important insights into reactor configuration and catalytic bed requirement for the engineering of the reactor. A pilot plant, integrated in an industrial gasification plant, will allow to investigate the reliability and profitability of the technology.

At the end of the chapter it is worth noting that a reactor model is only as good as the experimental data used. In the presented model, catalyst efficiency and membrane permeability are measured data. Although the qualitative trends resulting from the simulations remain valid, the quantitative evaluations strongly depend on the reliability of these data.

Acknowledgments I am grateful to Michele De Santis, Enrico Malfa, Stefano Martelli, Patrizia Miceli and Ali Smith for revisions and corrections. My special thanks to Paolo Granati for discussions and suggestions.

References

1. Holladay JD, Hu J, King DL, Wang Y (2009) An overview of hydrogen production technologies. Catalysis Today 139:244–260
2. Garcia Cortés C, Tzimas E, Peteves SD (2009) Technologies for coal based hydrogen and electricity co-production power plants with CO_2 capture. Joint Research Centre report JRC49206, EUR 23661 EN, doi:10.2790/23969. http://publications.jrc.ec.europa.eu/repository
3. Raggio G, Pettinau A, Orsini A, Fadda M, Cocco D, Deiana P, Pelizza ML, Marenco M (2005) Coal gasification pilot plant for hydrogen production. Part A: coal gasification and syngas desulphurization. Second international conference on clean coal technology for our future, Castiadis (Cagliari) Italy, 10–12 May. www.sotacarbo.it Accessed Oct 2010
4. Simbeck DR (2004) CO_2 capture and storage: the essential bridge to the hydrogen economy. Energy 29:1633–1641
5. Ladebeck JR, Wagner JR (2003) Catalyst development for water-gas shift. In: Vielstich W, Gasteiger HA, Lamm A (eds) Handbook of fuel cells: fundamentals technology and applications, Vol 3, Part 2. John Wiley & Sons, Chichester, pp 190–201
6. Ciferno JP, Fout TE, Jones AP, Murphy JT (2009) Capturing carbon from existing coal fired power plants. Chemical Engineering Progress, April, pp 33–41. www.aiche.org/cep Accessed Oct 2010
7. IPCC (2005) IPCC Intergovernmental Panel on Climate Change. In: Metz B, Davidson O, de Coninck HC, Loos M, Meyer LA (eds) Special report on carbon dioxide capture and storage,

Working group III of IPCC, Cambridge University Press, Cambridge, UK. http://www.ipcc.chAccessed Oct 2010

8. Ebner AD, Ritter JA (2005) Separation technology R&D needs for hydrogen production in the chemical and petrochemical industries. Chemical industry vision 2020 technology partnership www.chemicalvision2020.orgAccessed Oct 2010
9. Ebner AD, Ritter JA (2009) State-of-the-art adsorption and membrane separation processes for carbon dioxide production from carbon dioxide emitting industries. Separ Sci Technol 44:1273–1421. doi:10.1080/01496390902733314
10. Carbo MC, Jansen D, Haije WG, Verkooijen AWG (2006) Advanced membrane reactors for fuel decarbonisation in IGCC: H_2 or CO_2 separation? Fifth annual conference on carbon capture and sequestration, 8–11 May, Alexandria VA, USA. ECN-RX–06-084. http://www.ecn.nl/publicationsAccessed Oct 2010
11. HYDROSEP (2010) Hydrogen separation in advanced gasification processes. RFSC contract RFCR-CT-2006-00003
12. Barin I, Knacke O (1973) Thermochemical properties of inorganic substances. Springer-Verlag, Berlin
13. Favetta B (2007) Modellizzazione e simulazione di reattori a membrana per la produzione d'idrogeno da gas di sintesi. Tesi di laurea (PhD thesis), Università di roma La Sapienza
14. AGAPUTE (2010) Advanced gas purification technologies for co-gasification of coal, refinery by-products, biomass and waste, targeted to clean power produced from gas and steam turbine generator sets and fuel cells. RFCS Contract RFC-CR-04006
15. Piemonte P, De Falco M, Favetta B, Basile B (2010) Counter-current membrane reactor for WGS process: membrane design. Int J Hydrogen Energ. doi:10.1016/j.ijhydene.2010.07.158
16. Criscuoli A, Basile A, Drioli E (2000) An analysis of the performance of membrane reactors for the water–gas shift reaction using gas feed mixtures. Catal Today 56:53–64
17. Enick RM, Hill JW, Cugini AV, Rothenberger KS, McIlvried HG (1999) A model of a high temperature high pressure water-gas shift tubular membrane reactor. ACS Fuels Fall (New Orleans) 44:919–923
18. Raja LL, Kee RJ, Deutschmann O, Warnatz J, Schmidt LD (2000) A critical evaluation of Navier–Stokes boundary-layer and plug-flow models of the flow and chemistry in a catalytic-combustion monolith. Catal Today 59:47–60
19. Gosiewsky K, Warmuzinsky K, Tankzyk M (2010) Mathematical simulation of WGS membrane reactor for gas from gasification. Catal Today. doi:10.1016/j.cattod.2010.02.031
20. Alefeld G, Volkl J (1978) Hydrogen in metals, Vols I and II. Springer-Verlag, Berlin
21. Sonwane CG, Wilcox J, Ma YH (2006) Solubility of hydrogen in PdAg and PdAu binary alloys using density functional theory. J Phys Chem B 110(48):24549–24558
22. Jemaa N, Grandjean BPA, Kaliaguine S (1995) Diffusion coefficient of hydrogen in a Pd–Ag membrane effect of hydrogen solubility. Can J Chem Eng 73:405–410. doi:10.1002/cjce.5450730318
23. Whitaker S (1972) Forced convection heat transfer correlations for flow in pipes, past flat plates single cylinder single sphere and for flow in packed beds and tube bundles. AIChE J 18:361–371
24. Dullien FAL (1979) Porous media fluid transport and pore structure. Academic Press, London, UK
25. McCabe WE, Smith JC, Harriott P (2001) Unit operations of chemical engineering. McGraw-Hill, New York
26. Huang SC, Lin CH, Wang JH (2010) Trends of water gas shift reaction on close-packed transition metal surfaces. J Phys Chem 114:9826–9834
27. Smith RJB, Muruganandam L, Shantha MS (2010) A review of the water gas shift reaction kinetics. Int J Chem React Eng 8:R4
28. Callaghan CA (2006) Kinetics and catalysis of the water–gas-shift reaction: a microkinetic and graph theoretic approach. PhD thesis, Worcester Polytechnic Institute Electronic Theses & Dissertation. http://www.wpi.edu/Pubs/ETD/Accessed Oct 2010

29. Pex PPAC, van Delft YC, Correia LA, van Veen HM, Jansen D, Dijkstra JW (2004) Membranes for hydrogen production with CO_2 capture. Seventh international conference on greenhouse gas control technologies (GHGT-7), Vancouver, Canada, 5–9 September. http://www.ecn.nl/docs/library/reportAccessed Oct 2010
30. Kluiters SCA (2004) Status review on membrane systems for hydrogen separation. Intermediate report EU project MIGREYD NNE5-2001-670, ECN-C-04-102. http://www.ecn.nl/publicationsAccessed Oct 2010
31. Ockwig NW, Nenoff TM (2007) Membranes for hydrogen separation. Chem Rev 107:4078–4110. doi:10.1021/cr0501792
32. Gallucci F, Basile A, Iulianelli A, Kuipers HJAM (2009) A review on patents for hydrogen production using membrane reactors. Recent Patents on Chemical Engineering 2:207–222

Modelling Superstructure for Conceptual Design of Syngas Generation and Treatment

Aarón D. Bojarski, Mar Pérez-Fortes, José María Nougués and Luis Puigjaner

Abstract In this chapter a description of how the process synthesis problem can be casted as a superstructure optimisation problem is done. The first section draws on how the superstructure can be built, while the second section depicts the different techniques that can be used to reduce the computational time required to run a superstructure optimisation. Section 3 describes the different integration and control considerations embedded in a possible superstructure for analysing syngas generation and treatment. This last section also shows the results of different scenarios of this superstructure model.

Notation

ANN	Artificial neural network
ASU	Air separation unit
CC	Combined cycle
CGE	Cold gas efficiency
CSTR	Continuous stirred tank reactor
DOF	Degree of freedom
EO	Equation oriented
ER	Equivalence ratio

A. D. Bojarski (✉) · M. Pérez-Fortes · J. M. Nougués · L. Puigjaner
Universitat Politècnica de Catalunya, ETSEIB, Diagonal 647,
08028 Barcelona, Spain
e-mail: aaron.david.bojarski@upc.edu

M. Pérez-Fortes
e-mail: mar.perez-fortes@upc.edu

J. M. Nougués
e-mail: jose.maria.nougues@upc.edu

L. Puigjaner
e-mail: Luis.Puigjaner@upc.edu

L. Puigjaner (ed.), *Syngas from Waste*, Green Energy and Technology,
DOI: 10.1007/978-0-85729-540-8_8,

GA	Genetic algorithm
GT	Gas turbine
HRSG	Heat recovery steam generator
HP	High pressure
IGCC	Integrated gasification combined cycle
IP	Intermediate pressure
KPIs	Key performance indicators
LHV	Lower heating value
LP	Linear programming
LP	Low pressure
MILP	Mixed-integer linear programming
MINLP	Mixed-integer non-linear programming
MOO	Multi-objective optimisation
MSE	Mean square error
NLP	Non-linear programming
OF	Objective function
PFR	Plug flow reactor
PR	Pressure ratio
PSA	Pressure swing adsorption
PRENFLO	Pressurised entrained flow
RSM	Response surface methodology
SA	Sensitivity analysis
SM	Sequential modular
SQP	Sequential quadratic programming
ST	Steam turbine
TIT	Turbine inlet temperature
TOT	Turbine outlet temperature
VL	Vapour-liquid
WHB	Waste heat boiler
cg	Clean gas
rg	Raw gas

1 Introduction, Superstructure Building, Architecture and Functionalities

The design and modelling of chemical processes in general and syngas processes in particular consist of a series of steps where different refinements on a given process flowsheet are done aiming at generating a final design for the production of a given product.

The design of new processes has different stages associated to its *life cycle*, according to Cameron [1] different standards, such as ISO15288 [2] and ISO14040 [3], discuss the life cycle related to systems engineering. These standards introduce the following stages: concept, development, production, utilisation, support and retirement, which in the particular case of chemical processes involve the following (Sect. 2 in chapter "Modelling Syngas Generation" of [4]):

- *Strategic planning*: Initial ideas regarding resource utilisation or new product/service are generated; this phase is driven by new business opportunities.
- *Research and development* (R&D): Ideas are tested in laboratory; market research is done for promising products. From a process perspective, research covers areas such as product qualities, reaction kinetics, product yields and physicochemical prediction models.
- *Conceptual design*: Promising ideas are further developed and input–output process are generated. Initial process feasibility is assessed by means of general mass and energy balances. Simple models in steady state are used, some structural optimisation can be considered. Study of alternate reaction/production routes is performed.
- *Detailed design*: Here the final engineering flowsheet is obtained (piping, controls and instrumentation). Models used are more complex and unit specific, steady state assumptions are dropped and dynamic behaviour is modelled for start-up, shutdown, emergency response and regulatory control.
- Plant installation/construction and commissioning.
- *Operations*: It involves day-to-day process operations, problems associated to debottlenecking for retrofit, start-up or maintenance.
- *Decommissioning or close/pull down*: This is an important consideration in the life cycle given that most product and process have an "expiry" date and inevitably come to a natural end.
- *Remediation or rehabilitation*: This stage might involve significant financial resources and specialised chemical modelling and experimentation to consider ways of achieving remediation of land and environment.

In this book we are primarily concerned with *conceptual design*, given that during the conceptual design, modification costs are even lower than during R&D and detailed engineering design stages, as it was shown by Yang and Shi [5] and Rebitzer et al. [6]. In this sense the conceptual design phase lies between laboratory research and engineering design (detailed design), and serves as the connecting link between them.

Two main approaches are available for process design: one based on *mathematical programming* and the other centred on a *hierarchical decomposition of decisions*. In the latter the flowsheet is solved in layers, first the reaction steps, then separations, then heat integration and subsequently other layers as proposed by Douglas [7]. The former approach is based on the appropriate representation of all possible flowsheets for the production of a given product from different raw materials using different processing units by means of a process *superstructure*. This superstructure is commonly coded using a mathematical program which is subsequently optimised. The principal proponents of this approach are Biegler et al. [8].

1.1 Superstructure

In the approach proposed by Biegler et al. [8] the process synthesis problem is formulated as a mathematical programming problem. The whole superstructure, which is understood as the ensemble of all feasible flowsheets, of all possible combinations of equipment, raw material and products is programmed as a mixed-integer non-linear problem (MINLP). Integer (binary) variables are related to the presence or not of given equipment in the solution while real variables represent equipment parameters such as temperatures, pressures or flowrates.

In Eq. 1, f is a vector of economic and environmental *objective functions* (OF), or commonly known as *key performance indicators* (KPIs); $h(x, y) = 0$ and $g(x, y) < 0$ are equality and inequality *constraints*, and x and y are the vectors of *continuous* and *integer variables*, respectively. Mass and energy balances are considered in the case of equality constraints, while inequality rises from consideration of process operating constraints such as temperature restrictions or planning-scheduling considerations such as demand satisfaction. If the integer set Z is empty and the constraints and OFs are linear, then Eq. 1 becomes a linear programming (LP) problem; if the set of integer variables is nonempty and non-linear terms exist in the OFs or constraints, then Eq. 1 is a mixed-integer non-linear problem (MINLP), while mixed-integer linear problems (MILP) incorporate integrality and linear functions. The reader is referred to the work of Yeomans and Grossmann [9], for the consideration of superstructure modelling using general mathematical models, while the remaining of this chapter will deal with the superstructure problem representation using process simulation environments (see Sect. 1.3).

$$
\begin{aligned}
\min_{x,y} \quad & f(x,y) \\
S.T. \quad & h(x,y) = 0 \\
& g(x,y) < 0 \\
& x \in \mathrm{R}^n \\
& y \in \mathbb{Z}^q
\end{aligned}
\tag{1}
$$

The superstructure representation of any process will involve the appearance in the flowsheet of different unit operations performing the same process part. For example if a reaction step is considered, the modeller could be interested in which of the following options performs better so two or more options might be placed. For example if a reaction step is considered then a plug flow reactor (PFR) and a continuous stirred tank reactor (CSTR) might be sought. In process simulators the former strategy can be easily considered by adding those models and a set of stream mixers and splitters accordingly. See Fig. 1 for clarification.

The *optimisation* algorithm in the case of a process being represented as a superstructure ends in selecting the split fractions that are considered in the stream splitter models. While building such models special attention has to be given to the occurrence of stream recycles which will make the computation cumbersome.

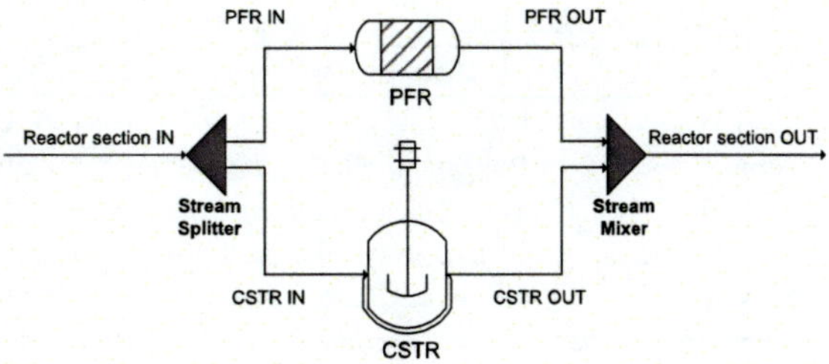

Fig. 1 Possible representation of a reaction step considering a PFR and a CSTR

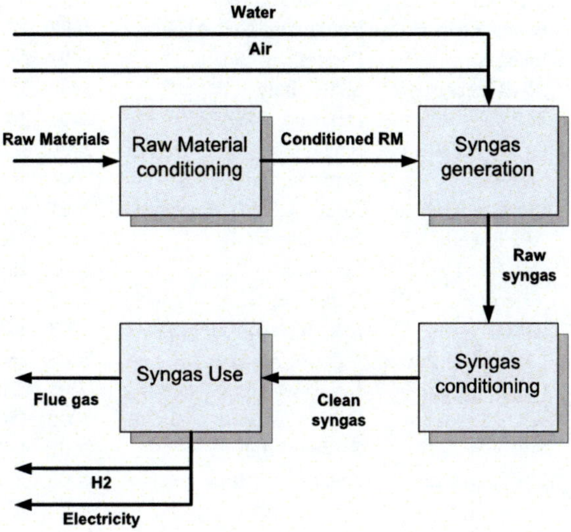

Fig. 2 Typical syngas production–consumption superstructure blocks

1.2 Syngas Superstructure Representation

Typically in the case of syngas production and its subsequent usage, the superstructure can be divided into the following blocks (see Fig. 2):

- Syngas raw material conditioning
- Syngas production
- Syngas conditioning, and
- Syngas usage

Given the broad amount of different raw materials that are available for syngas production, all possible combinations of pre-treatments should be included in the first block. Generally this stage will include unit operations for: particle sizing, raw material densification, mixing (if more than two sources are used), drying and movement from pile stocks. Further details on this stage can be found in the previous chapter "Modelling Syngas Generation". The case of syngas production by means of gasification was also discussed in chaper "Modelling Syngas Generation" and different strategies are presented which mainly differ in the reactor shape, type of bed, feeding and gasification agent. The last two blocks, syngas conditioning and usage are

Table 1 Worldwide IGCC power plants [33]

Owner	Location	Start	Net output (MWe)	Feed
Cool Water	USA	1984	120	Coal
Nuon	Buggenum, Netherlands	1994	250	Coal + biomass
Wabash	Terre Haute, Indiana, USA	1995	260	Coal + petcoke
Tampa Electric	Polk County, Florida, USA	1996	250	Coal + petcoke
SUV	Vresova, Czech Republic	1996	350	Lignite
Schwarze Pumpe	Lausitz, Germany	1996	40	Lignite + wastes
Texaco	El Dorado, Kansas, USA	1996	40	Petcoke
Sydkraft AB	Värnamo, Sweden	1996	6	Wood
Pernis Refinery	Rotterdam, Netherlands	1997	127	Visbreaker + tar
Elcogas	Puertollano, Spain	1997	285	Coal + petcoke
ISAB Energy	Sicily, Italy	1999	520	Asphalt
Motiva	Delaware, USA	2000	240	Petcoke
Exxon Mobil	Singapore	2000	180	Crude oil
Sarlux	Sardinia, Italy	2001	545	Visbreaker + tar
Chawan IGCC	Jurong Island, Singapore	2001	160	Tar
API Energia	Falconara, Italy	2002	280	Visbreaker + tar
Fife Energy	Scotland	2003	109	Coal + wastes
Valero	Delaware, USA	2003	160	Petcoke
NPRC	Sekiyu, Japan	2003	342	Petroleum residues
Clean Coal Power	Nakoso, Japan	2006	250	Coal
CITGO	LA, USA	2006	570	Petcoke
IOC	Orissa, India	2006	180	Petcoke
Sulcis	Sardinia, Italy	2006	450	Coal
Eni Sannazzaro	Sannazzaro, Italy	2006	250	Oil residues
Piñon Pine	Nevada, USA	–	100	Coal
Global Energy	Kentucky, USA	–	500	Coal + wastes

generally deeply interconnected given that syngas conditioning and its subsequent use have to be considered in parallel, as already further explained in chapter "Main Purification Operations". Mainly two uses are considered in our case of study: electricity generation and H_2 production. In the first case syngas conditioning for gas turbine use mainly requires desulphurisation, due to the production of sulphur oxides under the operating conditions of the gas turbine. Typically, syngas desulphurisation is done using similar equipment as the one used for natural gas desulphurisation which was described in chapter "Main Purification Operations". In order to produce H_2, additional units are required to separate CO_2 from the main stream. The H_2 stream can be used in a combined cycle (CC), or can be sold to the market. These different applications imply different degree of purity thus, different sequence of units.

In particular for the case of power production considered in this chapter, it follows the configuration of an IGCC power plant. In order to set out the different available options in the superstructure of concern, see as a reference in Table 1 a list of all the existing IGCC power plants at commercial scale, with its main feedstock. Gasification for electricity production in a CC is a process suggested approximately at the mid-term of the last century.

All these plants have been considered as *demonstration plants*, which have defined the current basis for future generation of IGCC power plants. The purpose of the R&D in the IGCC power plants field is to improve the environmental performance, decrease marginal costs and capital cost and technology availability/reliability. The idea that IGCC power plants are an opportunity is supported by the fact that in 2010, there are a lot of new projects envisaged around the world, mainly based on coal, and located (in order of starting projects) in USA, Canada, China and Europe [33]. On the other hand, new IGCC power plants with CO_2 capture technologies have also been a reference for the conception of this superstructure. Metz et al. [10] state that new projects are driven by Shell, Texaco and E-gas, during years 2002–2005. The main technology used is the Selexol capture system in pre-combustion configuration, as justified in chapter "Modelling Syngas Generation".

In summary, and dealing with the main features of the superstructure, the raw material that enters the system is firstly dried and crushed. The obtained feedstock dust is transformed into syngas in a PRENFLO gasifier, which was discussed extensively in chapter "Modelling Syngas Generation". Before entering the gas purification step, syngas is cooled down in a waste heat boiler (WHB) that allows for heat profit in the steam turbine (ST). Syngas is cleaned by removing solids, basic and acid trace components; and liquid sulphur is obtained as a by-product. The clean gas is finally sent to a CC. An important feature for this type of plants is the *integration* between the different blocks. Higher integration implies higher efficiencies, but with the drawback of higher dependencies, that is when some of the integrated units/flowrates does not work properly, the inter-connected units cannot work independently, this being not their proper work. The main integrated systems and integration flowrates that can be found in our type of syngas generation process deal with:

- *Oxygen consumption*. The oxygen that is required by the gasifier comes from the air separation unit (ASU), which is fed by a flowrate of air that is firstly compressed in the gas turbine (GT) system. Possibilities to this option are the partial integration with the GT compressor, which implies the installation of a compressor in the ASU itself, or non-integration, where the two streams of compressed air come to their respective compressors.
- *N_2 use*. After the ASU process, two N_2 streams are obtained: a relatively pure one, and the one called waste nitrogen. The first one can be used in the feeding system to transport and pressurise the feedstock dust, as well act as a temperature moderator in the gasifier. It is also further used for fly ashes discharge, filters cleaning and slag transportation. The second one is usually used in the GT combustor to control NO_x formation.
- *Steam network*. A net consumption of steam is usually counted for the gasifier, for the venturi scrubber and for the GT combustor (for clean gas saturation). These consumptions come from the produced steam in the plant, which is in the ST cycle. Moreover, heat from the imperative syngas cooling is profited to produce steam in the CC. Usually the heat exchangers are integrated in the water-steam net, with

purges and condensates recuperation. Heat exchangers network (HEN) is further discussed in chapter "Process Integration: HEN Synthesis, Exergy Opportunities".

The insights of H_2 production from syngas have been discussed in previous chapter "Main Purification Operations", while the production of electricity using syngas is discussed in Sect. 3.2. Next section deals with the different tools that can be used to tackle the design of IGCC flowsheets.

1.3 Software and Algorithms Used in Process Simulation

Process simulation in general is understood as the use of computer software resources to develop mathematical models for the construction of an accurate, representative model of a chemical process aiming at understanding its behaviour during regular plant operations and to explore other possible working conditions [11, 12]. The complexity of process simulation rises from the mathematical functions that are used in the model.

Process simulation environments are software tools constituted of mainly three parts: (a) a flowsheeting environment that allows for models connectivity, (b) a set of models that range from thermodynamic functions to process models and (c) a set of optimisation and solving algorithms that allow for solving the models connected altogether.

In the case that variables are not changing along time or position, a non-linear set of equations such as Eq. 1 appears in chemical problems, where both the number of functions in the $\boldsymbol{h}$ vector function plus design specifications equals the number of variables $\boldsymbol{x}$. If Eq. 1 is considered as a system of two equations and three variables: $h_1(x_1, x_2, x_3) = 0$ and $h_2(x_1, x_2, x_3) = 0$, by having fixed the value of one single x_i the system is square. In order to solve the set of equations in each particular case, two main approaches regarding process simulation are available: the *equation oriented* (EO) and the *sequential modular* (SM).

In the EO approach any x_i can be set freely and the system is solved altogether using different algorithms. Model's equations are solved altogether provided that the degree of freedom (DOF) of the system of equations is square; if this was the case two equations were available (h_1, h_2), while three variables were involved, consequently one x_i had to be fixed in order to make the system square. The techniques for solving non-linear equations overlap in their motivation, analysis and implementation with optimisation techniques [13]. In unconstrained optimisation, the objective function is the natural choice of merit function that measures progress towards the solution, but in non-linear equations various merit functions[1] can be used, all of which have some drawbacks. The sequence of estimates

[1] The merit function is a scalar-valued function that indicates whether a new iterate is better or worse than the current iterate, in the sense of making progress towards a root of the equation system. The most widely used merit function is the sum of squares.

converging to the optima can be generated using only first derivatives of the objective function (for example, steepest descent and conjugate gradient), or second order derivatives (Newton method, quasi-newton methods or sequential quadratic programming—SQP). The discussion of optimality conditions and optimality conditions regarding constrained optimisation can be found in different optimisation books such as: Steuer [14], Statnikov and Matusov [15], Nocedal and Wright [13] and Griva et al. [16], the reader is referred to those references. In all those books the constrained optimisation basics regarding Karusch–Kuhn and Tucker (KKT) conditions is discussed.

Contrary to the EO approach, in the SM approach a different scheme is adopted given that partial information is used. Lets say that x_1 is selected as fixed DOF, the SM approach solves the system using explicit expressions: $x_2 = H_1(x_1)$ and $x_3 = H_2(x_1)$, and uses custom made algorithms for the case of presence of cycles between variables and functions. The model's equations have to be especially suited and coded to manage the possibility of connecting results of some of them as inputs to others.

One important thing to emphasise when dealing with an *optimisation* considering SM simulation environments is the DOF available. In order to check if a given model variable is suitable for optimisation, it is convenient to perform a sensitivity analysis on such variable calculating the desired objective function and checking if changes in such variable affects appreciably the OF value. In the case of EO modelling, the former is not as difficult given that most simulation environments provide with different ways of knowing if a given variable has been fixed or if it is an available DOF suitable for optimisation. Process simulation environments provide the user with optimisation capabilities for NLP, the nonlinearities rising from the constraints and objective function. Constraints are required to enforce mass and energy balances, for which thermodynamic properties estimations are also required. Unit operation performance also introduces constraints as well as the calculation of objective function metrics.

Other set of optimisation algorithms are referred to as heuristics and are based on rules that do not involve first or second OF derivatives. These algorithms, such as Genetic Algorithm (GA), Simulated annealing or tabu search, provide with a robust optimised solution, but cannot guarantee any optimality condition as the ones based on function derivatives. They are based on oriented exhaustive searches of the solution space. The most simple of these algorithms is a random search, where, randomly generated solutions are tested and the feasibility/optimality of each solution is checked and the best solution is considered as the optimisation problem solution. GA, simulated annealing and tabu search use different heuristics that consider the generation of "better" solutions from the best solutions of previous solution sets. In the case of GA the heuristics use mutation, crossover and selection operators over good solutions to generate a new set of solutions to be tested, while in the case of SA different parts of the solution vector are left to be changed with decreasing probability aiming at mimicking the annealing effect produced in solidifying solutions that arrive to equilibrium conditions.

In Aspen Plus©, PRO/II© and Aspen Hysys©, the optimisation problem is solved first by calculating the process models and their respective variables before evaluating the constraints and objective function value. Due to its SM approach the optimisation problem is solved in an outer loop, while the model equations are converged in an inner loop. At least a single process model evaluation is required every time the objective and constraint functions are evaluated for optimisation [17]. Aspen Plus©, in the SM approach, has coded two algorithms, the complex algorithm which is a feasible path "black-box" pattern search, and a SQP method.[2] In the case of Aspen Hysys© the optimiser algorithms available are several, differing mainly in the ability in handling inequality and equality constraints, most of them are based on different quasi-Newton or SQP implementations [21].

Caballero et al. [17] points out that the process simulators capabilities involving integer variables or discontinuous domains for the equations are very limited. Moreover the optimisation capability for process topology changes is rather small and the usage of complex objective functions, such as complex cost models or detailed units sizing models involving discontinuities, can only be done "a posteriori" after the simulation has converged. In this sense, the combined use of commercial simulation coupled with stand alone optimisation algorithms has been proposed by several authors. The combined use of Aspen Hysys© together with MS Excel optimiser has been done by Alexander et al. [22], while its connection to GA is exemplified by Chen et al. [23]. While the former authors dealt with NLP, Caballero et al. [17, 24] proposed different algorithms for MINLP, where they combined Aspen Hysys© with Matlab[3] using different decomposition strategies for tackling with integer variables. In the case of Aspen Plus©, Diwekar et al. [25], Chaundhuri and Diwekar [26] and Fu et al. [27] proposed the use of simulated annealing included as a calculation block within the simulator, which requires using the input language of Aspen and custom made FORTRAN, to implement the simulated annealing algorithm. In all the former cases the authors emphasise the flexibility that is attained when connecting the process simulator to an external optimiser, this flexibility arises from the different algorithms that can be applied. This last part is highly important with regard to the implementation of multi-criteria optimisation. None of the commercial simulation environments provide with the capabilities to solve multi-objective optimisation (MOO) problems. Consequently in order to solve such problems the user is required to combine the process simulation environment with other tool for dealing with multiple objectives.

Summarising, the former optimisation methods are used to calculate the appropriate values for the splits in a superstructure, consequently the splits value define which structure will be used.

[2] It provides with three different implementations one of them is based on the work of Biegler and Cuthrell [18] and Lang and Biegler [19], while other implements the Broyden-Fletcher-Goldfarb-Shanno (BFGS) approximation to the Hessian of the Lagrangian [20].

[3] Matlab© has already a set of optimiser codes for solving NLP problems but it can also access other stand alone solvers easily.

2 Meta-Modelling or Surrogate Modelling

One important aspect when performing process simulation in SM mode is that in many cases the whole process simulation takes too long to run, and that an optimisation of the process considering very exhaustive models may be impossible due to computational time constraints. In this sense one important method is the replacement of complex models by means of meta-models, which are the subject of this subsection. Please note that in the case of EO mode surrogate models can also be used, but the use of surrogate models to replace computationally expensive models is specially suited for the SM mode.

Any *meta-model* or *surrogate model* methodology consists in building a mathematical function (say $g(x)$), which is cheaper from the computational point of view, and which approximates the behaviour of the pre-existing model (i.e., $f(x)$) over the domain of variation of its inputs [28].

The primary goal of meta-modelling is to predict the true model output (y) behaviour ($y = f(x)$) at an untried point x by using $g(x)$. In general a meta-model is built on a pre-existing computer experiment sample, with a set of pairs (x_i, y_i), $i = 1, \ldots, n$. Intuitively, it is desired to have the residual or approximate error, defined simply as $f(x) - g(x)$, as small as possible over the whole experimental region T. In order to do that the mean square error (MSE) defined as in Eq. 2 is minimised.

$$\mathrm{MSE}(g) = \int_T f(x) - g(x)^2 \mathrm{d}x \tag{2}$$

Most meta-models can be written as in Eq. 3, where the set of $\{B_0(x), \ldots, B_L(x)\}$ is a set of basis functions which depend on the type of meta-model selected, while the β_j factors correspond to DOF to be fixed when fitting the meta-model to the model.

$$g(x) = \sum_{j=1}^{L} B_j \beta_j \tag{3}$$

Fang et al. [28] state that since outputs of computer experiments are deterministic, the construction of a meta-model is in fact an interpolation problem. To interpolate the observed set of outputs $\{y_1, \ldots, y_n\}$, over the observed inputs $\{x_1, \ldots, x_n\}$ using the basis $\{B_1(x), \ldots, B_L(x)\}$an L value is taken large enough such that Eq. 4 has a solution.

$$\begin{aligned} &Y = BB \cdot \beta_G \\ &Y = (y_1, \ldots, y_n)^T \\ &\beta_G = (\beta_1, \ldots, \beta_n) \\ &BB_{ij} = B_j(x_i) \quad i = 1, \ldots, n, \quad j = 1, \ldots, L \end{aligned} \tag{4}$$

Diverse basis functions are available for usage, but the most commonly used are polynomials and splines. Other methods are kriging and artificial neural networks (ANN): Fang et al. [28] make the following recommendations:

- *Polynomial models* are primarily intended for regression with random error. Polynomial modelling is the best established meta-modelling technique, and it is probably the easiest to implement. They are recommended for exploration in deterministic applications with a few fairly well-behaved factors.
- *Kriging* may be the best choice in the situation in which the underlying functions to be modelled are deterministic and highly non-linear in a moderate number of factors (less than 50).
- *Multi-layer perceptron networks* may be the best choice (despite their tendency to be computationally expensive to create) in the presence of many factors to be modelled in a deterministic application.

Other methodologies rise from the design of experiments and response surface techniques. In these cases the models to be fitted are similar to the ones used in the Analysis of Variance (ANOVA). Examples of using response surface methods (RSM) in the context of optimisation are the works of Chen and Frey [29] while a brief consistent review is done by Almeida-Bezerra et al. [30].

An ANN is formed by simple processing elements called neurons, which are activated as soon as their inputs exceed certain thresholds. Neurons are arranged in several layers, which are inter-connected in such a way that input signals are propagated through the complete network to the output. Thus, they provide a way of correlating complex relationships between input and output responses in a model. The choice of the transfer function of each neuron (e.g., a sigmoidal function) contributes to the overall non-linear behaviour of the network. In general four characteristics define an ANN (chapter "Modelling Syngas Generation" of [31]): type of neurons/nodes, architecture of the connections between neurons (presence of loops, separates feedforward and feedback architectures) and learning algorithm.

In previous works [32, 33] some unit operation models have been replaced with meta-models. These meta-models have been used in the context of process simulation, since two different process simulators, Aspen Plus© and Aspen Hysys©, are used to simulate different parts of the same process and it is required to use the results from one in the other. In this case a multi-layer perceptron network is used. Data fitting to the ANN was done using the Matlab's© toolbox for ANNs. Specifically, Aspen Hysys© has a proprietary interface which accepts COM objects called Aspen Hysys© Extensions, while Aspen Plus© allows for user models, coded in FORTRAN to be directly linked to its model library. One possible situation that is considered is the case of using Aspen Plus© results inside Aspen Hysys©, to use Aspen Plus© results in Aspen Hysys©, due to the fact that the whole superstructure was constructed in Aspen Hysys©. The implementation of this approach requires three steps:

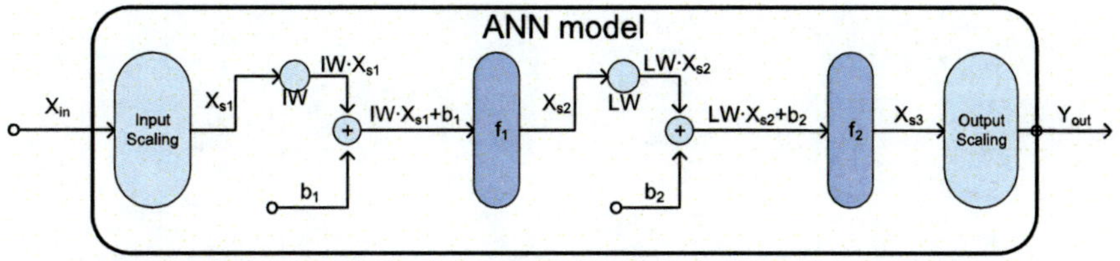

Fig. 3 ANN structure used

(i) Generating representative data in Aspen Plus©
(ii) Training the ANN
(iii) Using the trained ANN in Aspen Hysys©

Step (i) is carried out in Aspen Plus© using its sensitivity analysis tool. Step (ii), which encompasses the ANN training task, can be carried out using the ANN toolkit provided with Matlab©, taking into consideration different sets for training and validation. Step (iii) requires a model that uses the ANN results and provides with the appropriate results. The algorithm has been implemented as an Aspen Hysys© Unit Operation Extension. The ANN structure used is shown in Fig. 3. Initially, input values (X_{in}) are scaled to $[-1;1]$ interval (X_s^0). The first level of neuron response is obtained by performing the function evaluation of the first level over the result of multiplying the input matrix *IW* and adding the corresponding bias (b_1). This result is multiplied by a middle layer matrix *LW*, and other bias is added (b_2) together with a last function evaluation. The number of neurons in the first level has been fixed to a given number.

The number of neurons (*nNeu*) in the middle level fixes the sizes of all matrix and vectors used in the ANN, given that only one level is considered. *IW* is a matrix of size [*nNeu, nIn*], while *LW* is a matrix of [*nOut, nNeu*]. The functions used are "*tansig*" for first level and "*purelin*" for the second level. Results of the second function evaluation are scaled back to real values. The use of ANNs instead of polynomials or other meta-modelling techniques such as krigging is based on the ANNs ability to cope with multi-output models straightforward, while other techniques require one meta-model for each output variable.

A more simple approach can be the use of surrogate modeling, consisting in the substitution of a complex model by a more simplified version of it which is easier to compute although may have lower predictability. One example of surrogate modelling approach is the utilization use of a component splitter model; this model divides the input flows in as many streams as desired with user predefined species flows. For example, this model can easily replace absorption towers if the component splits a calculated previously. This approach is followed in this chapter, where most RadFrac units have been substituted by component splitters, based on the base case simulation results. This fact has implied to change the property package from ELECTNRTL to Peng-Robinson. The use of these surrogate models allowed us to perform the different sensitivity analysis (SA) described in Sect. 3.3, while decreasing the computational time required.

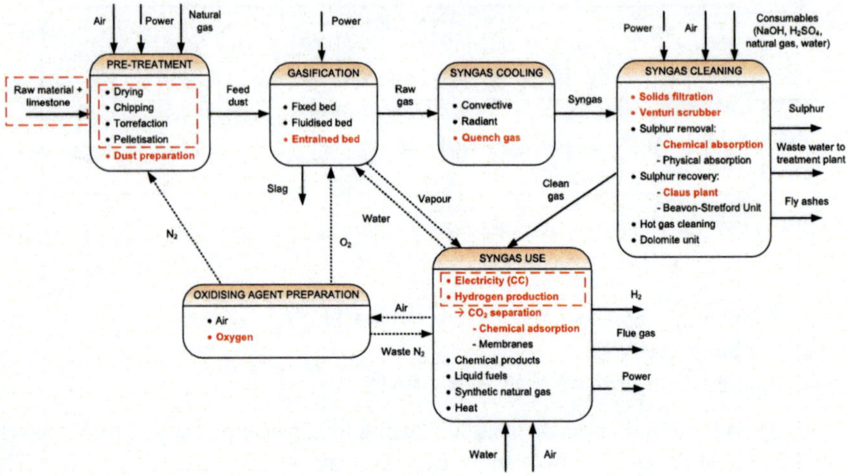

Fig. 4 Process superstructure; *discontinuous* indicate options considered in this book. The considered modelled flowsheet is in *red*. Integration flows are represented by *dotted lines*

3 Integrated Design and Control Considerations

As discussed in Sect. 1, the objective of syngas could be the production of H_2, or electricity. The first use was discussed in chapter "Main Purification Operations" and summarised in Sect. 3.1, while the production of electricity is discussed next, in Sect. 3.2. In a superstructure context, both applications have to be considered simultaneously by providing with appropriate models for each case.

A common IGCC process superstructure is shown in Fig. 4. This flow diagram assembles the different technology possibilities. Raw materials can be of different origins, and the addition of limestone is recommended, as described in chapter "Modelling Syngas Generation". Pre-treatment options contemplate energy densification and matter drying, as further discussed in chapter "Raw Materials Supply". A fixed unit is here, that is, the dust preparation before the chosen gasification technology, the entrained bed gasifier, which uses pure oxygen as main gasification agent (see chapter "Modelling Syngas Generation"). The raw gas from the gasifier is cooled down before cleaning through syngas recycling (the quench gas option). Other possibilities contemplate cooling through heat exchange with steam. Syngas cleaning use contemplates amines absorption and Claus plant for sulphur removal and recovery (see chapter "Main Purification Operations"). The Beavon-Stretford unit is a complement to the Claus plant that treats tail gas. Hot gas cleaning counts with specific beds, and the dolomite unit is an example of process intensification as described in chapter "Emerging Technologies on Syngas Purification: Process Intensification". Finally, in syngas usage a list of the syngas possibilities makes reference to the explanation of chapter "Main Purification

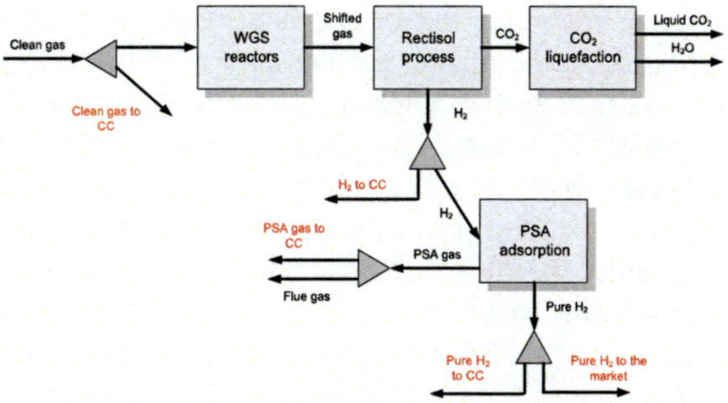

Fig. 5 CO_2 capture and H_2 production process superstructure

Operations". The use of membranes as CO_2 capture units is an emerging option, as described in chapter "H_2 Production and CO_2 Separation".

Among all the possibilities that a general gasification plant presents to be optimised, the dashed lines show the considered design choices here, thereby, the considered options for superstructure optimisation. In red they have marked the units modelled in Aspen Plus©. Pre-treatment and feedstock-syngas final use options elections are treated separately and through different methods.

The pre-treatment step optimisation is developed in chapter "Raw Materials Supply" where, depending on the feedstock characteristics (mainly moisture content and LHV) the pre-treatment units to be used should be elected in order to adapt to the raw material and to the required inlet conditions by the power or by the hydrogen plant. Feedstock mixture and final syngas usage election is treated through multi-objective optimisation in chapter "Selection of Best Designs for Specific Applications". The possibilities of the superstructure concerning this last subject are shown in the last sections of this chapter.

Finally, and concerning the levels of integration mentioned in Sect. 1.2, in our superstructure approach we are not considering a complete and integrated HEN; the integrate heat exchange considered here makes reference to the WHB and to the heat recovery steam generator (HRSG); and to the steam consumption that reduces the ST power generation. Consequently, all the heat streams from cooling and heating are set free. See an analysis of these heat streams in Sect. 3.4. Moreover, the N_2 net contemplates its use in NO_x emissions reduction and in feedstock transportation.

3.1 Hydrogen Production

Hydrogen, once separated from CO_2, could be exploited in different ways: it can be sold as a product, or if highly purified can be used in fuel cells; otherwise it could be sent to a gas turbine which is what is usually done with synthesis gas. Figure 5 shows the superstructure implemented to meet with these objectives. Looking at the splitters, several decisions should be taken concerning the separation factors. Firstly, the choice whether sending the clean gas to a Combined

Cycle or else to produce H_2; then, the purity of the H_2 to be sent to the turbine or to be sold to the market through the pressure swing adsorption selection.

Therefore it is deduced that the possibility of co-generation of power and H_2 is one of the possible choices in the superstructure. The base case contemplates the use of the whole clean gas stream to produce power. When adding the possibility of CO_2 capture (as in the last section of this chapter), the basic splitter fractions consider the "extreme" case where on the one hand, clean gas is separated to produce power in the CC, and on the other, the H_2 stream is being sold in the market. The PSA gas is released as flue gas.

3.2 Power Generation

The global energy balance in the syngas production for Combined Cycle application deals with the energy that enters via feedstock (LHV) and the final energy obtained in the CC. The net power production contemplates as main consumers the GT cycle compression and the ASU operation. In terms of heat, all steam consumed in the plant comes from the steam cycle, influencing the ST power production. As mentioned before, oxygen and nitrogen flows are integrated within the system formed by the CC-ASU-gasifier: air to the ASU is fed from the gas turbine compressor; waste nitrogen produced in the ASU is sent to the GT to diminish the flame temperature, thus reducing NO_x emissions, and the other nitrogen stream is used in the feeding system of the gasifier.

A CC has several mechanical and operating limitations whose consideration might require several modelling assumptions:

- Air flow to the compressor is limited by an upper and lower bound, to avoid choke and surge phenomena [34]. Nevertheless, in this modelling approach we do not considered any mass flow restriction.
- Gas turbine inlet temperature (TIT) is limited by turbine design conditions. This temperature is not controlled in our approach.
- A flash tank is used in the CC to collect feed water and produce steam. This tank plays a main role in the steam net for heat exchange. It has not been modelled, since the steam net for heat exchange has not been modelled.
- The HRSG constitutes in real plants a very complex HEN which considers different stream temperatures, and combines and intercalates different heat exchangers for producing steam at different temperatures and degrees of superheating. In the conceptual (preliminary) design, a very simple approach has been considered. Simple heater models are used, where heat transfer area and heat transfer coefficients are disregarded when calculating the heat produced from the flue gas temperature changes, which are integrated with the GT cycle to produce steam. In this sense the only thing that we consider is that outlet gas temperature has to be higher than the steam temperature desired.

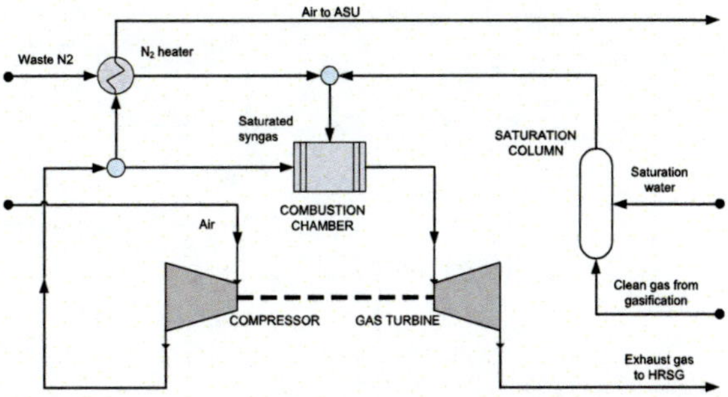

Fig. 6 Simplified flowsheet of the GT cycle

Regarding the overall CC modelling strategy, it is customarily to follow the approach found in Zhu [35] and in Ongiro et al. [36]. The *GT cycle* is composed of the system of compressor—combustion chamber—turbine that uses the cleaned syngas. A saturation column, before the clean gas combustion, saturates this stream with steam and nitrogen. A simplified flowsheet of the process is seen in Fig. 6.

Turbine and compressor stages can be modelled using turbine and compressor models from the simulation software model libraries. In general, simulation suites provide with models which include single or multiple effect compressors/turbines, and also provide with different algorithms for estimating their behaviour (e.g., polytropic and isentropic); typically these models require defining a given efficiency. The saturation column can be modelled as a two-phase flash model that considers the addition of water until the relative syngas humidity reaches the value of 100% [37].

In general gas turbines have air cooling, which draws cool air from compression stages into turbine stage inlets. In the case that we consider, we assume four stages gas turbine, thus the compressor has been modelled as a four-step process as well, where part of its compressed air can be sent directly to the turbine without passing through the combustion chamber. Each corresponding stage (i) has the same pressure loss or gain ratio (*PR*), according to Eq. 5, and it has been introduced into the model by FORTRAN code using a calculator block. PR is the pressure ratio of the turbine or compressor, P_0 is the inlet pressure and n_{st} is the number of compressor/turbine stages.

$$P_{i_{loss}} = P_0 \mathrm{PR}^{\frac{i}{n_{st}}} \tag{5}$$

The tuning of these block parameters includes calibration of cooling air split fractions and of the model's efficiencies of each of the compressor and turbine stages. The combustor is modelled using a Gibbs reactor. Turbine outlet temperature (TOT), which is a critical value since it influences the HRSG heat recovery, is controlled by the air mole flow that goes into the combustor by using a calculator block that, similar to the one present in the gasifier, sets the air mole rate as a proportional stream of the stoichiometric air needed to burn the clean gas mixture. Again, the code is introduced in FORTRAN. In this base case, the proportional factor is of 2.45, but for other cases the amount of air that enters the

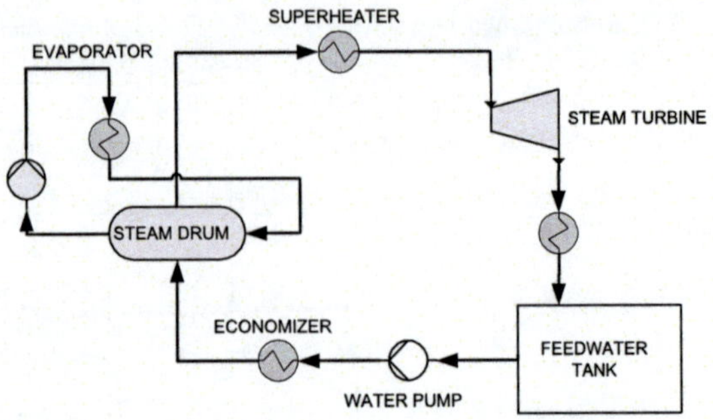

Fig. 7 Simplified flowsheet of a ST cycle

system destined to the GT is defined by a design specification that controls this flowrate in order to obtain a predefined TOT.

The HRSG profits the temperature drop from 540 to 100°C. This way, four heat exchangers have been contemplated to generate HP (high pressure), IP (intermediate pressure) and LP (low pressure) steam and one more to condensate water that returns to the feed water tank. In order to maximise the generated power in the ST, the outlet temperatures for HP, IP and LP steam side in the heat exchangers have been estimated as 300, 180 and 170°C.

The *ST cycle* considers three cycles, which are composed of a water pump, economiser, evaporator, super heater and the turbine itself; using models from the Aspen Plus© simulator. A general flowsheet is shown in Fig. 7.

Water is available at different pressures and it flows from their corresponding cycle, down to the LP cycle, thus being the LP turbine the one that produces the highest amount of power, due to the large amount of steam that is depressurised. The water mass flow of each cycle is calculated considering the total heat that is recovered from the WHB and the HRSG. Water consumption has to be considered due to different net consumptions, such as the steam fed to the gasifier, the net steam consumed in the venturi scrubber to clean the raw gas, the steam flow that is used in the saturator to saturate the clean gas before its combustion in the GT and the water consumption due to the WGS reactor to produce H_2. These consumptions penalise the heat flow from WHB and HRSG to the ST cycle.

With regard to the ASU, this unit produces enriched air by performing an oxygen and nitrogen separation to accomplish the requirements of the gasification and the CC blocks. The gasifier can use air enriched at different amounts that range from 85 to 99%. Two different ways of enriching air are available: one is the typical cryogenic procedure while the other is based on the use of selective permeating membranes, which is the subject of current research [see 38]. In general most IGCC plants use an oxygen stream of 85% of purity in molar basis. The ASU products are: enriched oxygen, pure nitrogen and waste nitrogen which are pressurised externally, to adapt them to the plant conditions, since the pressures present in the ASU process are governed by the cryogenic and distillation conditions of the air. We can distinguish several steps in this procedure. They can be modelled in a simplified way with a component separator, compressors and heater units:

- *Air pre-cooling*. In this step, air is taken to the purification unit temperature. Generally, this step is done in fact by a refrigeration fluid.
- *Air purification*. The main objective here is to remove impurities, such as water, from the air inlet stream. It is based on solid–gas absorption, for instance alumina. As any other adsorption process, the efficiency increases with high pressures and low temperatures. As the pressure is determined by the GT compressor, the previous air pre-cooler is crucial to obtain the desired efficiency. A component separator block is used in our approach.
- *Distillation*. It is a cryogenic process where nitrogen and oxygen can be separated through a distillation column. It occurs at about 13 bars and −165°C; oxygen is released as liquid, and nitrogen as gas. Waste N_2 has a composition of approximately 98 mol%, and pure N_2 of about 99.9 mol%.
- *Gas supply*. Adaptation of the outlet streams to the desired conditions of pressure and temperature for the gasifier and the CC. These are about 30 bars for the oxygen, 20 bars for the waste N_2 and 50 bars for the IP N_2. Isentropic efficiencies of the compressors are assumed to be 0.72.

Finally, the CC net power is obtained by considering the gross power resulting from the ST and the GT (considering the air compressor consumption) and subtracting the consumption from the ASU, the crusher, and from the compressors and pumps from syngas cleaning unit blocks, including the CO_2 capture system when co-production is considered.

3.3 Overall Modelling Data

Major technical data considered in the superstructure is summarised in Table 2. In the same way, see in Table 3 a summary of all modelled units and its representation in Aspen Plus©.

3.4 Calibration and Validation

Typically the model's response is compared against experimental or plant data available. In the case of IGCC models, this step involves the comparison of several models against available data. First and most important is the validation of thermodynamic properties estimation. Discussion of the appropriate thermodynamic representation was done in chapters "Modelling Syngas Generation" and "Main Purification Operations" and the reader is referred to those chapters. The comparison of the model against data is customarily done in two ways: (a) comparison of model against data, point by point, or (b) comparison of model tendencies against plant behaviour (e.g., if this variable increases then the other should increase or decrease in a given percentage). For the first case different models are

Table 2 Technical data for the IGCC plant model

Gasification	
Feedstock	2,600 tons day^{-1}
Crusher fineness	50–60 μm
Feedstock MC	2%
P of lock hoppers	30 bars
T_{gasif}	1,400–1,500°C
P_{gasif}	25 bars
O_2/O_2 stoich (mole basis)	Raw material dependant
H_2O/C_{in} (mass basis)	Raw material dependant
T before the WHB	850°C
T before gas cleaning	235°C
Max. P before CC	3 bars
Combined cycle (CC)	
Air_{in}	Conditioned by gasifier and gas turbine cycle requirements
HP/IP/LP	127/35/6.5 bars
HP/IP/LP T before ST	510/520/265°C
Compressor pressure ratio	15.7
T_{cg} before combustor	300°C
TOT	540°C
P of the saturation column	20.8 bars
T of the flue gas	100°C
T of the air to air separation unit (ASU)	130°C
Consumables	
NaOH (15 mass%)	Syngas and cleaning water dependant
H_2SO_4 (96 mass%)	Cleaning water dependant
Natural gas	Feedstock dependant
H_2O	Syngas and raw material dependant
CH_3OH	Proportional to the amount of CO_2 in the shifted gas
Air separation unit (ASU)	
Oxygen purity	85% volume basis
Radiation losses	2% adiabatic reactor
T_o	15°C
P_o	0.9309 bars

run at the desired plant conditions and model outputs are compared, while for the other type of comparisons typically a sensitivity analysis is performed.

Also the accuracy of the model is checked in terms of the net power, by-products and clean gas characteristics, as shown in Table 4. Measured data come from ELCOGAS power plant.

The biggest discrepancy is in the estimation of the ASU consumption and in the amount of sulphur recovery, which is overestimated. It can be deduced that the underestimation of the LHV of the clean gas is compensated in the CC with a bigger efficiency. Nevertheless, by considering the results as a whole, it can be said

Table 3 Modelled units in Aspen Plus©

Block	Aspen Plus model©
Feedstock dust preparation	Dust preparation: crusher Combustion chamber: Gibbs reactor Dryer: stoichiometric reactor Bag filters: two phase (VL) flash separator Lock hopper: mixer
Gasification and WHB	Gasifier: yield and Gibbs reactors Solid removal: component splitter WHB: heating units and a compressor for gas recycling
Syngas cleaning	Venturi scrubber: RadFrac column and two-phase (VL) flash separator Sour water stripper: RadFrac columns COS hydrolyser: stoichiometric reactor MDEA absorber: RadFrac columns Claus plant: stoichiometric and Gibbs reactors and two-phase (VL) flash separators WGS reactor: two equilibrium reactors Rectisol process: RadFrac columns PSA unit: Component splitter Liquefactor: flash separator and compressor
CC and HRSG	Turbines, compressors and heating units Combustor: Gibbs reactor Saturator: two-phase (VL) flash separator HRSG: heating units
ASU	Compressors and heating units, and component splitter

Table 4 Overall IGCC plant validation data

	Measured	Model	Error %
Raw gas flow (kg h^{-1})	202,500	192,272	−5.05
LHV clean gas (kJ kg^{-1})	9,911	9,547	−3.68
Gross power (MW)	308	315	2.27
ASU consumption (MW)	29	27.4	−5.52
Net power (MW)	283	285	0.71
Sulfur (kg h^{-1})	3,113	3,318	6.59

that in a conceptual stage this modelling approach represents an accurate enough approximation of an IGCC power plant, with an error in the range of preliminary design, where error within 15–20% are acceptable [39].

Many different metrics can be defined to prove the validity of the IGCC overall model checking model tendencies. Some of them are the raw gas composition or the LHV; others such as cold gas efficiency (CGE, defined in Eq. 31 of chapter "Modelling Syngas Generation"), or the $Eff_{CCapture}$ (defined in Eq. 28 of chapter "Main Purification Operations" and revised here as well in Eq. 11) previously defined, are also used. The following paragraphs define other efficiency terms usually referred.

The energy efficiency of the combined cycle (Eff_{CC}) is defined as the ratio between the net power of the CC (considering the energy consumed by the compressor), and the total chemical energy contained in the clean gas (cg) which is fed to the gas turbine, see Eq. 6. Even if the most usual stream is clean gas, hydrogen or PSA purge gas can also be fed to the GT. Note that even if Eq. 6 refers it as cg, depending on the scenario characteristics, the denominator can have either of the abovementioned natures.

$$\mathrm{Eff}_{\mathrm{CC}} = \frac{\mathrm{Power}_{\mathrm{net}}}{m_{\mathrm{cg}}\mathrm{LHV}_{\mathrm{cg}}} \tag{6}$$

By considering the plant as a whole, the total profitable energy that goes into the system is related with the final electricity to the grid. This energy efficiency ($\mathrm{Eff}_{\mathrm{global}}$) is calculated as in Eq. 7.

$$\mathrm{Eff}_{\mathrm{global}} = \frac{\mathrm{Power}_{\mathrm{net}}}{m_{\mathrm{feed}}\mathrm{LHV}_{\mathrm{feed}}} \tag{7}$$

In an analogous way, in the case of H_2 production, the global efficiency is calculated as in Eq. 8.

$$\mathrm{Eff}_{\mathrm{global}_{\mathrm{H_2}}} = \frac{m_{\mathrm{H_2}}\mathrm{LHV}_{\mathrm{H_2}}}{m_{\mathrm{feed}}\mathrm{LHV}_{\mathrm{feed}}} \tag{8}$$

Equation 31 of chapter "Modelling Syngas Generation" and (6) evaluate the performance of the gasifier and of the CC separately, while the $\mathrm{Eff}_{\mathrm{global}}$ provides a metric of the overall installation. In order to relate the former definitions the efficiency associated to the gas cleaning operations ($\mathrm{Eff}_{\mathrm{cl}}$) can be defined as in Eq. 9, where m_{rg} and m_{cg} refer to the raw and clean gas mass flows.

$$\mathrm{Eff}_{\mathrm{cl}} = \frac{m_{\mathrm{rg}}\mathrm{LHV}_{\mathrm{rg}}}{m_{\mathrm{cg}}\mathrm{LHV}_{\mathrm{cg}}} \tag{9}$$

Consequently, $\mathrm{Eff}_{\mathrm{global}}$ can be calculated from the partial efficiencies as in Eq. 10.

$$\mathrm{Eff}_{\mathrm{global}} = \mathrm{Eff}_{\mathrm{CC}}\mathrm{Eff}_{\mathrm{cl}}\mathrm{CGE} \tag{10}$$

In an analogous way, for the H_2 production the global efficiency can be calculated by considering in this case the partial efficiencies, including the efficiency of the carbon capture train ($\mathrm{Eff}_{\mathrm{CCapture}}$), where this is defined last as in Eq. 12.

$$\mathrm{Eff}_{\mathrm{CCapture}} = \frac{m_{\mathrm{H_2}}\mathrm{LHV}_{\mathrm{H_2}}}{m_{\mathrm{cg}}\mathrm{LHV}_{\mathrm{cg}}} \tag{11}$$

$$\mathrm{Eff}_{\mathrm{global}_{\mathrm{H_2}}} = \mathrm{Eff}_{\mathrm{CCapture}}\mathrm{Eff}_{\mathrm{cl}}\mathrm{CGE} \tag{12}$$

The validation of model trends is done by computing the former metrics, by means of sensitivity analysis (SA). In every SA the LHV of each stream, raw gas mole flowrates, emissions, by-products, power and water consumption are considered and taken directly from model results. While the LHV of fuels is calculated

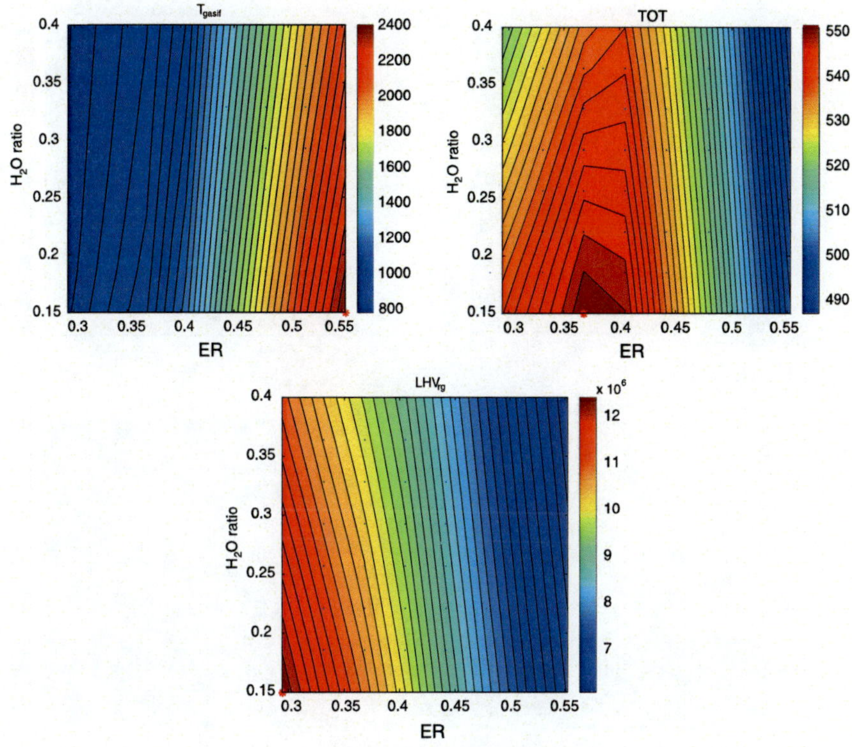

Fig. 8 Iso-lines for T_{gasif}, TOT, and LHV_{rg} in °C and J/kg, respectively, for the base case

using Eq. 32 in chapter "Modelling Syngas Generation", the estimation of LHV_{rg} and LHV_{cg} is performed using the Aspen Plus© property sets (QVALNET) [20]. Differences between each case study lie on raw material composition, and they consider the oxygen ratios to the gasifier and to the combustion chamber in the CC with an adjustment of the gasifier temperature (T_{gasif}), to remain close to the 1,400°C and the TOT, to be as close as possible to 540°C.

The sensitivity analysis (SA) performed changes considering the gasifier feedstocks: the equivalence ratio, ER, is changed in an interval of ±30% its base case value (i.e., a variation from 0.29 to 0.55), while the steam ratio is varied between 0.15 and 0.4 for all case studies. To point out that due to the oxygen ratio value into each case study, the ratio, defined as proportionate to the stoichiometric oxygen of the mixture, is always below the complete combustion value. Eight values are considered in both ranges; accordingly 64 points are resulted for each analysis. The CC fed oxygen ratio is maintained constant.

The SA results are represented in two-dimensional graphs, where the axis corresponds to the input variables and the output variable changes are plotted using contour lines. Blue points in the graphs are the SAs results, while the red star point is the highest value, for the output variable in the SA.

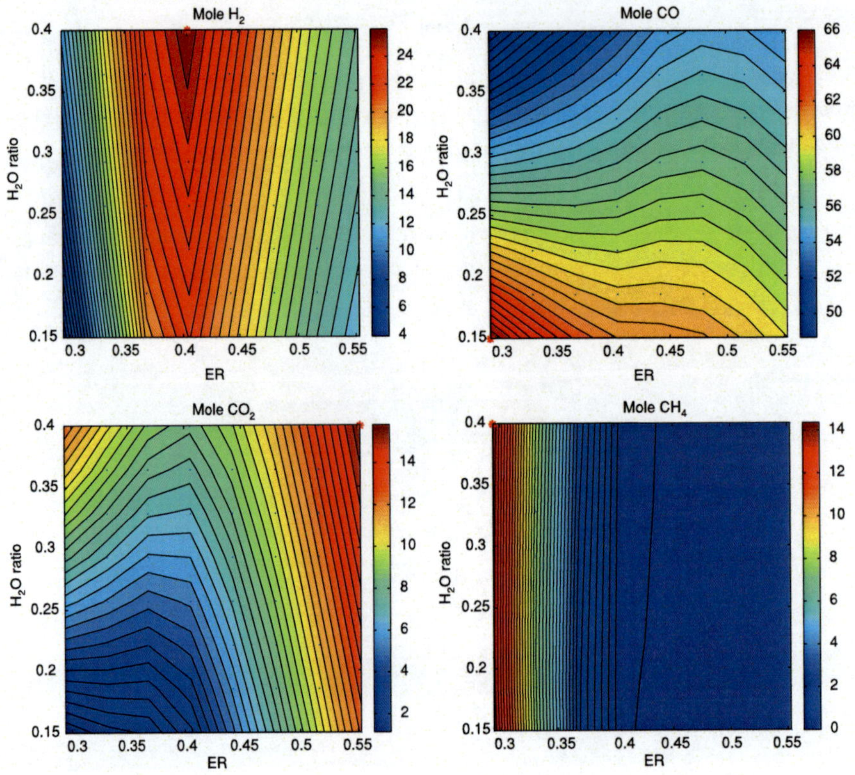

Fig. 9 Iso-lines representing the molar fraction (%) of raw gas components in dry basis for the base case

Figure 8 reports the variation observed for T_{gasif}, TOT and the LHV_{rg}. It is clearly seen that in the role as a moderator of the steam in the gasifier higher steam ratios are followed by lower temperatures, and that more inlet oxygen implies higher temperatures.

The considered SA leads to gasification temperatures in the range of 800–2,400°C, even if the *gasification zone* is in the middle, limited by CH_4 and sulphur oxides production. TOT is between 550 and 490°C. In the modelled installation and regarding the HRSG, lower temperatures than 540°C are not feasible operating regions due to the steam temperature requirements. With regard to the LHV_{rg}, it can be seen that the highest values are obtained for the combination of lowest H_2O and ER.

Syngas compositions (H_2, CO, CO_2 and CH_4) are represented in Fig. 9. In the case of CH_4 its occurrence is mainly driven for the O_2 amount coupled with high gasifier temperatures (see Fig. 8 *left*, and Fig. 9 *bottom right*). For the case of H_2, CO and CO_2 non-linear shapes are found. In the case of H_2, the highest H_2 production is found for an ER close to 0.4, and for the highest ratio of steam fed,

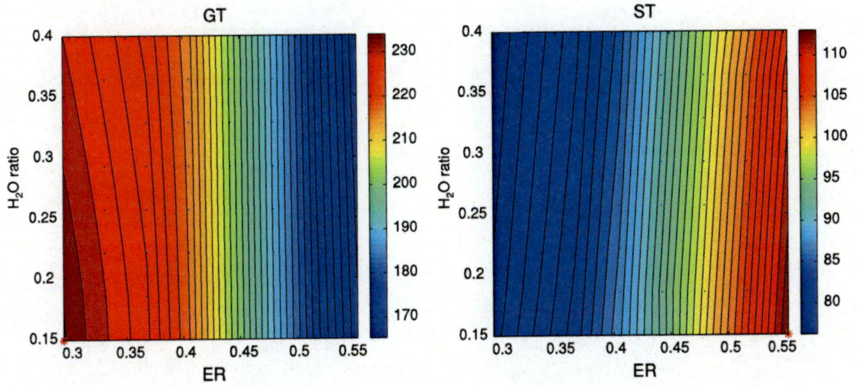

Fig. 10 Iso-lines for gas and steam cycles power production (MW) for the base case

which is in clear concordance with Eqs. 22 and 23 in chapter "Modelling Syngas Generation". In the case of CO_2, a clear relationship with H_2 composition is found but its highest values are found for the maximum amount of O_2 and steam. As expected, the CO behaviour is opposite to the one obtained for CO_2.

Figures 10 and 11 are highly interrelated; firstly, steam and gas turbines power production are shown in Fig. 10. The amount of power produced by the gas turbine reaches a maximum when ER and steam ratio are the lowest. On the contrary, steam turbine production is maximum at higher oxygen ratios but minimum at lower steam ratios. In both cases the influence of O_2 ratio is more important than the one of steam ratio.

In the case of CGE, it is found in Fig. 11 (*top left*) that the highest value coincides with the one obtained for the maximum gas turbine power production, and is in concordance with the higher LHV_{rg} values found in Fig. 8, given that the fuel LHV remains constant. Similarly, the Eff_{global} shows the same behaviour due to the fact that most of power comes from GT, and the fact that fuel LHV remains constant. In the case of the highest value for Eff_{CC}, it coincides with the highest steam turbine power production, therefore with highest gasification temperatures the WHB recovers more heat for steam production.

It is demonstrated here that oxygen ratio has higher influence in gasification than the steam one. The CGE is the main parameter that influences in the final global efficiency of the plant, since it is the GT that produces more power.

3.5 Results

Let us compare the base case, where all the syngas goes to the CC, with the extreme cases where all the clean gas is used to produce: (a) pure H_2 to be sold to the market, (b) pure H_2 to be sold to the market with PSA purge gas used in the CC or (c) H_2 to be used in a hydrogen turbine in a CC. The main difference between

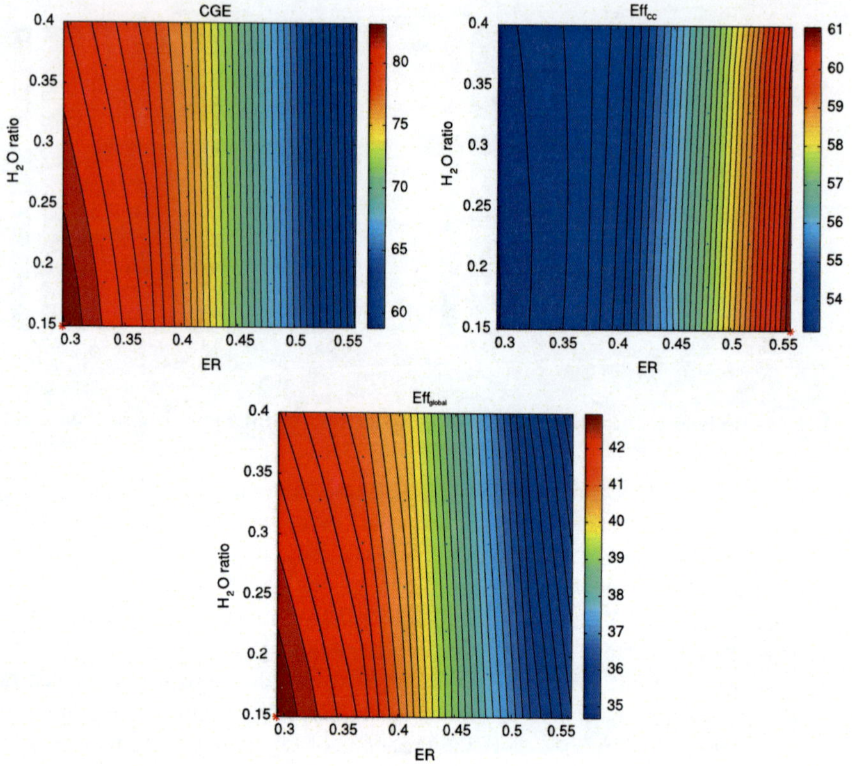

Fig. 11 Iso-lines for unitary and global efficiencies (%) for the base case

Table 5 Comparison between electricity and H_2 production

	Syngas to CC	H_2 to market	H_2 to market – PSA to CC	H_2 to CC
Eff_{global} (%)	42	–	5.4	33.3
$Eff_{global_{H_2}}$ (%)	–	51	51	–
Gross power MW	315	0	85	273
Net power MW	285	−46	37	226

the (a) and (b) with the (c) scenarios is the use or not of the PSA unit, since if the H_2 is not going to the market, its required purity is not restrictive. Moreover, in all cases the Rectisol block is used to separate H_2 from the remaining gases, which could be profited by a carbon capture scheme, and it is not profited in the GT.

The main advantage of using H_2 in GT combustion is the lesser level of emissions to the atmosphere, which hypothetically will involve only water. In these cases, we are assuming that the GT conceived for syngas (operating and main characteristics) can be used for H_2. Please note that this requires the HRSG structure to be defined for generating steam from GT flue gas works similarly when being fed with steam from H_2 combustion. See in Table 5 the most important parameters reported for each case study: the efficiency calculated as in Eqs. 7 and 8 depending on the final output, and the gross and net power for each situation. Note that even if the compressor in the GT cycle is not working for the GT itself, it is needed to compress the air that goes to the ASU.

The first column corresponds to the base case where all syngas is combusted in the CC. The other columns correspond to the following scenarios that imply several changes in the flowsheet layout. In all the scenarios performed, the amount of air that goes to the GT is regulated with a design specification in Aspen Plus© in order to adjust TOT to a temperature around the 540°C:

- *H_2 to market*: In this scenario, the HRSG has been deactivated. We are assuming that no air for the GT combustor is entering the system, and that no refrigeration is needed for the turbine. The waste N_2 from the ASU can be sent to the feeding system, since it is not needed in the GT system.
- *H_2 to market–PSA to CC*: It is similar to the previous one. In this case, an amount of "residual" power is produced. It checks the validity of this PSA residual gas as feed for the gas turbine, but it does not justify the presence of a turbine, as seen in Table 5. The HRSG is again activated, as well as the refrigeration system of the gas turbine.
- *H_2 to CC*: In this scenario, as this differs from the two previous ones, the PSA unit is not used; thus, the hydrogen produced in the WGS reactors is directly sent to the CC.

Table 5 shows that the efficiency penalty of a combined cycle with syngas or with H_2 is around 7%. This is due to the extra power consumption of the carbon removal train, to obtain the final H_2 stream. On the other hand, the increase of feedstock that would supply the same amount of power in a CC moved by hydrogen instead of syngas, according to this data is about 21%.

Analogous to chapter "Main Purification Operations", the outputs of the superstructure (power and H_2 calorific value) have been analysed versus the percentage of syngas stream that goes to the CC. The electricity consumed by the carbon capture technology has been considered. In the case of H_2 production, the consideration here is the production of pure H_2 to be sold in the market. It is interesting to appreciate that at the value of 50%, the calorific value of the H_2 is the same than the gross power produced by the CC, thus deducing that both applications have similar efficiencies. Figure 12 shows the lineal dependencies of the abovementioned parameters. It is interesting to mention that the production of hydrogen has an almost constant efficiency of 68% in all the scenarios. In the case of the gross power, two different behaviours have been plotted: in dashed black the one that considers all waste N_2 available from the ASU being fed to the combustor of the GT, and in colour the one considering a fraction of the waste N_2 stream being separated according to the same fraction that of syngas. This waste N_2 stream has a flowrate of 246,000 kg h^{-1} and a temperature of 384°C at a pressure of 20 bars, and in general is profited to increase the GT overall flowrate, while the syngas is fed to the GT at 147°C. The efficiency of the CC is different in the former two situations: in the first one, the efficiency increases while the proportion of

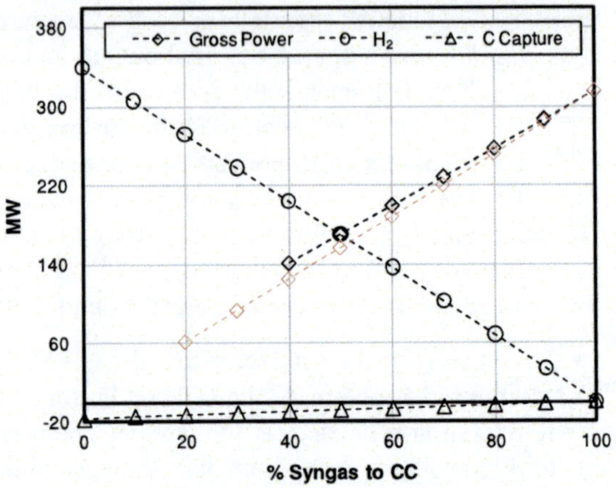

Fig. 12 H_2, carbon capture technology consumption and gross power variation with the percentage of syngas that goes to the CC. In colour, again the gross power but with waste N_2 separation

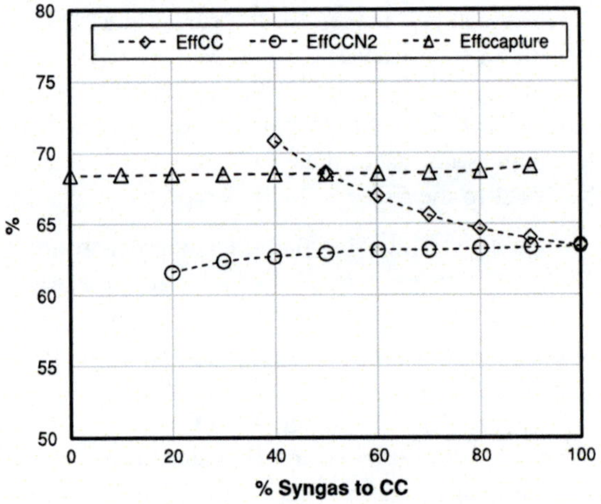

Fig. 13 Efficiencies for the CC and the carbon capture sections. Eff_{CCN2} contemplates N_2 separation

syngas to the CC decreases, ranging from 63 to 71% for values of the splitter between 1 and 0.4. In the second situation the efficiency is approximately constant around the 63%. It is due to the importance of the total flowrate that goes to the GT, which is crucial for the final power produced. See Fig. 13 where the efficiencies comparison is plotted.

In both Figs. 12 and 13, note that the lines related with the CC operation are not plotted for all the split fractions, given that there is a operative limitation of reaching a TOT value around 540°C, in order to make the GT work properly. It is found that for some split fraction scenarios it is impossible to obtain a value of TOT close to its requirement, even with the stoichiometric combustion of the syngas. This means that there is a certain amount of waste N_2 that refrigerates the combustion mixture too much, which can also be interpreted that the combustion mixture does not have the appropriate flowrate to produce the high temperature gases adequate for the GT operation. Note that the gross power produced with waste N_2 constant is higher than the one that separates it, even if this last option allows for an operation with a smaller amount of syngas.

4 Conclusion

This chapter has provided an overall view of the possible ways of handling the IGCC design problem from a flowsheet optimisation point of view (preliminary design). Clearly, it has shown the importance of the notion of *superstructure*, which is introduced in the first section. It provides the necessary flexibility and versatility in the selection of the best alternative design during the synthesis procedure. Moreover, different algorithms introduced for tackling the design optimisation are readily accessed by the superstructure. Model validation is emphasised by means of plant data comparison and by the use of sensitivity analysis. Finally, the efficiency penalty because of CO_2 capture methods is demonstrated through four case studies. The limitation in poly-generation associated to the value of TOT in the flue gas stream is analysed in depth. In this sense the superstructure and its versatility have been proved and ready for scenarios optimisation.

References

1. Cameron I (2005) Modelling across the process life cycle: a risk management perspective. Eur Symp Comput Aided Process Eng 15:3–19
2. ISO15288 (2008) Systems and software engineering—system life cycle processes. Technical report ISO, Genève, Switzerland
3. ISO14040 (1997) Environmental management—life cycle assessment—principles and framework. Technical report ISO. Genève, Switzerland
4. Cameron IT, Newell RB (2006) Modeling in the process life cycle. In: Puigjaner L, Heyen G (eds) Computer aided process and product engineering, Sect. 4.2. Wiley, New York, pp 667–693
5. Yang Y, Shi L (2000) Integrating environmental impact minimization into conceptual chemical process design: a process systems engineering review. Comput Chem Eng 24:1409–1419
6. Rebitzer G, Ekvall T, Frischknecht R, Hunkeler D, Norris G, Rydberg G, Schmidt W, Suh S, Weidema B, Pennington D (2004) Life cycle assessment Part 1: framework, goal and scope definition, inventory analysis, and applications. Environ Int 30:701–720
7. Douglas JA (1985) Hierarchical decision procedure for process synthesis. AIChE J 31:353–362
8. Biegler L, Grossmann I, Westerberg A (1997) Systematic methods of chemical process design. Prentice-Hall, Englewood Cliffs
9. Yeomans H, Grossmann IE (1999) A systematic modeling framework of superstructure optimization in process synthesis. Comput Chem Eng 23:709–731
10. Metz B, Davidson O, Coninck H, Loos M, Meyer L (2005) Special report on carbon dioxide capture and storage, Chap. 3. IPCC, Switzerland

11. Diwekar U (2005) Green process design, industrial ecology, and sustainability: a systems perspective. Resour Conserv Recy 44:215–235
12. Diwekar U, Small M (2002) Process analysis approach to industrial ecology. In: Ayres RU, Ayres LW (eds) A handbook of industrial ecology. Edward Elgar, Cheltenham, UK, pp 114–137
13. Nocedal J, Wright S (2006) Numerical optimization, 2nd edn. Springer, Berlin
14. Steuer R (1986) Multiple criteria optimization: theory computation and application. Wiley series in probability and mathematical statistics–applied, 1st edn. Wiley, New York
15. Statnikov R, Matusov J (1995) Multicriteria optimization and engineering, 1st edn. Chapman and Hall, New York
16. Griva I, Nash G, Sofer A (2009) Linear and nonlinear optimization, 2nd edn. Society for Industrial and Applied Mathematics (SIAM), Philadelphia
17. Caballero J, Odjo A, Grossmann I (2007) Flowsheet optimization with complex cost and size functions using process simulators. AIChE J 53:2351–2366
18. Biegler L, Cuthrell J (1985) Improved infeasible path optimization for sequential modular simulators, Part II: the optimization algorithm. Comput Chem Eng 9:257–267
19. Lang Y-D, Biegler L (1987) A unified algorithm for flowsheet optimization. Comput Chem Eng 11:143–158
20. AspenTech INC (2005) Aspen plus system reference. Aspen Tech, Cambrige
21. AspenTech INC (2005) Aspen HYSYS system reference. Aspen Tech, Cambrige
22. Alexander B, Barton G, Petrie J, Romagnoli J (2000) Process synthesis and optimisation tools for environmental design: methodology and structure. Comput Chem Eng 24:1195–1200
23. Chen H, Rogers T, Barna B, Shonnard D (2003) Automating hierarchical environmentally-conscious design using integrated software: VOC recovery case study. Environ Prog 22:147–160
24. Caballero J, Milan-Yanez D, Grossmann I (2005) Rigorous design of distillation columns. Integration of disjunctive programming and process simulators. Ind Eng Chem Res 44(200):6760–6775
25. Diwekar U, Grossmann I, Rubin E (1992) An MINLP process synthesizer for a sequential modular simulator. Ind Eng Chem Res 31:313–322
26. Chaudhuri P, Diwekar U (1997) Synthesis under uncertainty with simulators. Comput Chem Eng 21:733–738
27. Fu Y, Diwekar U, Young D, Cabezas H (2001) Designing processes for environmental problems. In: Sikdar SK, El-Halwagi M (eds) Process design tools for the environment, vol 1, Chap. 12. Taylor & Francis, New York, pp 295–317
28. Fang K, Li R, Sudjianto A (2006) Design and modeling for computer experiments, 1st edn. Chapman & Hall/CRC, Taylor & Francis, Boca Raton, FL
29. Chen J, Frey C (2004) Optimization under variability and uncertainty: a case study for NO*x* emissions control for a gasification system. Environ Sci Technol 38:6741–6747
30. Almeida-Bezerra M, Santelli R, Padua-Oliveira E, Escaleira L (2008) Response surface methodology (RSM) as a tool for optimization in analytical chemistry. Talanta 76:965–977
31. Kasabov N (1998) Foundations of neural networks, fuzzy systems, and knowledge engineering, 2nd edn, a Bradford book. MIT Press, Cambridge, Massachusetts
32. Pérez-Fortes M, Bojarski AD, Velo E, Nougués JM, Puigjaner L (2009) Conceptual model and evaluation of generated power and emissions in an IGCC plant. Energy 34:1721–1732
33. Pérez-Fortes M, Bojarski AD, Velo E, Puigjaner L (2010) IGCC power plants: Conceptual design and techno-economic optimization. In: Harris AM (ed) Clean energy: resources, production and developments. NOVA, Hauppauge, NY
34. Giampaolo T (2006) The gas turbine handbook, principles and practice, 3rd edn. Fairmont Press, Lilburn, Georgia
35. Zhu Y (2004) Evaluation of gas turbine and gasifier-based power generation system. PhD thesis, Graduate Faculty, North Carolina State University
36. Ongiro AO, Ugursal VI, Altaweel AM, Blamire DK (1995) Simulation of combined-cycle power-plants using the Aspen Plus shell. Heat Recovery Syst CHP 15:105–113

37. Coca MT (2003) Tecnología de gasificación integrada en ciclo combinado: GICC. Aplicación real en España: Elcogas. Puertollano. Technical report, ENERCLUB
38. Highman C, van der Burght M (2003) Gasification. Elsevier Science, Amsterdam
39. Wells G, Rose L (1986) The art of chemical process design, Vol 1–2 of computer-aided chemical engineering. Elsevier, Amsterdam

Process Integration: HEN Synthesis, Exergy Opportunities

Zdravko Kravanja, Miloš Bogataj and Aleksandr Soršak

Abstract This chapter provides a brief description of thermodynamic analysis, the maximization of heat-recovery and power generation and the synthesis of IGCC's heat exchanger network. Trade-offs between the income from power generation and utility costs plus the investment in HEN are studied with respect to the generation of different pressure-levels of steam and temperature driving force losses within the heat-recovery network. A combined pinch analysis/mathematical programming approach is applied and the optimization models are described for (i) the maximization of heat-recovery and power generation, and (ii) a synthesis of the heat-recovery and steam/power generation network. A sensitivity analysis for the synthesis is performed in order to show how optima are sensible for the expected increase in future electricity and utility prices, and the project's lifetime. The results obtained from the studied IGCC process indicate that detailed optimization has to be performed during the network synthesis step, otherwise optimal trade-offs are missed that may result either in serious power generation losses or in obtained over-designed networks.

Z. Kravanja (✉) · M. Bogataj
Faculty of Chemistry and Chemical Engineering, University of Maribor,
Smetanova ulica 17, 2000 Maribor, Slovenia
e-mail: kravanja@uni-mb.si

M. Bogataj
e-mail: milos.bogataj@uni-mb.si

A. Soršak
Javni zdravstveni zavod Mariborske lekarne Maribor,
Minarikova ulica 6, 2000 Maribor, Slovenia
e-mail: sandi.sorsak@mb-lekarne.si

L. Puigjaner (ed.), *Syngas from Waste*, Green Energy and Technology,
DOI: 10.1007/978-0-85729-540-8_9,

Notation

ASU	Air separator unit
CP	Claus plant
GCC	Grand composite curve
GT	Gas turbine
HE	Heat exchanger
HEN	Heat exchange network
HPS	High-pressure steam
HRAT	Heat-recovery approach temperature
HRSG	Heat-recovery steam generator
LP	Linear programming
LPS	Low-pressure steam
MINLP	Mixed-integer non-linear programming
MPS	Medium pressure steam
OA/ER	Outer-approximation/equality-relaxation
ST	Steam turbine

1 Introduction

In this chapter we demonstrate on the integrated gasification combine cycle (IGCC) case-study how to perform energy efficiency enhancement analysis and synthesis of heat/energy-recovery networks. Either heuristic, pinch analysis or mathematical programming can be applied to fulfil the task. In our case, a combined pinch analysis/mathematical programming approach is described. The enhancement analysis and synthesis are based on the assumption of considering process streams with fixed flows, temperatures, pressures, phase conditions and composition. If so, then, at this stage, we skip the gas turbine (GT) system by focusing only on the enhancement of IGCC's process heat-recovery network for steam/power generation.

Basic features of the IGCC process in regard to heat integration and the production of steam for power generation are the following:

- We are dealing with a large number of process hot and cold streams (about 40) within a wide range of temperatures (100–1,270 K).
- Low-pressure (LP-), medium-pressure (MP-) and high-pressure steam (HPS) can be produced for power generation, presented as additional hot and cold streams (16) with variable heat capacity flowrates.
- The heat to be transferred between hot and cold streams is extensive (about 400 MW), giving rise to very large areas of heat exchangers.
- Approximately one third of all the process and utility streams are isothermal.
- On the one hand we have a large heat-recovery steam generator (HRSG) for the production of various steams, and on the other hand, we have the generation of power within a three-stage steam turbine system (vapour turbine—VT) operating at a high, medium and low pressure.

It should be noted that the exergy content of steam, and, hence its potential to produce power, increases with pressure level and temperature. Thus, with HPS we can produce more power than with MPS, and with MPS more than with LPS. It is then apparent that, for the maximization of power generation, both the amount of heat recovered from process hot streams (the load) and steam levels are equally important. However, there is a general trade-off between the load and the level since the higher the level, the less heat can be recovered, and vice versa. Another very important trade-off is the one between the investment in HEN and the amount and level of heat recovered. A smaller heat-recovery approach temperature (HRAT) enables better integration; however, it is achieved at a higher investment cost. On the other hand, a larger HRAT would recover less heat into steam at a lower exergy level, at the same time wasting more heat and using more cold utility. Another important question relates to the selection of appropriate types of heat exchanger (HE) units. In order to exploit the temperature potentials of process hot streams as much as possible, in such a load versus level problem, it is very important to consider those HE types that exhibit counter-flow or a very near counter-flow arrangement of hot and cold streams. The selection of different hot and cold utilities is also an important issue. The higher the level of a hot utility and the lower the level of a cold utility below the ambient temperature, the higher are the costs. And lastly, due to the economy-of-scale effect, investment costs for equipment are generally lower with a smaller number of process units, so that it is also necessary to design a HEN with a smaller number of HE units. It should also be noted that one has to consider a parallel arrangement of HE units when a calculated area exceeds the maximal allowable areas for selected types of HE units.

With respect to the above-mentioned trade-offs, it is apparent that we are dealing with a very complex optimization problem. Therefore, the main goal is to obtain a solution where optimal trade-offs are established between the income from power generation and the outcome from the consumption of utilities plus investment. Such an optimal HEN design should exhibit an optimal generation of different quality steam (load and levels) at optimal temperature driving forces in the optimal arrangement, number and types of HE units. In order to accomplish this task, the following step-wise procedure is applied:

1. Thermodynamic analysis of the IGCC process flowsheet in order to assess the opportunity for heat and exergy integration for heat-recovery and power generation.
2. Maximization and sensitivity analysis of heat-recovery and power generation versus HRAT.
3. Optimal synthesis of a heat-recovery and steam/power generation network.

Whilst in the first step a thermodynamic approach with pinch analysis is applied, the second and third steps rely on mathematical programming: the second step on linear programming (LP) applies an extended model of the simultaneous heat integration model by Duran and Grossmann [1], and the third HEN synthesis step on mixed-integer non-linear programming (MINLP) is based on an extension of the simultaneous model for the synthesis of HEN by Yee and Grossmann [2].

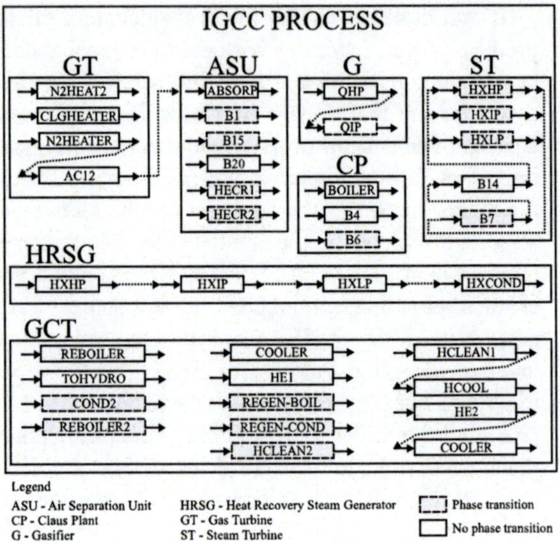

Fig. 1 Schematic representation of enthalpy streams in IGCC process

2 Thermodynamic Analysis of the IGCC Process

A very important step when performing thermodynamic analysis of any process is its data collection and acquisition. Let us first describe how data is extracted from the IGCC flowsheet.

2.1 Data Extraction

The extraction of process streams' thermodynamic properties plays an important role in the heat integration analysis of a given process. Regardless of whether the analysis is performed using the pinch methodology or any of the approaches based on mathematical programming, great care must be taken to extract the data properly. By this we mean that the data should be extracted in a way that imposes no unnecessary constraints on the problem being analyzed and yet ensures feasibility of the final design.

Let us now consider the schematic representation of an IGCC process in Fig. 1. The figure is a condensed representation of a converged Aspen Plus flowsheet [3]. In each of the blocks of the flowsheet (GT, ASU, G, CP, ST and GCT), the represented streams were identified as those with the need for cooling or heating in order to satisfy process constraints. Stream names correspond to the names of the process units in the flowsheet.

As shown in Fig. 1, some of the streams undergo a phase change. These streams are denoted with shaded rectangles, for example, condensation of the top product in the distillation column (REGEN-COND in GCT). On the other hand, some

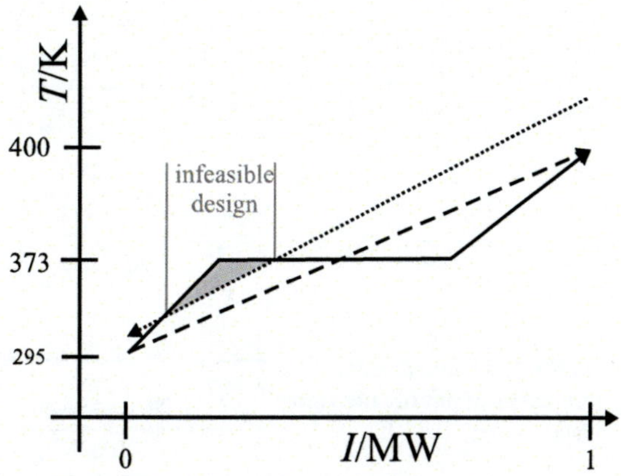

Fig. 2 Temperature-enthalpy profiles of a stream subjected to phase transition

streams actually represent a single stream, although extracted as multiple streams with linked target and supply temperatures. Such a stream may even cross the borders of blocks. These streams are denoted with dotted lines. For example, stream N2HEATER in GT is linked to stream AC12 in the same block, which is then linked to stream ABSORP in ASU.

The above two cases are the most common examples which, unless the stream temperature-enthalpy profiles are represented correctly, can lead to errors and missed opportunities. Firstly, let us consider the case when a phase transition takes place on a generic cold stream, as depicted with a solid line in Fig. 2. The stream with a supply temperature of 295 K is being heated up to boiling temperature (373 K), evaporated, and then superheated to a target temperature of 400 K. The total enthalpy flowrate equals 1 MW. Please note that since we are dealing with a continuous process, enthalpy flowrates (*I*/W) are used rather than enthalpies (*H*/J). Obviously, we are dealing with three segments, each of them having a unique heat capacity—slope of the temperature-enthalpy profile. Now, suppose that the stream is extracted as a single segment stream i.e., having constant heat capacity through the entire temperature-enthalpy profile (*dashed line* in Fig. 2).

Clearly, as it can be seen from this figure, the latter case causes thermodynamic infeasibilities due to negative temperature differences in a portion of the profile where the hot stream (*dotted line*) is colder than the properly represented cold stream. For this reason, whenever a phase transition occurs in a given stream, its temperature—enthalpy profile must be represented by segments. The number of segments of each stream, their target and supply temperatures and enthalpy flowrates can be obtained by simulating each such stream, as shown in Fig. 3.

Next, we must address those streams extracted as multiple streams, but are actually a single physical stream. Unlike the prior example, where improper representation (extraction of thermodynamic properties) of streams can lead to serious errors, this is not the case here. Nonetheless, extracting the data in this way does not open up additional opportunities for improving the design. The latter comes from the fact that each such stream imposes additional constraints (fixed supply and target temperatures, and enthalpy flowrate) on the problem being

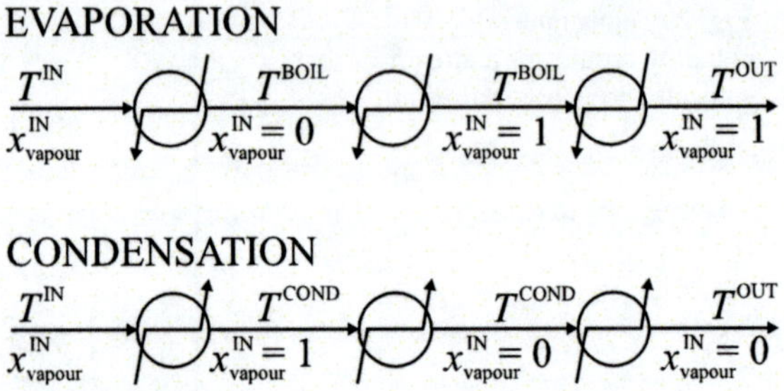

Fig. 3 Flowsheet for extracting the thermodynamic properties of streams undergoing phase transition

Table 1 Thermodynamic properties of hot streams (raw data)

Block	Process unit/stream	T^{in}/(K)	T^{out}/(K)	*I*/(MW)
1	ABSORP	400.1	287.1	16.397
1	HECR1	287.1	110.6	33.578
1	HECR2	370.1	123.3	16.268
2	QHP	1,108.1	658.1	72.210
2	QIP	658.1	508.1	22.797
3	COND2	383.1	382.1	0.559
3	COOLER	384.7	313.1	0.536
3	HCLEAN1	415.4	382.0	6.721
3	HCOOL	382.0	306.1	12.865
3	HE2	389.1	344.0	5.668
3	COOLER	344.0	306.1	4.720
3	$\text{REGEN}^{\text{COND}}$	357.5	356.5	16.785
4	BOILER	1,273.1	543.1	7.602
4	B4	573.1	349.2	1.529
4	B6	777.6	414.1	4.728
5	HXHP	849.3	599.3	146.475
5	HXIP	599.3	449.3	83.740
5	HXLP	449.3	443.1	3.392
5	HXCOND	443.1	373.1	38.143
6	N2HEATER	623.9	563.1	9.050
6	AC12	563.1	400.1	23.818
7	B7	307.2	306.2	276.700

Blocks: *1* ASU, *2* G, *3* GCT, *4* CP, *5* HRSG, *6* GT, *7* ST

analyzed, thus reducing the degrees of freedom. Obviously, it is a must that such streams are merged into a single stream having a supply temperature equal to the supply temperature of the first stream in a series, a target temperature equal to the target temperature of the last stream in a series and an enthalpy flowrate equal to the sum of the enthalpy flowrates for all the streams in a series.

Having discussed the above issues, let us take a closer look at the raw data in Tables 1 and 2. In Table 1, the basic thermodynamic properties of hot streams are given as extracted from the flowsheet. The properties of cold streams are given in Table 2. In total, 22 hot and 15 cold streams were identified. Among the hot and cold steams, the streams found in the GT block (pre-heating of condensate in B14,

Table 2 Thermodynamic properties of cold streams (raw data)

Block	Process unit/stream	T^{in}/(K)	T^{out}/(K)	I/(MW)
1	B20	110.6	510.6	33.502
1	B15	121.6	661.4	15.985
1	B1	110.6	111.6	10.403
3	REBOILER	408.1	409.1	0.597
3	TOHYDRO	397.6	414.1	1.271
3	REBOILER2	408.1	409.1	0.814
3	HE1	328.9	371.1	5.668
3	$REGEN^{BOIL}$	389.0	390.0	25.195
3	HCLEAN2	342.4	406.8	6.938
6	N2HEAT2	846.6	966.0	9.050
6	CLGHEATER	397.7	533.1	10.778
7	HXHP	413.1	764.1	218.959
7	HXIP	413.1	567.8	106.579
7	HXLP	413.1	518.0	3.392
7	B14	306.2	413.1	56.413

Blocks: *1* ASU, *2* G, *3* GCT, *4* CP, *5* HRSG, *6* GT, *7* ST

production of steam at different pressures in HXHP, HXIP and HXLP and condensation of turbine exhaust in B7) are also given, although the steam is generally treated as a utility and should, therefore, be left out of the analysis. However, in this case, it may be advantageous to treat these streams as hot and cold streams, since they directly contribute to the production of the final product (electrical power).

The raw data given in Tables 1 and 2 carries all the above-discussed pitfalls. Therefore, the data should be refined further in accordance with the above discussion. All of the streams which undergo a phase transition should be checked as to whether they should be broken up into multiple segments. In addition, some streams should be merged and represented as a single stream. The refined data is given in Tables 3 and 4. Besides the information on the supply and target temperatures and enthalpy flowrates, additional data needed for the HEN synthesis approach, which is described and discussed in subsequent sections, is also presented in these tables. As a result of the merging and breaking-up of streams, we end-up with 19 hot streams (16 process and 3 steam-condensing streams) and 22 cold streams (13 process and 9 steam-producing streams). Note that for the sake of further analysis steam-condensing and steam-producing streams will be treated with variable heat capacity flowrates. Therefore, specific capacity flowrate, defined in MW per mass production rate of steam in ton h^{-1}, is introduced for those streams.

Last, but not least, it must be pointed out here that the streams given in Tables 3 and 4 are not all the process streams, found in the flowsheet, which could also be considered in the analysis. For practical reasons, half of dozen or so streams having enthalpy flowrates well below 0.5 MW were neglected in this study.

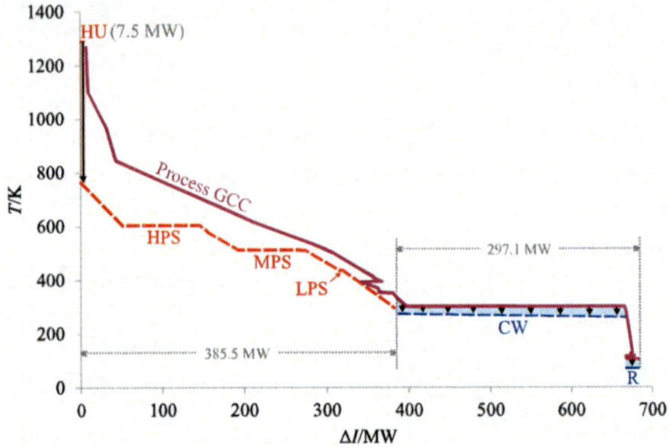

Fig. 4 GCC of an IGCC process' current state

3 Analysis of the Current State

One of the fastest and most practical ways to analyze how energy is being utilized within a given process is to construct a Grand Composite Curve (GCC), a fundamental concept of the pinch analysis. The usefulness of the GCC lies in its unique feature, which enables a representation of utility requirements both in enthalpy and temperature terms against the profile of the process background. For additional information on the theory and applications of the pinch analysis, the reader is referred to Smith [4] and Klemeš et al. [5].

In this work, the GCC shown in Fig. 4 was constructed from the data given in Tables 3 and 4. The profile of the process background was constructed on the basis of all the streams, with the exception of the cold steam-producing streams in the GT block. These streams were used to plot a heat-recovery profile, plotted below the process background profile. Note that the length (enthalpy content) of the heat-recovery profile (385.5 MW) was fixed according to the data in Table 4. In other words, the production of steam was fixed according to the one obtained from the converged flowsheet. Fixing the enthalpy content of the heat-recovery profile, causes the process background profile to shift to the right, as this is the only way to prevent negative temperature differences, thus creating a gap of approximately 7.5 MW. This gap corresponds to the hot utility deficit and points to a slight overdesign of steam production and, consequently, an overoptimistic production of electrical power (107.9 MW). Analyzing the GCC further, we find that the total cold utility requirement equals 297.1 MW, of which 95% could be cooled with cooling water or cooling tower system. The rest must be cooled with a suitable refrigerant.

Apart from a slight overdesign of steam production, GCC raises some additional questions. Is the heat-recovery profile constructed optimally, as clearly the temperature differences between the process background and heat-recovery profile are, in some parts, well greater than 300 K? If so, then how can we improve

Table 3 Thermodynamic properties of hot streams (refined data)

Block	Stream	Stream No.	T^{in} (K)	T^{out} (K)	I (MW)	fc (MW K^{-1})	fc^{spec} (MWh ton^{-1})	h (W m^{-2} K)	p (bar)	Tox
	Process streams									
1	$HECR1^{COOL}$	1	287.1	112.8	19.101	0.1096		300	13.0	0
1	$HECR1^{COND}$	2	112.2	110.6	14.477	9.0481		4,500	13.0	0
1	$HECR2^{COOL}$	3	370.1	124.3	12.166	0.0495		600	25.0	0
1	$HECR2^{COND}$	4	124.3	123.3	4.102	4.1020		4,500	25.0	0
2	QHP, QIP	5	1,108.1	508.1	95.006	0.1583		600	25.0	1
3	COND2	6	383.1	382.1	0.559	0.5590		2,500	22.0	0
3	COOLER	7	384.7	313.1	0.536	0.0075		1,500	22.0	0
3	HCLEAN1, HCOOL	8	415.4	306.1	19.586	0.1792		500	22.0	1
3	HE2, COOLER	9	389.1	306.1	10.389	0.1252		1,500	22.0	1
3	$REGEN^{COND}$	10	357.5	356.5	16.785	16.7850		2,500	22.0	1
4	BOILER	11	1,273.1	543.1	7.602	0.0104		100	0.9	1
4	B4	12	573.1	349.2	1.529	0.0068		100	0.9	1
4	$B6^{COOL}$	13	777.6	447.6	2.443	0.0074		100	0.9	1
4	$B6^{COND}$	14	447.6	414.1	2.285	0.0682		100	0.9	1
5	HXHP, HXIP, HXLP, HXCOND	15	849.3	373.1	271.750	0.5707		100	1.0	1
6	N2HEATER, AC12, ABSORP	16	623.9	287.1	49.265	0.1463		300	13.0	1
	Steam-condensing streams									
7	$B7^{HPS}$	17	307.2	306.2	Total:		0.6047	4,500	0.05	0
7	$B7^{MPS}$	18	307.2	306.2	276.700		0.6047	4,500	0.05	0
7	$B7^{LPS}$	19	307.2	306.2			0.6475	4,500	0.05	0

Blocks: *1* ASU, *2* G, *3* GCT, *4* CP, *5* HRSG, *6* GT, *7* ST

Table 4 Thermodynamic properties of cold streams (refined data)

Block	Stream	Stream No.	T^{in} (K)	T^{out} (K)	I (MW)	fc (MW K^{-1})	fc^{spec} (MWh ton^{-1})	h (W m^{-2} K)	p (bar)	Tox
	Process streams									
1	B20	1	110.6	510.6	33.502	0.0838		300	13.0	1
1	$B15^{EVAP}$	2	121.6	122.3	2.229	3.1843		1500	13.0	0
1	$B15^{HEAT}$	3	122.3	661.4	13.756	0.0255		300	13.0	0
1	$B1^{EVAP}$	4	110.6	111.6	10.403	10.403		1500	13.0	1
3	REBOILER	5	408.1	409.1	0.597	0.5970		2000	22.0	0
3	TOHYDRO	6	397.6	414.1	1.271	0.0770		500	22.0	1
3	REBOILER2	7	408.1	409.1	0.814	0.8140		2000	22.0	0
3	HE1	8	328.9	371.1	5.668	0.1343		1500	22.0	1
3	$REGEN^{BOIL}$	9	389.0	390.0	25.195	25.1950		2000	22.0	1
3	$HCLEAN2^{EVAP}$	10	342.4	366.2	4.018	0.1688		500	22.0	1
3	$HCLEAN2^{HEAT}$	11	366.2	406.8	2.920	0.0719		500	22.0	1
6	N2HEAT2	12	846.6	966.0	9.050	0.0758		500	20.0	1
6	CLGHEATER	13	397.7	533.1	10.778	0.0796		500	20.0	1
	Steam-producing streams									
7	$HXHP^{PREHEAT}$	14	307.3	599.0	108.374		0.3727	1500	122.6	0
7	$HXHP^{EVAP}$	15	599.0	600.0	94.832		0.3293	2000	122.6	0
7	$HXIP^{SUPERHEAT}$	16	600.0	764.1	51.709		0.1771	600	122.6	0
7	$HXIP^{PREHEAT}$	17	307.3	511.0	39.232		0.2446	1500	32.0	0
7	$HXIP^{EVAP}$	18	511.0	512.0	79.331		0.4947	2000	32.0	0
7	$HXIP^{SUPERHEAT}$	19	512.0	567.8	7.555		0.0456	600	32.0	0
7	$HXLP^{PREHEAT}$	20	307.3	425.0	0.715		0.1380	1500	5.0	0
7	$HXLP^{EVAP}$	21	425.0	426.0	3.036		0.5863	2000	5.0	0
7	$HXLP^{SUPERHEAT}$	22	426.0	518.0	0.289		0.0550	600	5.0	0

Blocks: *1* ASU, *2* G, *3* GCT, *4* CP, *5* HRSG, *6* GT, *7* ST the exergy efficiency of the IGCC plant? How does improving the exergy efficiency affect the profitability of the process? The answers to these questions will be provided in subsequent sections.

4 Maximization of Power Generation

In the previous section we described how data from flowsheet simulation can be used to define hot and cold process streams. In this section, we will show how it is possible, in a simple way, to maximize the generation of power from steam whilst recovering the maximum amount of heat from the IGCC process hot composite curve. Firstly, we will briefly describe an optimization model and then perform a simple sensitivity analysis to assess the maximum amount of power generated as a function of a given temperature driving force loss (HRAT), within the heat-recovery network.

4.1 Optimization Model for Maximal Energy Recovery

Several indexes have to be specified before presenting the model:

- Index i for hot process streams and hot steam-condensing streams, given by set H
- Index j for cold process streams and cold steam-producing streams, given by set C

In addition we define a subset of cold process streams C^{P} and a subset of cold steam-producing streams C^{S}, and similarly, a subset of hot process streams H^{P}, and a subset of hot steam-condensing streams H^{S}.

We can apply either a variation of the minimum utility transhipment model by Papoulias and Grossmann [6] based on a heat cascade diagram or a variation of the simultaneous heat integration model by Duran and Grossmann [1] based on the pinch location method. Since the latter needs less data handing, we chose the pinch location method. Only a short description of the model and its adaptation to the IGCC problem is given here. For further details, please refer to the book by Biegler et al. [7]. The basic idea of the pinch location method relies on an insight that the pinch can only occur at one of the inlet temperatures for hot and cold streams. Thus, a heat balance is posed at every inlet temperature T_i^{in} and T_j^{out} to calculate any heat deficit above it as the difference between the heat content of a cold composite curve (CCC) above this temperature, and the heat content of a hot composite curve (HCC) above this temperature with the addition of the minimum recovery approach temperature (HRAT). It can be shown that the maximal heat deficit corresponds to the minimum consumption of the hot utility $Q_{\mathrm{min}}^{\mathrm{HU}}$ whilst the minimum consumption of cold utility $Q_{\min}^{CU}$ can be calculated from the overall heat balance. In our case, all inlet and outlet temperatures, and the heat capacity flowrates of the process streams are fixed, only the flowrates of the cold steam-producing streams are variable, since we would like to obtain the optimum production of steam at different pressures. Thus, the variable heat capacity flowrate of steam-producing streams can be represented as the product of its specific heat capacity flowrate (the heat capacity flowrate of the steam at a production rate of one ton per hour), fc_j^{spec}, $j \in C^S$ and a steam mass flowrate, $q_{m_{js}}$ (please note that this is our optimization variable). Note that fc_j^{spec} is a constant and can be defined from steam tables or by a flowsheet simulator based on a given pressure, temperature range and phase change. Typically, any steam can be represented by three cold steam-producing streams: one for pre-heating, one for evaporation and one for superheating and, in addition, with one hot steam–condensing stream, all having the same q_m. In order to define the mapping of hot and cold streams to different steams, a steam set and multi-dimensional (process stream, steam) sets for hot and cold streams are defined, respectively:

- $S = \{s|\text{steam } s \text{ produced in a process}\}$
- $\text{HS} = \{(i, s)|\text{hot stream } i \text{ used as a condensing stream for steam } s\}$
- $\text{CS} = \{(j, s)|\text{cold stream } j \text{ representing the production of steam } s\}$

Finally, for every steam we also need a specific power P_s^{spec} which is defined as power generated at the steam production rate of one ton per hour. The objective of the optimization is the maximization of the annual profit as the difference between income from the generation of electricity, and the utility cost at a given HRAT. The following optimization model can now be defined:

$$\max \text{ Profit} = c^{el} \sum_{s \in S} q_m P_m^{\text{spec}} - c^{hu} Q_{\min}^{HU} - c^{cu} Q_{\min}^{CU}$$

$s.t.$

$$\begin{aligned} Q_{\min}^{HU} \geq & \sum_{j \in C} FC_j \times \left[\max\left\{0, T_j^{\text{out}} - \left(T_{ii}^{\text{in}} - \Delta T_{\text{HRAT}}\right)\right\} - \max\left\{0, T_j^{\text{in}} - \left(T_{ii}^{\text{in}} - \Delta T_{\text{HRAT}}\right)\right\}\right] - \forall ii \in H \\ & \sum_{i \in H} FC_i \times \left[\max\{0, T_i^{\text{in}} - T_{ii}^{\text{in}}\} - \max\{0, T_i^{\text{out}} - T_{ii}^{\text{in}}\}\right] \\[2ex] Q_{\min}^{HU} \geq & \sum_{j \in C} FC_j \times \left[\max\left\{0, T_j^{\text{out}} - T_{jj}^{\text{in}}\right\} - \max\left\{0, T_j^{\text{in}} - T_{jj}^{\text{in}}\right\}\right] - \forall jj \in C \\ & \sum_{i \in H} FC_i \times \left[\max\left\{0, T_i^{\text{in}} - \left(T_{jj}^{\text{in}} + \Delta T_{\text{HRAT}}\right)\right\} - \max\left\{0, T_i^{\text{out}} - \left(T_{jj}^{\text{in}} + \Delta T_{\text{HRAT}}\right)\right\}\right] \end{aligned} \tag{1}$$

$$Q_{\min}^{CU} = \sum_{i \in H} FC_i \times \left(T_i^{\text{in}} - T_i^{\text{out}}\right) + Q_{\min}^{HU} - \sum_{j \in C} FC_j \times \left(T_j^{\text{out}} - T_j^{\text{in}}\right) \tag{2}$$

where the heat capacity flowrates of the process streams are fixed:

$$FC_i = fc_i,\ \forall i \in H^P \text{ and } FC_j = fc_j,\ \forall j \in C^P \tag{3}$$

whilst the heat capacity flowrates of steam-producing streams are variables:

$$FC_i = \sum_{s \in S, (i,s) \in HS} fc_i^{\text{spec}} \times q_{m_s}, \forall i \in H^S \tag{4}$$

$$FC_j = \sum_{s \in S, (j,s) \in CS} fc_j^{\text{spec}} \times q_{m_s}, \forall j \in C^S \tag{5}$$

Note that, since the inlet and outlet temperatures are fixed, the above model is linear and gives global optimal solutions. The max operator is used to calculate the portion of every stream above pinch candidates $T_{ii}^{\text{in}}, \forall ii \in H$ and $T_{jj}^{\text{in}}, \forall jj \in C$ at a given HRAT.

4.2 Sensitivity Analysis

A simple, yet very effective sensitivity analysis can now be carried out by performing a sequence of the above optimization problems, each time increasing HRAT by a suitable ΔHRAT step-size. Figure 5 shows the dependence of annual gross profit, power production from LPS, MPS and HPS and the consumption of cold utility on HRAT. It is interesting to note that at small values for HRAT, the temperature potential of the IGCC process is high enough to enable the complete production of steam at a high pressure. As the values for HRAT increase, more MPS and, finally, more LPS are produced, whilst the total production of steam and, hence, power generation is decreasing. Since the total production of steam is decreasing, less heat is recovered from the process and more cold utility has to be used. For example, increasing HRAT from 2 to 100 K would result in a reduction of power generation from 118 to 88 MW and an increase in the consumption of the cold utility from 287 to 318 MW. 441 tons h^{-1} of HPS can be produced at a HRAT of 2 K compared with the production of 283 tons h^{-1} HPS, 58 tons h^{-1} MPS and 0.4 tons h^{-1} LPS or 341 tons h^{-1} totals at a HRAT of 100 K. Note that, besides lowering the level of steam produced, the total production of steam is also significantly decreased, resulting in a 25% reduction of power generation and more than 40% reduction of annual gross profit (53 M€ annum^{-1} vs. 31 M€ annum^{-1}). It is evident that choosing an optimal HRAT should have the highest priority. Since the optimal HRAT is an approach temperature where appropriate trade-offs are established between the income from the power generation and the cost of utilities plus investment, a detailed optimization of the IGCC heat-recovery network, steam and power generation has to be performed.

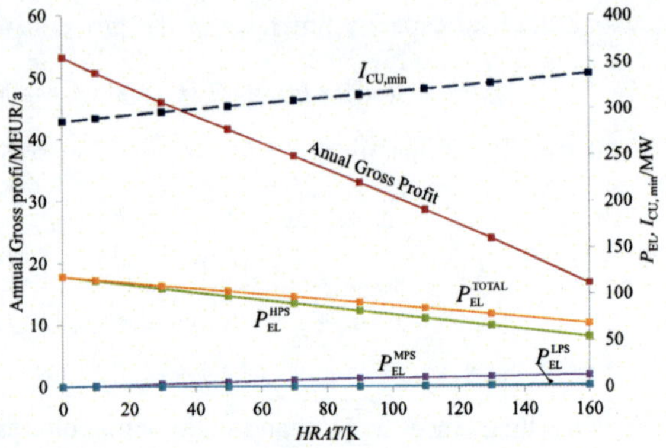

Fig. 5 Sensitivity analysis results

5 Optimal Synthesis of a Heat-Recovery Network for Effective Power Generation

In order to achieve optimal trade-offs between the income from the generated power, and investment plus utility cost, any optimization model we would like to apply has to account for the production of different pressure-levels of steam, the consumption of different utilities (e.g., cooling water, refrigeration, etc.) and the investment in HEN, which depends heavily on an arrangement of heaters, coolers and HE units within the network and their optimal temperature distributions and driving forces. With respect to obtaining optimal temperature driving forces, such model should enable the consideration of driving forces as optimization variables individually in each exchanger. In addition, since temperature distributions and, hence, driving forces depend strongly on the selection of HE types, this model should provide the possibility for a simultaneous selection of HE types during the course of optimization. It should be noted that the stage-wise model for the synthesis of HEN by Yee and Grossmann [2] has the capability of optimizing the driving forces individually within each unit. However, it is based on single-type pure counter-flow exchangers. In order to accomplish our task, the Yee and Grossmann's model has been extended to:

- Alternative exchanger types, similar to that proposed by Soršak and Kravanja [8]
- Multi-utility configurations
- Alternative pressure-level steam production, and
- Accounting for the parallel arrangement of HE units when a total transfer area exceeds an upper limit for a selected exchanger type

5.1 Superstructure

Since we are dealing with a stage-wise superstructure, alternative hot and cold utilities and alternative exchanger types, the following indexes are now introduced:

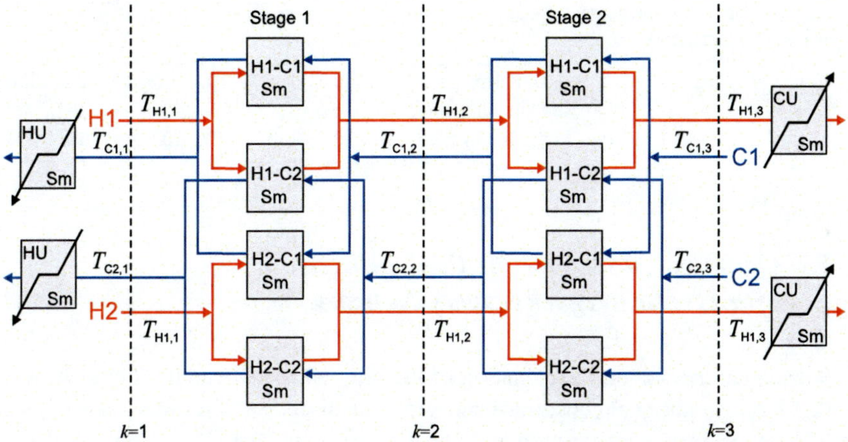

Fig. 6 Two-stage superstructure of HEN with two hot and two cold streams (Sm represents the match superstructure)

- Index k for the stages given by set K
- Index l for the alternative exchanger types given by set L
- Index hu for hot utilities given by set HU, and
- Index cu for cold utilities given by set CU

Within the superstructure, each cold stream (j) can be potentially matched with each hot stream (i) over several stages (k) (Fig. 6), and each potential match is now represented by the match superstructure comprising the following exchanger types:

- Double-pipe heat exchanger (index for exchanger types $l = 1$)
- Plate and frame heat exchanger ($l = 2$)
- Fixed plate shell and tube heat exchanger ($l = 3$), and
- Shell and tube heat exchanger with U-tubes arranged by an even number of passes ($l = 4$)

Note that the heaters and coolers are represented by the heater superstructure and the cooler superstructure in order to consider multi-utility configurations.

HEN topology and the selection of exchanger types is specified by the vectors of binary variables determined during MINLP optimization. The selection of the HE types is modelled by disjunctions based on operating limitations. Since different types of heat exchangers involve different design geometries, which influence the inlet and outlet temperatures of heat exchangers, additional constraints are specified to provide a feasible temperature distribution in HEN. The basic usage recommendations for different areas, pressures and temperature ranges that hold for different exchanger types [9] are shown in Table 5. In addition, due to the leakage problem, the use of a plate and frame heat exchanger is not recommended, when one of the involved streams in the exchanger is toxic.

Table 5 Data of considered heat exchanger types

Type	p_{max} (MPa)	T(°C) range	A(m^2) range
Double pipe	30.7	−100–600	0.25–200
Plate and frame	1.6	−25–250	1–1,200
Shell and tube	30.7	−200–600	10–1,000

5.2 Optimization Model for the Synthesis of a Heat/Energy-Recovery Network

Since a description and explanation of the original model can be found elsewhere [2, 7], only additional constraints and terms are discussed. This model is composed of heat balance constraints and logical constraints that determine selection of the matches and HE types, and the profit objective function. In order to simplify the task, coolers and heaters are only represented by fixed plate shell and tube exchangers. The following binary variables are introduced for the discrete decision:

- $y_{hu,j}^{HU}$ for the selection for hot utility hu for cold stream j,
- $y_{cu,i}^{CU}$ for the selection for cold utility cu for hot stream i, and
- $y_{i,j,k,l}$ for the selection of exchanger l in (i, j, k) match

Heat balances:

For every hot and every cold stream:

$$\left(T_i^{in} - T_i^{out}\right) \times FC_i = \sum_{j \in C} \sum_{k \in K} Q_{ijk} + \sum_{cu \in CU} Q_{cu,i}^{CU} \quad \forall i \in H \tag{6}$$

$$\left(T_j^{out} - T_j^{in}\right) \times FC_j = \sum_{i \in H} \sum_{k \in K} Q_{ijk} + \sum_{hu \in HU} Q_{hu,j}^{HU} \quad \forall j \in C \tag{7}$$

Note that, by the summation of utilities over their corresponding sets, we allow for the designing of multi-utility configurations.

- For every segment of hot and cold stream in every stage:

$$\left(T_{i,k} - T_{i,k+1}\right) \times FC_i = \sum_{j \in C} Q_{ijk} \quad \forall i \in H, k \in K \tag{8}$$

$$\left(T_{j,k} - T_{j,k+1}\right) \times FC_j = \sum_{i \in H} Q_{ijk} \quad \forall j \in C, k \in K \tag{9}$$

- Inlet temperature assignment:

$$T_i^{in} = T_{i,1} \quad \forall i \in C \text{ and } \quad T_j^{out} = T_{j,N+1} \quad \forall j \in C \tag{10}$$

where N is the number of stages.

• A monotonic decrease in temperatures from the hot-left down to the cold-right part of the HEN superstructure:

$$T_{i,k} \geq T_{i,k+1} \quad \forall i \in H,\, k \in K \quad \text{and} \quad T_{j,k} \geq T_{j,k+1} \quad \forall j \in C,\, k \in K \tag{11}$$

• Outlet temperature:

$$T_i^{\text{out}} \leq T_{i,N+1} \quad \forall i \in H \quad \text{and} \quad T_j^{\text{out}} \geq T_{j,1} \quad \forall j \in C \tag{12}$$

• Cold utility consumption for every hot stream:

$$\left(T_{i,N+1} - T_i^{\text{out}}\right) \times FC_i = \sum_{cu \in \text{CU}} Q_{cu,i}^{\text{CU}} \quad \forall i \in H \tag{13}$$

• Hot utility consumption for every cold stream:

$$\left(T_j^{\text{out}} - T_{j,1}\right) \times FC_j = \sum_{hu \in \text{HU}} Q_{hu,j}^{\text{HU}} \quad \forall j \in C \tag{14}$$

Convex hull logical constraints for exchanger types:

• For every exchanger type heat load:

$$Q_{i,j,k,l}^{L} \leq Q^{\text{UP}} \times y_{i,j,k,l} \quad \forall i \in H, j \in C,\, k \in K,\, l \in L \tag{15}$$

• Match heat balance:

$$Q_{i,j,k} = \sum_{l \in L} Q_{i,j,k,l}^{L} \quad \forall i \in H, j \in C,\, k \in K \tag{16}$$

• Only one HE type can be selected if a match is selected:

$$\sum_{l \in L} y_{i,j,k,l} \leq 1 \tag{17}$$

Note that if none of the HE types is selected, then all heat loads for exchanger types are zero (Eq. 15), and the match heat load becomes zero, too (Eq. 16).

• Both temperature differences on the left and right sides of the match should be positive if HE is selected:

$$\left.\begin{aligned} \Delta T_{i,j,k} &\leq \left(T_{i,k} - T_{j,k}\right) + \Delta T_{i,j}^{\text{UP}} \sum_{l \in L} y_{i,j,k,l} \\ \Delta T_{i,j,k+1} &\leq \left(T_{i,k+1} - T_{j,k+1}\right) + \Delta T_{i,j}^{\text{UP}} \sum_{l \in L} y_{i,j,k,l} \end{aligned}\right\} \quad \forall i \in H, j \in C,\, k \in K \tag{18}$$

• Feasible temperature distribution

The original model (Yee and Grossmann [2]) contains constraints for feasible temperature distribution only for pure counter-flow exchangers. Since the extended model comprises different exchanger types, the temperature distribution in HEN that holds for a pure counter-flow exchanger may become infeasible and even

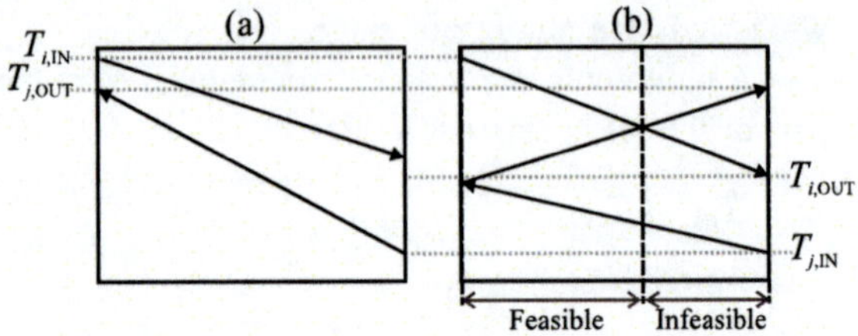

Fig. 7 Temperature arrangement in **a** counter-flow HE and **b** shell and tube HE with U-tubes

unattainable if the shell and tube exchanger is selected. The problem can be observed if we take a closer look at the temperature distribution of counter-flow and the shell and tube heat exchangers (Fig. 7).

In counter-flow heat exchangers, the outlet temperature of the cold stream can be higher (Fig. 7a) because of the transfer area's geometry. When U-tubes are used, the flow arrangement combines the counter and co-current flows. The temperature distribution becomes infeasible when a shell and tube exchanger is selected and $T_j^{\mathrm{out}} \geq T_i^{\mathrm{out}}$ (Fig. 7b). To overcome this problem, an additional constraint has to be specified for the shell and tube with U-tubes exchangers ($l = 4$):

$$T_{i,k+1} - T_{j,k} + \Delta T^{\mathrm{UP}} \cdot (1 - y_{i,j,k,4}) \geq \Delta T_{\min} \quad (i \in H, j \in C, k \in K) \tag{19}$$

Note that this constraint is a complicated one, since it tends to increase temperature differences in the shell and tube exchangers causing some additional loss of temperature potential for heat-recovery. This loss can be significant, especially when streams are defined within wide ranges between inlet and outlet temperatures. On the other hand, fixed plate shell and tube HE exhibit counter-flow arrangement, so its temperature distribution should disobey the above constraint. This means that by using a fixed plate shell and tube exchanger we may efficiently recover a considerable fraction of otherwise lost temperature potentials. However, at higher pressure, this type of HE is considerably more expensive than shell and tube with U-tubes (see Table 6). Thus, there is a trade-off between the potential losses and the selection of different types of exchangers.

5.3 *Disjunctive Modelling of Operating Limitations for Selected Exchanger Types*

The selection of exchanger types depends on the operating temperatures and pressures of the involved streams (Table 5). As stream pressures are fixed, they can be taken into account in a pre-screening procedure before optimization. However, all stage temperatures are optimization variables and, therefore, affect the selection of exchanger types. The temperature ranges of the comprised exchanger types have been modelled as scalar products between the vector for binary variable regards exchanger types and the vector for lower and upper temperature operating limits:

Table 6 Cost data

Exchanger type	Design pressure (bar)	Fixed charge coefficient (k€)	Area cost coefficient (k€ m^{-2})
Double pipe	0.01–125	46.0	2.742
Plate and frame	0.01–125	129.8	0.347
Fixed plate shell and tube	0.01–10	121.4	0.193
Fixed plate shell and tube	10–25	105.6	0.304
Fixed plate shell and tube	25–125	174.4	0.919
Shell and tube with U-tubes	0.01–125	100.9	0.272
Utility heat exchangers	0.01–125	105.6	0.304
Utilities	Cost (€ kWh^{-1})		
High-pressure steam		0.04	
Cooling water		0.01	
Refrigeration		0.08	
Electricity		0.08	
Other data			
Expected project's lifetime, a		15	
Interest rate (%)		10	
Yearly working fraction (h $annum^{-1}$)		8,000	

$$\left.\begin{aligned} T_{i,k} &\leq \sum_{l\in L} T_l^{\mathrm{UP}} y_{i,j,k,l} + T_i^{\mathrm{UP}}\left(1-\sum_{l\in L} y_{i,j,k,l}\right) \\ T_{i,k} &\geq \sum_{l\in L} T_l^{\mathrm{LO}} y_{i,j,k,l} + T_i^{\mathrm{LO}}\left(1-\sum_{l\in L} y_{i,j,k,l}\right) \end{aligned}\right\}, \forall i\in H, j\in C, k\in K \tag{20}$$

$$\left.\begin{aligned} T_{j,k} &\leq \sum_{l\in L} T_l^{\mathrm{UP}} y_{i,j,k,l} + T_j^{\mathrm{UP}}\left(1-\sum_{l\in L} y_{i,j,k,l}\right) \\ T_{j,k} &\geq \sum_{l\in L} T_l^{\mathrm{LO}} y_{i,j,k,l} + T_j^{\mathrm{LO}}\left(1-\sum_{l\in L} y_{i,j,k,l}\right) \end{aligned}\right\}, \forall i\in H, j\in C, k\in K \tag{21}$$

If an exchanger type in a match is selected then the last terms in the upper constraints become zero and stage temperatures are enforced within the corresponding operating limits; otherwise the constraints are redundant.

Logical constraints for utilities:

- Selection of hot and cold utilities, respectively:

$$Q_{hu,j}^{\mathrm{HU}} \leq Q_j^{\mathrm{UP}} \times y_{hu,j}^{\mathrm{HU}}, \quad \forall hu \in \mathrm{HU}, \forall j \in C \tag{22}$$

$$Q_{cu,i}^{\mathrm{CU}} \leq Q_i^{\mathrm{UP}} \times y_{cu,i}^{\mathrm{CU}}, \quad \forall hu \in \mathrm{CU}, \forall i \in H \tag{23}$$

- Positive temperature differences:

$$\left.\begin{aligned}\Delta T^1_{hu,j} &\le \left(T^{\text{in}}_{hu} - T^{\text{out}}_{j}\right) + \Delta T_{hu,j}\left(1 - y^{\text{HU}}_{hu,j}\right)\\ \Delta T^2_{hu,j} &\le \left(T^{\text{out}}_{hu} - T_{j,1}\right) + \Delta T_{hu,j}\left(1 - y^{\text{HU}}_{hu,j}\right)\end{aligned}\right\}, \forall hu \in \text{HU}, \forall j \in C \tag{24}$$

$$\left.\begin{aligned}\Delta T^1_{cu,i} &\le \left(T_{i,N+1} - T^{\text{out}}_{cu}\right) + \Delta T_{cu,i}\left(1 - y^{\text{CU}}_{cu,i}\right)\\ \Delta T^2_{cu,1} &\le \left(T^{\text{out}}_{i} - T^{\text{in}}_{cu}\right) + \Delta T_{cu,i}\left(1 - y^{\text{CU}}_{cu,i}\right)\end{aligned}\right\}, \forall cu \in \text{CU}, \forall i \in H \tag{25}$$

Generation of steam:

As in the previous section, we have to define variable heat capacity flowrates for hot steam producing steam-condensing $i \in C^S$ and cold steam-producing streams $j \in C^S$ proportional to a steam mass production flowrate q_{m_s}, whilst the heat capacity flowrates of process streams $i \in H^P$ and $j \in H^P$ are fixed. Thus, the same constraints as the presented in Sect. 4.1 are applied (Eqs. 3, 4 and 5).

Objective function:

The objective function of the extended model defines the annual profit of heat-recovery network and steam/power generation and comprises the following terms:

- Income from selling electricity, Eq. 26a
- Cost for utility consumption, Eq. 26b
- Annualized investment for coolers and heaters, Eq. 26c, and
- Annualized investment for heat exchangers of different types, Eq. 26d

$$\text{Profit} = c^{el} \times \sum_{s \in S} q_{m_s} P^{\text{spec}}_{s} \tag{26a}$$

$$-c^{\text{HU}}_{hu} \times \sum_{hu \in \text{HU}} \sum_{j \in C} Q^{\text{HU}}_{hu,j} - c^{\text{CU}}_{cu} \times \sum_{cu \in \text{CU}} \sum_{i \in H} Q^{\text{CU}}_{cu,i} \tag{26b}$$

$$\begin{aligned}&-f_d \times \sum_{i \in H} \sum_{cu \in \text{CU}} \left[c^{cu}_{fix} \times y^{cu}_{cu,i} + c^{cu}_{\text{var}} \times \frac{Q^{\text{CU}}_{cu,i}}{U_{cu,i} \times \Delta_{\ln} T_{cu,i} \times Ft^{cu}}\right]\\ &-f_d \times \sum_{j \in C} \sum_{hu \in \text{HU}} \left[c^{hu}_{fix} \times y^{hu}_{hu,j} + c^{hu}_{\text{var}} \times \frac{Q^{\text{HU}}_{hu,j}}{U_{hu,j} \times \Delta_{\ln} T_{hu,j} \times Ft^{hu}}\right]\end{aligned} \tag{26c}$$

$$-f_d \times \sum_{i \in H} \sum_{j \in C} \sum_{k \in K} \sum_{l \in L} \left[c_{fix} \times y_{i,j,k,l} + c_{\text{var}} \times \frac{Q^l_{i,j,k,l}}{U_{i,j} \times \Delta_{\ln} T_{i,j,k} \times Ft_l}\right] \tag{26d}$$

where f_d is an annualized factor depending on the given project's lifetime, and the selected interest rate. In order to avoid the numerical difficulties of using the above logarithmic mean temperature differences when $\Delta T^1 = \Delta T^2$, the temperature driving forces are specified by Chen's approximation [10]:

$$\Delta_{\ln} T_{cu,i} = \left[\Delta T^1_{cu,i} \times \Delta T^2_{cu,i} \times \left(\frac{\Delta T^1_{cu,i} + \Delta T^2_{cu,i}}{2}\right)\right]^{(1/3)} \tag{27}$$

$$\Delta_{\ln} T_{hu,j} = \left[\Delta T^1_{hu,j} \times \Delta T^2_{hu,h} \times \left(\frac{\Delta T^1_{hu,j} + \Delta T^2_{hu,j}}{2}\right)\right]^{(1/3)}$$

$$\Delta_{\ln} T_{i,j,k} = \left[\Delta T_{i,j,k} \times \Delta T_{i,j,k+1} \times \left(\frac{\Delta T_{i,j,k} + \Delta T_{i,j,k+1}}{2}\right)\right]^{(1/3)}$$

Note that *Ft* denotes the correction factor for the temperature driving forces within different exchanger types. It is equal to one for counter-flow and is less than one (0.8–0.95) for exchanger types with combination of counter and co-current flows as is the case with shell and tube HE with U-tubes.

5.4 Computational Feature of the Model

It should be noted that, in order to preserve the linearity of the constraints, all the non-linear cost terms are presented in the objective function. However, the consideration of different exchanger types drastically increases the combinatorics, size and, hence, the computation effort needed to solve the problem. Problems, such as IGCC with 20 hot and 20 cold streams easily have 2,000 or more binary variables and ten thousands of continuous variables and constraints. Efficient strategies have to be applied in order to decrease the computational burden and to find a good solution:

- Pre-screening of alternatives for some exchanger operating limitations can be done ahead of the optimization. The space for discrete decisions can thus be drastically reduced.
- In many cases, an outer-approximation/equality-relaxation (OA/ER) algorithm applied to solve this MINLP problem may not converge, since the sequence of major iteration produces infeasible NLP subproblems. The problem is very time consuming especially when MILP master problems take long CPU times for the needed integrality gap. In such a case integer infeasible path optimization [8] can be applied, thus preventing the OA/ER algorithm from obtaining infeasible NLPs. Considerably faster convergence of the algorithm can thus be achieved.

5.5 Results and Discussion

In the IGCC problem we dealt with 16 hot and 13 cold process streams, and 3 hot and 9 cold steam-producing streams (Tables 3 and 4), for which we assume that the power is generated in a cascade of HP, MP and LP turbines operating at

Table 7 Sensitivity analysis results

Electricity/utility cost factor	1	1	1	1	1.5	2	3
Estimated lifetime (annum)	5	10	15	30	30	30	30
HPS production (tons h^{-1})	378	384	398	401	405	409	409
LPS production (tons h^{-1})	0	0	0	4	0	0	0
Recovered enthalpy flowrate HPS (MW)	332	338	350	353	356	360	360
Recovered enthalpy flowrate LPS (MW)	0	0	0	3	0	0	0
Power HPS (MW)	101	103	107	108	107	110	110
Power LPS (MW)	0	0	0	1	0	0	0
Cold utility load (MW)	304	302	298	296	296	295	295
Annualized HEN investment (M€ $annum^{-1}$)	17.7	15.2	10.2	10.2	11.6	13.1	18.0
HEN investment (M€)	67.2	93.2	101.7	100.5	109.7	123.1	161.0

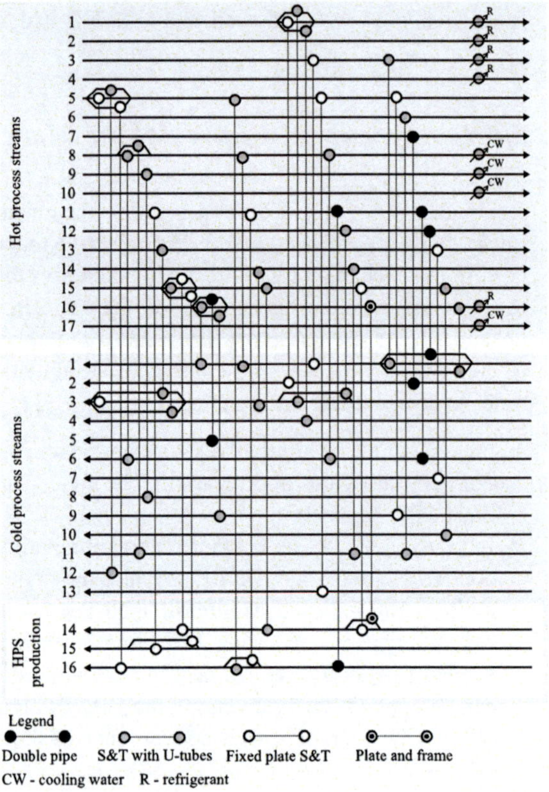

Fig. 8 Heat exchanger network for actual prices of utilities and energy

122, 32 and 5 bar, respectively. Note that the present cost data are given in Table 6. When applying the above MINLP model for simultaneous heat-recovery and steam/power generation network synthesis, optimal trade-offs between electricity income and utility cost plus investment can be obtained over different lifetime periods and the prediction of future utility and electricity prices. Table 7

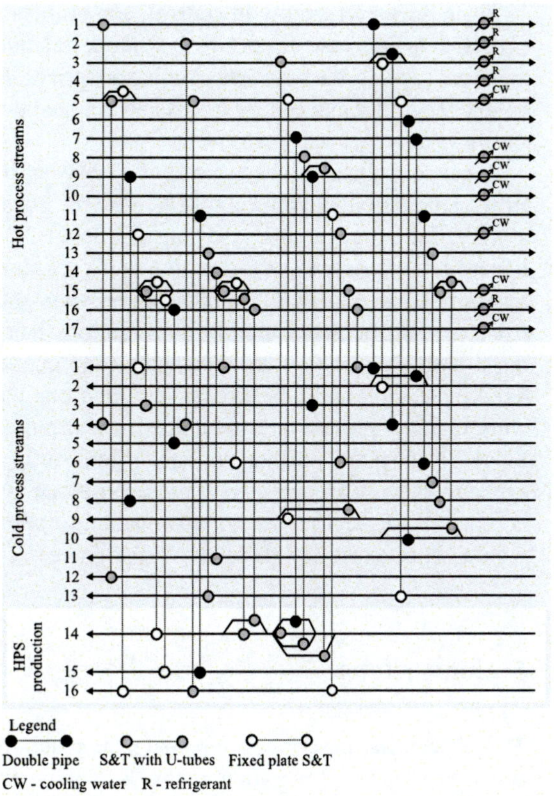

Fig. 9 Heat exchanger network for double the utility and energy prices

shows the amount of generated steam and power, cold utility consumption, annualized and total investment for IGCC HEN. From Table 7 it can be seen how, at lower energy prices and shorter depreciation periods, the generation of power decreases, when higher values are obtained for the actual HRAT. Note that a similar trend was observed in the previous section, where it was seen that by increasing HRAT the generation of power was reduced and the consumption of cold utility increased. In all cases, complex heat-recovery networks are obtained. The network, where actual prices for utilities and electricity, and a 15 year depreciation period are considered, is shown in Fig. 8. It can be seen that about 35 MW of energy is wasted: 2/3 heat content of HCLEAN1-HCOOL, ¾ of HE2-COOLER and complete heat from the REGEN-COND condenser, whilst 350 MW of heat is recovered for the production of 398.1 tons h^{-1} HPS and 107 MW power. HEN comprises 39 HE units, 4 coolers at ambient temperature and 5 refrigeration units at very low temperatures and no heater. Due to integration, more than half (28.2 MW) of a very low temperature cooling requirement could be saved. This solution yields 68.5 M€ annum^{-1} of electricity income, 36 M€ annum^{-1} cooling cost and 13.4 M€ annum^{-1} annualized investment for HEN, giving 19.1 M€ annum^{-1} of annual profit before tax.

Since prices for energy in general, and utilities and electricity in particular, are expected to increase faster in the future relative to other prices, we also, as mentioned, performed synthesis for higher prices. HEN obtained when considering double the utility and energy prices and doubled the project's lifetime (30 years) is shown in Fig. 9.

A quick analysis showed that about 7 MW of heat is additionally recovered by steam production enabling 2.9 MW more of power generation. Also, integration at very low temperatures saves an additional 2.0 MW of the refrigeration load. It can again be seen that this solution is in accordance with the observation about a trade-off between HRAT, and heat-recovery and power generation efficiency. It also should be noted that, on average, one half of those HE units selected are those with counter-flow arrangements (one quarter double-pipe and one quarter fixed plate shell and tube exchangers) and one half is shell and tube exchangers with U-tubes providing a combined co-current and counter-flow arrangement. Since fixed plate exchangers are significantly more expensive than U-tube exchangers, they are selected at those positions in HEN where a higher efficiency is required for heat recovery, thus enabling optimal steam and power generation with a better exergy exploitation of the process.

6 Conclusions

This chapter has described a three-step approach for the synthesis of heat-recovery and steam/energy generation network. Firstly, we start with a thermodynamic analysis of a studied process flowsheet. Detailed data extraction and acquisition should be performed in order to thoroughly undertake the pinch analysis. The results from the pinch analysis indicate possible heat-recovery and exergy opportunities, which can be elaborated in more details over the next two mathematical programming steps, where in the first next step a simple, yet effective, maximization of steam/power generation can be performed, with respect to temperature potential losses within a heat-recovery network. A detailed synthesis of the network is then performed in the last step where appropriate trade-offs are obtained between the income from power generation and utility costs plus investment. The synthesis comprises a heat-recovery network and variable steam production at different pressure-levels for power generation. Optimal arrangement, number, loads, temperature driving forces and types of heat exchanger units, as well as optimal mass flowrates for the production of LPS, MPS and HPS are obtained during this step. An optimal solution with current energy prices and 15 years of an expected project's lifetime yields 107 MW power productions, whilst the one with double prices and doubled lifetime enables 3 MW more power, which indicates that when considering future prices, and lifetime periods more realistically, additional increase in the heat/energy-recovery efficiency can be achieved. Whilst applying this three-step procedure we have also demonstrated how basic optimization models can be adapted and customized for solving large-size industrial problems.

Finally, it should be noted that only complete heat-recovery network was analyzed, together with steam and power generation at fixed process streams and fixed levels of steam. In order to obtain a global solution the whole process, including a gas turbine system with variable flows and their temperatures has to be optimized simultaneously, giving rise to an extensive optimization problem. Serious research work would be needed to accomplish this task.

References

1. Duran MA, Grossmann IE (1986) Simultaneous optimization and heat integration of chemical processes. AIChE J 32:123–138
2. Yee TF, Grossmann IE (1990) Optimization models for heat integration-II, heat exchanger network synthesis. Comput Chem Eng 14:1165–1184
3. Pérez-Fortes M, Bojarski AD, Velo E, Nougués JM, Puigjaner L (2009) Conceptual model and evaluation of generated power and emissions in an IGCC plant. Energy 34:1721–1732
4. Smith R (2005) Chemical process design and integration. John Wiley & Sons, Chichester, West Sussex
5. Klemeš J, Friedler F, Bulatov I, Varbanov P (2010) Sustainability in the process industry: integration and optimization. McGraw-Hill, Maidenherd, New York
6. Papoulias SA, Grossmann IE (1983) A structural optimization approach to process synthesis—II, heat recovery networks. Comput Chem Eng 7:707–722
7. Biegler LT, Grossmann IE, Westerberg AW (1997) Systematic method of chemical engineering. Prentice Hall PTR, New Jersey
8. Soršak A, Kravanja Z (2002) Simultaneous MINLP synthesis of heat exchanger networks comprising different exchanger types. Comput Chem Eng 26:599–615
9. Hewitt GF, Shires GL, Bott TR (1994) Process heat transfer. CRC Press, Boca Raton, Florida
10. Chen JJJ (1987) Letter to the editors: comments on improvement on a replacement for the logarithmic mean. Chem Engng Sci 42:2488–2489

Global Clean Gas Process Synthesis and Optimisation

Mar Pérez-Fortes and Aarón D. Bojarski

Abstract This chapter begins with an introduction to the different possible metrics related to clean gas process synthesis and its subsequent usage. Latter, the different techniques for tackling with multiple criteria are presented, emphasising the use of multi-criteria decision analysis (MCDA) and multi-objective optimisation (MOO). The different criteria elected here for optimisation are described and later used as key performance indicators (KPI) for the proposed scenarios, in chapter "Selection of Best Designs for Specific Applications". Finally, a case study related to the operation of an IGCC plant considering coal–petcoke or natural gas as a fuel is assessed applying the optimization concepts introduced here and taking into account the operation considerations developed in this and in previous chapters.

Notation

ADP	Abiotic depletion potentials
AHP	Analytical hierarchy process
AoPs	Areas of protection
AP	Acidification potentials
bc	Base case
CBA	Cost benefit analysis
CCS	Carbon capture and storage
CED	Cumulative energy demand
CExD	Cumulative exergy demand
CF	Characterisation factor
COE	Cost of energy
CSTR	Continuous-stirred tank reactor

M. Pérez-Fortes (✉) · A. D. Bojarski
Universitat Politècnica de Catalunya, ETSEIB,
Diagonal 647, 08028 Barcelona, Spain
e-mail: mar.perez-fortes@upc.edu

A. D. Bojarski
e-mail: aaron.david.bojarski@upc.edu

L. Puigjaner (ed.), *Syngas from Waste*, Green Energy and Technology,
DOI: 10.1007/978-0-85729-540-8_10,

E	Electricity produced
EF	Ecological footprint
EF_{ij}	Emission factor for pollutant i associate to activity j
EM	Environmental mechanism
EPRI	Electric Power Research Institute
ERA	Environmental risk assessment
FU	Functional unit
GWP	Global warming potential
IGCC	Integrated gasification combined cycle
IPA	Impact pathway analysis
IPCC	International Panel on Climate Change
ISO	International Standards Organisation
KPI	Key performance indicator
LC	Life cycle
LCA	Life cycle assessment
LCI	Life cycle inventory
LCIA	Life cycle impact assessment
MAUT	Multi-attribute utility theory
MAVT	Multi-attribute value theory
MCDA	Multi-criteria decision analysis
MCM	Multi-media compartment models
MOO	Multi-objective optimisation
NBI	Normal boundary intersection
NC	Normal constraint method
NGCC	Natural gas combined cycle
NPV	Net present value
NPW	Net present worth
OF	Objective function
PF	Pareto fronts
r_d	Discount rate
SC	Supply chain
t	Each individual year, during the project life
TAC	Total annualised cost
TCR	Total capital requirement
TOC	Total operating costs
TOPSIS	Technique for order by similarity to ideal solution

1 Introduction

At this point, several design decisions should be taken in the conceptual design of a gasification plant using the already described and built superstructure. This constitutes a problem of *decision-making* under *multiple objectives* with difficult trade-offs and uncertain outcomes, which requires an iterative and complex procedure. In these situations, methods of decision analysis can help decision makers to set out better decisions [1]. The methods to cope with this type of problems are generally known as multiple-criteria decision analysis (MCDA). *Decision analysis* is a merger of decision theory and systems analysis, where *decision theory* provides a foundation for a logical and rational approach to decision-making, whereas *systems analysis* provides methodologies for systems representation and modelling to capture the interactions and dynamics of complex problems.

In general, any decision-making process usually involves three general stages [2, 3] further divided in smaller tasks as follows:

- *Problem structuring*: It aims at defining the problem in terms of
  - Identification and involvement of stakeholders, and elicitation of preferences;
  - Identification of appropriate indicators; and
  - Generation and identification of alternatives.
- *Problem analysis*: It aims at checking the solution quality
  - According to the preference modelling;
  - Comparing and evaluating alternatives; and
  - Performing robustness, sensitivity and uncertainty analyses.
- *Problem resolution:* It counts with the choice of the most suitable alternative.

Regarding problem structuring, in Sects. 1.1 and 1.2 in chapter "Modelling Superstructure for Conceptual Design of Syngas Generation and Treatment", related to superstructure and syngas superstructure representation, the overall framework for analysing different syngas process alternatives has been outlined.

The most important aspects are those related to the selection and identification of appropriate indicators and alternatives. With regard to metrics, the next Sect. 2 discusses around this. Regarding generation and identification of alternatives, in chapter "Modelling Superstructure for Conceptual Design of Syngas Generation and Treatment" the representation of the whole process as a superstructure allows for easily selecting and generating different process alternatives, regarding electricity and/or hydrogen production. However, the generation could be done using different methodologies: single objective optimisation (efficiency or MW produced), which was used in chapter "Modelling Superstructure for Conceptual Design of Syngas Generation and Treatment" to exemplify the possibility of the superstructure results, or multi-objective optimisation described in next sections and developed in chapter "Selection of Best Designs for Specific Applications". Other widely used generation technique, which was exploited in chapter "Modelling Superstructure for Conceptual Design of Syngas Generation and Treatment", is the use of *scenario analysis*, where different superstructure inputs and topologies are tested and different criteria are gathered on their results. Regardless of the methodology to tackle the multi-criteria and preferences issues, the generation of different solutions is of paramount importance. Three possible methodologies are identified to tackle with this task: mathematical programming, the usage of some heuristic based on process knowledge and thermodynamic insight to generate flowsheet alternatives. The last one is used in this book approach.

2 Key Performance Indicators Associated to Syngas Generation

Key performance indicators (KPIs) are the metrics used for optimisation. They allow to value or quantify the proposed scenarios performance. There is consensus on the requirements for effective metrics; in general, they should satisfy the following criteria [4, 5]:

- Simple and understandable to a variety of audiences
- Reproducible and consistent in comparing different time periods, business units or decision alternatives
- Robust, unbiased and non-perverse (i.e., good metrics must indicate better performance)
- Relevant and complementary to existing regulatory programs
- Cost-effective in terms of data collection, making use of data already collected or available for other purposes, while minimising the effort of gathering new data sets
- Stackable along the supply chain (SC) or the product/process life cycle (LC) stages
- Scalable for multiple boundaries of analysis, and
- Protective of proprietary information

Sharratt [4] states that this list should be understood as an unachievable ideal, and some compromise is inevitable. Metrics should be able to reproduce changes at all levels in the system, it would be a fallacy to have a set of metrics that does not take into consideration the closely knit network of cause–effect relationships that comprise chemical processes [6]. This systems-wide approach requires the collection of more than one single metric, which implies a multivariate view of the system. The selection of one set of metrics in favour of the others rely on the agreement between the decision makers and in the underlying principles of each of the metrics calculation methodologies.

Typically, metrics tend to represent different process design aspects: It is common to see economic related metrics, efficiency metrics and environmental impact indices. It has often been advocated that quantitative indicators should be normalised to a unique measure of performance across different sectors to be comparable and used in weighting decision alternatives and comparing operational units [6, 7]. In this sense, all MCDA techniques require that alternative's attributes are normalised before weighting. Some of the examples include normalisation to the physical flows in the system (e.g., per tonne of product output), to a measure of economic performance (e.g., turnover of sales, shipment value, value added, operating profit, number of employees or total investments), or to a defined *functional unit* (FU) of the system under study, for instance, the MW produced, or the flowrate of the final product of concern.

The metrics considered here include economic, plant performance and environmental points of view. In the first metric, the cost of energy (COE) is of concern. In the second metric, the different efficiencies as described in chapter " Modelling Superstructure for Conceptual Design of Syngas Generation and Treatment" are considered. And last but no least is the environmental point of view through a life cycle assessment (LCA). An exhaustive calculation procedure of these parameters is described in Sects. 2.1, 2.2 and 2.3. It has been remarked the necessary trade-off involved in selecting best candidate among competing scenarios. In this sense, the calculated KPIs are used in Pareto curves, which is a graphical method to select the best candidate according to a pre-fixed criterion. This method does not need for normalisation.

2.1 Economic Point of View

Economic aspects have travelled side by side to chemical engineering since its very beginning, and different indicators are used to check the economic viability of different processing options. In both retrofit and starting up (green-field) projects, standard industrial practise calls for estimation of potential investment, working capital, sales revenue and operating expenses, to assess long-term economic impact of the project. The financial evaluation of a project, known also as cost benefit analysis (CBA) comprises basically three major steps:

- *Estimation of capital costs*: these represent discrete expenditures comprising a fixed capital (also known as investment costs) and working capital. Fixed capital can be estimated using factored methods while working capital is associated to inventories, cash and accounts receivables. Capital costs are expressed in monetary units.
- *Estimation of cash flows*: these represent the surplus of incomes over expenditures for all periods; calculation of these cash flows requires estimation of expected revenues and operating costs. Cash flows are expressed as monetary time flows (for instance, per year, month or week).
- *Evaluation of economic indicators*: this last step comprises the use of cash flows for the calculation of the selected metric. Besides cash flows, other parameters such as interest rate, depreciation and savage costs are also required.

Surprisingly, a recent survey by Pintaric and Kravanja [8] of economic objective functions (OFs) used in optimisation problems related to chemical process design, revealed that the most common OFs are different types of costs (such as the total annualised cost—TAC). Optimisation of profit or economic potential is found less common, while the usage of net present worth or value (NPW, NPV) or monetary value added are found rarely. However, other process

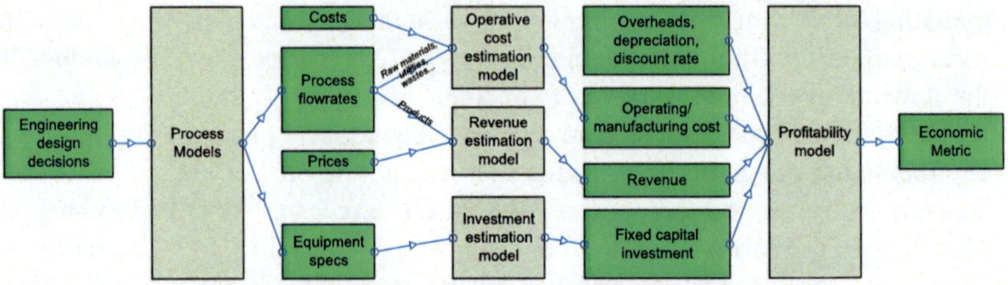

Fig. 1 Variables and models required for the calculation of an economic metric

design books emphasise on the use of metrics where the time value of money is taken into account (Chap. 5 of Biegler et al. [9]). In some cases the application of NPV and discounted cash flow for profitability evaluation and the economic comparison of alternatives are the most acceptable, as recommended in Chap. 10 of Peters and Timmerhaus [10]. A global view of how an economic metric can be calculated can be grasped from Fig. 1.

The selection of the discount rate (r_d) for any time discounted metric is also subject of controversy, given that it represents the trade-off between the enjoyment of present and future benefits and affects directly intergenerational aspects of sustainable development. Higher r_d's devaluate future impacts and consequently they count little on long time horizon projects, which could be perceived as contrary to the interest of future generations.

This study considers the economic point of view of electricity, hydrogen or electricity–hydrogen production plant. The economic parameter to evaluate the global performance of the plant taken into account is the COE, which in the case of electricity production, the term is referred to the net power produced, and in the case of the hydrogen production it is referred to final product sold to the market, minus the electricity consumed in the plant, as Eff_{total} equation already contemplates (see Sect. 2.2). It is defined as the final price per kWh that considers the return of investment, and the total operating costs, fixing a plant economic life [11, 12]. Therefore, the methodology counts with the calculation of the total capital requirement (TCR), the total operating costs (TOC) per year and the consequent final production cost.

The economic evaluation in this book takes into account two approaches:

- The first one follows the work of Frey [13] and Frey and Akunuri [14], which are based on an EPRI Technical Assessment Guide [15]. Probabilistic expressions, with their ranges of application for costs calculation are found, based on real IGCC power plants. Additionally, the general methodology for chemical projects evaluation described in Ulrich and Vasudevan [16] is considered for equipments that are not contemplated in the former sources, for instance, the carbon capture and storage (CCS) train. Globally, the dimensions variability is taken into account in the costs expressions considering the *sizing variable*. Stream values are mainly retrieved from the Aspen Plus simulation results. See in Table 1 the breakdown of the total cost, as it

Table 1 Breakdown of the COE calculation following the approach of Frey [13], Frey and Akunuri [14], and Ulrich and Vasudevan [16]. Each term counts with sizing variables expressions based on real experiences

Total capital requirement (TCR) (€)	Total direct costs	Equipment and general facilities
	Total indirect costs	Indirect construction costs, sales tax, fees, permits, etc.
	Contingencies	Process and project contingencies
	Others	Royalties, inventories, initial chemicals charge, land, etc.
Total operating costs (TOC) (€/year)	Fixed operating costs	Operating work
		Maintenance work
		Maintenance material
		Administrative and support work
	Variable operating costs	Total fuel cost
		Total consumable cost
		Ash disposal
		By-product credit (sulphur and slag)
COE (€/kWh)	f(TCR, TOC, kWh)	

calculated in Sect. 4, which applies this approach to evaluate the combined cycle performance in a plant of similar dimensions powered by natural gas or by solid fuel–syngas, in natural gas combined cycle (NGCC) and IGCC configurations. The COE is calculated as in Eq. 1, where t is each individual year, during the project life, thus the interest rate is considered for the calculation of each year inputs and outputs in monetary terms. E is the electricity produced (case study Sect. 4), in all cases such cash or energy flows should be properly discounted using a given interest rate. Nevertheless, in co-production scenarios, E and H_2 should be taken into account for the plant revenues (see chapter "Selection of Best Designs for Specific Applications"). It is possible to say that the COE is the energy price that should be used to pay back the power plant investment. For further detail about the methodology and about input values, see Pérez-Fortes et al. [17].

$$\mathrm{COE} = \frac{\sum \mathrm{TCR}_t + \mathrm{TOC}_t}{\sum E_t + \mathrm{H}_{2t}}. \tag{1}$$

- The second one is based on the works of Hamelinck and Faaij [18] and Vliet et al. [19]. In this case, the dimensions variability between one case study and another is based on *economies of scale*. The reason why we have followed another methodology for scenarios evaluation in chapter "Selection of Best Designs for Specific Applications" is the big number of scenarios and variability considered (16), where it is not necessary to obtain the level of detail of the previous methodology. Also the computing time for this approach is lower. The annualised total costs, as in the previous case, takes into account investment and variable costs. Instead, in this case, feedstock costs are considered separately. Thus, the final breakdown takes into account:

(a) TCR
(b) Annual costs (AC), which are the previous TOC minus fuel costs
(c) Feedstock costs (FC)
(d) Electricity supply, electricity demand and H_2 supply

In the second approach we are also considering specific costs related with the CCS methodology, as the cost of CO_2 avoided, or the cost of CO_2 captured [20], defined as in Eqs. 3 and 4, respectively, and taking into account only electricity production. The first value reflects the average cost of CO_2 to atmosphere reduction (total mass of CO_2 emitted, in tonnes is used in the formula) comparing both plants, with and without CCS technology (of the same type and dimensions). The second one reflects the economic-viability of a CO_2 capture system giving a price for CO_2: it means that if the CO_2 can be sold to the market at the calculated price (for instance, for the food industry), as a by-product in the plant, the COE of both plants, would be the same. In this case the reference plant having a disadvantage since its CO_2 emissions are higher. The CO_2 value in tonnes corresponds to the tonnes of CO_2 captured. Note that mass balances require that CO_2 is or captured or emitted as in Eq. 2. Notice that in Eqs. 2, 3 and 4, bc means base case, while CCS makes reference to the scenario with carbon capture units.

$$(CO_2)_{captured} = (CO_2)^{bc}_{emitted} - (CO_2)^{CCS}_{emitted} \tag{2}$$

$$\text{Cost of } CO_{2,avoided}[€/tonCO_2] = \frac{COE_{CCS} - COE_{bc}}{(CO_2, \text{emitted}/kWh)_{bc} - (CO_2, \text{emitted}/kWh)_{CCS}} \tag{3}$$

$$\text{Cost of } CO_{2,captured}[€/tonCO_2] = \frac{COE_{CCS} - COE_{bc}}{(CO_{2,captured}/kWh)_{CCS}} \tag{4}$$

Strictly speaking the cost of avoiding CO_2 emissions should include the costs associated to its transportation and pumping to final disposal, however, is common practise to address the capture scenario only, where costs are considered up to the liquefaction, disregarding the disposal.

2.2 Plant Engineering Point of View

In the case that syngas is used for co-production; the efficiency calculation cannot be based solely on the contribution of electricity or hydrogen. By combining the efficiencies related to electricity (Eff_{global}) and hydrogen ($Eff_{globalH2}$) production defined in Eqs. 7 and 8 in chapter "Modelling Superstructure for Conceptual Design of Synsas Generation and Treatment", the total co-production efficiency can be calculated as follows:

$$Eff_{total} = Eff_{global} + Eff_{global_{H_2}} = \frac{Power_{net} + m_{H_2} LHV_{H_2}}{m_{feed} LHV_{feed}}. \tag{5}$$

This Eff_{total} considers the possible destinies of syngas production, which in some cases might require the consumption of electricity instead of a positive contribution, see chapter "Selection of Best Designs for Specific Applications".

2.3 Environmental Point of View

In the case of environmental metrics no information is easily available to chemical process designers for its computation. There are two main reasons for this:

- Relevant properties of chemicals (e.g., toxicity, environmental degradation constants) are not readily available in the tools commonly used by chemical engineers (e.g., process simulators, chemical process design handbooks[1]).
- Location-specific knowledge is needed to estimate environmental impacts, with the exception of environmental problems that are global in nature (e.g., ozone layer depletion and increase of greenhouse gas concentration).

Sharratt [4] states that all environmental effects can in principle be linked to the concentration, dispersion and persistence of materials in the environment. Most chemicals in recent years have been categorised according to their potential for *persistence*,[2] *bioaccumulation*[3] and *toxicity*.[4] However, the environment is compromised by industry in two ways: emissions and the consumption of raw materials, this fact broadly separates typical environmental metrics into two categories [21]:

- *Pollution categories associated to system's output flows such as*: ozone depletion, global warming, human toxicology, eco-toxicology, smog formation, acidification, eutrophication, odour, noise, radiation and waste heat.
- *Depletion categories associated to system's input flows*: abiotic resource depletion, biotic resource depletion, land use and water use.
- The calculation of *environmental metrics* requires the estimation of *environmental interventions* (inputs and outputs) from the system. While inputs to the

---

[1] In many cases, the properties have not been measured for a large number of chemicals, and the measurements that have been made frequently show wide ranges of variation.

[2] Persistence is related to what extent materials will accumulate; at one extreme of behaviour are materials that are not degradable and thus accumulate while at the other extreme are highly degradable materials that will quickly reach an essentially steady level in the environment as their rate of release is balanced by their destruction. In this sense, persistence is associated to the substance resistance to chemical (hydrolysis, photolysis, etc.) or biological (biodegradation, metabolism, etc.) degradation or breakdown.

[3] Bioaccumulation is related to the chemical tendency to become increasingly concentrated (in fat tissues) as one moves up along the food chain from microorganism to mammals.

[4] Toxicity is the most contentious/disagreeable area of concern where multiple tests are available depending on the endpoint (lethality, fecundity, endocrine disruption, etc.), each chemical has different effects and consequently different toxicity.

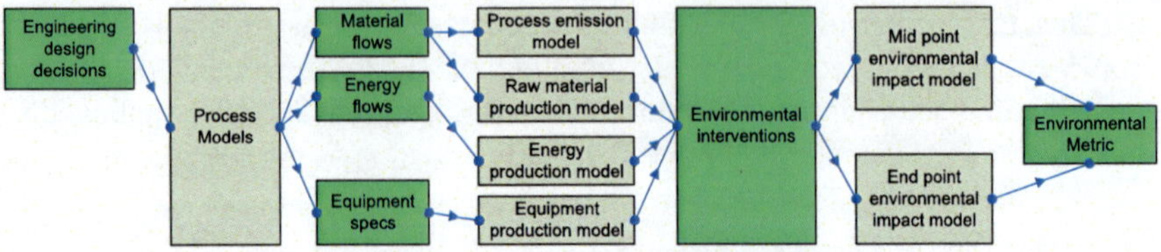

Fig. 2 Variables and models required for the calculation of an environmental metric

system can be easily gathered from the raw material and energy consumption, the estimation of emissions is not straightforward and several authors propose different ways to assess them.

Once environmental interventions (understood as extractions, emissions from and to the environment, or different types of land use) are estimated, it is important to know how the chemical compound will distribute along the different environmental compartments, this requires the use of environmental models. Some impacts are not directly related to the chemical concentration on a given environmental compartment, but to the exposure of this chemical to the subjects of impact, consequently another layer of modelling is required, namely the impact model. The calculation of an environmental metric can be summarised in Fig. 2, which shows the different models required to calculate an environmental metric.

Most methods differ on the way mid- or endpoint impacts are measured and in the way that weights are assessed for each impact. Moreover not all methods consider the same environmental areas of protection, or how each mid-point indicator affects the endpoint. However, there is a tendency to define indicators at common mid-points to ensure simplicity in their definitions and to minimise perceived uncertainty [22]. While reliable endpoint modelling seems within reach for some categories such as acidification, cancer effects and photochemical ozone formation, it is still under development for climate change (a mid-point indicator is still used early along the environmental mechanism, i.e., increase in radiative force), and the endpoint modelling is encumbered with large uncertainties because of many unknowns of the global climate system and because of the long time horizon of some of the involved balances [22]. The selection between one of the impact assessment methods or the usage of different mid-point models from different methods is a matter of the decision maker and the goal that the study follows.

2.3.1 Emission Estimation

The chemical process emits directly and indirectly: direct emissions are associated with the process and are known as foreground process emissions, while indirect are associated with other parts of the process life cycle and are known as background emissions. Releases may be further classified as intended (such as stacks and flares) or accidental (such as leaks and spills). Stefanis and Pistikopoulos [23] classify direct emissions in four groups as follows:

- Accidental releases mainly because of the occurrence of scenarios such as leakage, equipment failure and human error.
- Fugitive emissions that involve small leaks or spills from pumps or flanges which are generally tolerated in industry.
- Releases from normal process operations such as: start-up, shutdown, maintenance/cleaning procedures and from operation conditions changes.
- Episode releases as a result of sudden weather changes or other occurrences.

Typical sources of fugitive emissions are valves, flanges, pump and compressor seals, process drains and open-ended lines. In this sense, Chap. 8 of Allen et al. [24] state that common sources of releases that are overlooked in flowsheet are fugitive emissions (leaks) and venting of equipment (breathing and displacement losses), periodic equipment cleaning and transport container residuals.

One typical way of estimating emissions is the usage of emission factors. An *emission factor* (EF_{ij}) is a representative value that attempts to relate the quantity of a pollutant (i) released to the environment sink (air, water, soil, . . . , j) with an activity associated with the release of that pollutant. Several lists of emission factors are available for different activities, being the usual environmental sink air, examples are available: UN (International Panel on Climate Change), Europe (European Environmental Agency), Australia (Environment Australia), United Kingdom (UK National Atmospheric Emissions Inventory) and United States (US-Environmental Protection Agency EPA-AP42). The actual pollutant i emission (EM_{ij}) is calculated as in Eq. 6.

$$EM_{ij} = A_R EF_{ij} \tag{6}$$

where A_R represents the activity rate, usually a mass flow.

2.3.2 Environmental Models and Emission Fate Estimation

Once emission has been estimated, via process models, emission factors are measured; the question of the fate of the compound must be addressed. Chemical environmental fate is highly component dependant and is modelled by means of environmental fate models. Sinclair-Rosselot and Allen [25] describe the appearance of two types of environmental model approaches: (i) focusing on a single compartment and (ii) taking into account *multimedia compartment models* (MCMs). In the first case typical examples are: prediction of air concentrations downwind from a stationary source, or the estimation of concentration using ground water dispersion models, their main disadvantage is that they provide concentration in only one compartment.

The complexity in MCMs rises from characteristics such as: number of environmental compartments considered, homogeneity and heterogeneity of each one of them and steady or unsteady conditions. In Mackay [26], environmental models taxonomy is provided in levels of increasing complexity:

1. *Level I*: Corresponds to multiple phases closed systems, where pollutants do not react, i.e., are conserved in their chemical form. Each phase is considered as a closed vessel that attains thermodynamical equilibrium, see Chap. 2 of Mackay [25].
2. *Level II*: Corresponds to steady state multiple phase open systems, where pollutants are subject to advective flows (related to '*the direct movement of a chemical by virtue of its presence in a medium that happens to be flowing*'), chemical reactions (biodegradation, hydrolysis, oxidation and photolysis) and attain physicochemical equilibrium. Each phase is considered as a CSTR where outlet concentrations equal phase concentrations, see Chap. 2 of Mackay [26].
3. *Level III*: Corresponds to steady state multiple phase open systems, where pollutants are subject to advective flows, chemical reactions and diffusive flows between environmental compartments, so chemical equilibrium is used but not attained, see Mackay [26].
4. *Level IV*: Corresponds to level III models where some compartments are taken into non-steady state conditions.

In all MCMs where equilibrium is hypothesised the partitioning of a chemical between environmental phases is described using the concept of fugacity for the description of mass transfer and reaction phenomena.

2.3.3 Environmental Impact Estimation

The concept of *environmental impact* is closely related to the concept of risk, which in many cases is embedded in the way fate, dose and impact of a chemical compound are calculated. In the case of risk there are two analytical tools available for such analysis: *Environmental risk assessment* (ERA) and *Impact pathway analysis* (IPA). Both tools put emphasis on impacts to humans, in the case of ERA emphasis is put on ingested dose, while in the case of IPA the focus is on air concentration, see Sonnemann [27, p. 27].

Risk in the environmental sense is defined by Allen and Shonnard [28] as '*the probability that a substance or situation will produce harm under specific conditions*'. This risk will be the combination of two factors (Chap. 9 of Cameron and Raman [29]): (i) the probability that the adverse event will occur and (ii) the consequences/effects of such event. It is generally accepted that risk is a function of a given hazard and the exposure to such hazard; considering that hazard is the potential of a given substance/situation to produce harm or adverse effects in people or the environment, while exposure is the contact time or exposition to such hazard. To assess the risk, the following items have to be addressed properly:

- *Hazard assessment*, which addresses the question of which are the adverse effects that a given substance or situation produces (mortality, shortened lifespan or impairment).
- *Dose response* is the mathematical relationship between the dose of a given substance and the appearance of negative effects.

- *Exposure assessment* is linked to dose measurement and it studies how much and which subjects are 'exposed' to the substance or situation.
- *Risk characterisation* addresses how big the adverse impact of the chemical/ situation is.

Most of the environmental metric methodologies consider a given set of emissions into some compartments which are modelled using a given environmental model. These emissions are assessed in terms of hazard/dose/exposure/risk and a given *characterisation factor* (*CF*) is obtained which relates the emission to its impact. Several environmental metrics have been developed within the LCA context for LCIA (*life cycle impact assessment*), where two important terms are crucial to be defined appropriately, these are: impact category and *environmental mechanism* (EM). An impact category represents environmental issues of concern to which some LCI (*life cycle inventory*) results may be assigned. According to de Haes et al. [30], all physical process and variables starting from extractions, emissions or other types of interaction between the product/process system and the environment, which are connected with a given impact category, are called the EM of that impact category. Within and connected to a given EM it can be distinguished:

- *Areas of protection* (AoPs), these are variables of direct societal concern, also known as classes of endpoints which have some well recognisable value for society. Each impact assessment methodology has a predefined set. Common AoPs are: *human health, natural resources, natural environment and man-made environment* [30].
- *Category mid-points*: these variables which appear within the EM of an impact category fit between environmental interventions and the impact category endpoints. Examples are: concentration of toxic substances, deposition of acidifying substances, global temperature or sea level.

Regarding *environmental impact assessment* two schools of methods have evolved [22, 31]:

- Problem oriented or mid-point methods like CML [32, 33], EDIP [34, 35] and TRACI [21, 36], which restrict quantitative modelling to relatively early stages in the EM to limit uncertainties and classify and characterise emission results in mid-point categories. Themes are common mechanisms (e.g., climate change) or commonly accepted grouping (e.g., aquatic ecotoxicity).
- Damage oriented or endpoint methods such as Eco-indicator 99 [37] or EPS [38] try to model the EM up to the damage to a given area of protection, sometimes with high uncertainties. These methods differ on the way endpoint impacts are measured and in the way that weights are assessed for each impact. Moreover not all methods consider the same AoPs, or how each mid-point indicator affects the endpoint.

The actual environmental impact can then be interpreted in terms of AoPs (EI^{AoP}) or category midpoints (EI^{cat}), and its calculation can be performed using Eqs. 7–9.

$$\mathrm{EI}^{\mathrm{cat}} = \sum_{j}^{\text{all sinks}} \sum_{i}^{\text{all species}} m_{ij}\mathrm{CF}_{ij}^{\mathrm{cat}} \tag{7}$$

$$\mathrm{EI}^{\mathrm{AoP}} = \sum_{j}^{\text{all sinks}} \sum_{i}^{\text{all species}} m_{ij}\mathrm{CF}_{ij}^{\mathrm{AoP}} \tag{8}$$

$$\mathrm{EI}^{\mathrm{AoP}} = \sum_{c}^{\text{all cats}} W_{\mathrm{AoP,cat}}\mathrm{EI}^{\mathrm{cat}} \tag{9}$$

where characterisation factors could be available for all different environmental interventions (m_{ij}) in terms of mid-point impact categories ($\mathrm{CF}_{ij}^{\mathrm{cat}}$) or in terms of areas of protection ($\mathrm{CF}_{ij}^{\mathrm{AoP}}$). Some impact assessment methods also provide with a set of weights ($W_{\mathrm{cat,AoP}}$) to transform mid-point environmental impacts into environmental impacts at the areas of protection level as in Eq. 9. In the context of LCIA, *normalisation* is conducted to obtain a comprehensive view of impact category indicator results. Normalisation values in LCIA are calculated on the basis of chosen reference systems, e.g., all society's activities in a given area and over a specified period of time, or the interventions of the world as a whole in a certain year [39, 40].

Besides emission factors, the most common way of estimating an emission is to measure it or to model it using process simulation. The use of both methods has been shown extensively in the literature.

The former sections discussed the different available metrics for measuring the key parameters of a given design, the next sections discuss how to deal with such information in two different ways: for generating different options by using multi-objective optimisation and how to decide among different optimised and non-dominated solutions.

3 Generation and Identification of Alternatives

Different objective functions could be used according to the decision maker's criteria. Multiple objective programming methods aim at finding suitable solutions of mathematical problems with multiple conflicting objective functions, and different alternative strategies can be applied to solve a multiobjective problem [41, 42].

One typical approach consists of optimisation of alternating objectives, that is, solving the problem for one objective, and next an additional objective function subject to constraints for the objectives already optimised. The main drawback of the former consists of the optimisation process usually leading to different solutions depending on the order in which the objectives are selected. Another approach consists of aggregating the different objectives in a single objective function with varying numerical weights. Unfortunately, these coefficients usually lack of physical meaning, and entail an arbitrary assignment of values, see Sect. 3.1. Thus, there is no unique optimal solution for multiobjective problems, but rather a set of feasible solutions which may be suitable. The preferred approach consists of providing a set of Pareto optimal solutions: a Pareto solution is one for which any improvement in one objective can only take place if at least another objective worsens. Pareto optimal solutions are also termed dominating solutions, while the remaining possible optimisation solutions are dominated. This latter approach implies that the decision maker is interested in all possible trade-off solutions resulting from no previous prioritisation of the objective functions. Particularly in the case of objective functions related to the environment, economic metrics are always prioritised in companies and constraints on the environmental interventions (emissions, concentrations and others) are given by stringent environmental policies. However, a view of process operation that sees environment as an objective and not just as a constraint on operations can lead to the discovery of operating policies or plant designs that improve both environmental and economic performance [43].

3.1 Techniques Applicable for MOO

The techniques for generating a set of Pareto optimal solutions should have some desirable properties. Namely, they should be able to find all available Pareto points, generate them evenly along the possible solutions in the feasible region (understood as the collection of points that satisfy all problem constraints), and they should not generate and explore dominated solutions [44]. However, all the available techniques present deficiencies in some of the former aspects. For example, the weighted sum must be carefully applied since it does not generate all available Pareto points, and the Pareto frontier does not represent an evenly set of solutions of the feasible region [45]. Finally, normal boundary intersection (NBI) [46] and normal constraint method (NC) [44] generate points that are not in the Pareto frontier, but NBI is more prone to generate dominated solutions. In general, all previous procedures require of a filtering step to distinguish and classify dominated from non-dominated solutions. In general the preferred approach for generating Pareto fronts (PF) is the use of a normalised constraint method.

3.2 Problem Analysis from Pareto Efficient Solutions and MCDA

Once a certain number of alternatives are generated the selection of the '*best compromise*' alternative from the set of non-dominated alternatives requires input about the values and preferences of the people responsible for making the decision. Thus, design teams working on a problem with multiple objectives are faced with the need to apply multi-attribute decision analysis (MADA) or MCDA techniques. These are briefly outlined in this section based on a review given by Seppala et al. [1].

- *Elementary methods*: these methods do not require any inter criteria weighting, and some cases nor a relative ranking.
  - Maxi-min: it assigns total importance to the attribute with respect to which alternative performs worst, its equivalent to assess the strength of an alternative by its weakest link.
  - Maxi-max: it assigns total importance to the attribute with respect to which alternative performs best; assessing the strength of an alternative by its strongest link.
  - Conjunctive and disjunctive methods are screening methods that select different alternatives given that attributes are exceeding threshold values for all alternatives (conjunctive) or for some (disjunctive). In general they allow for selection of satisfactory alternatives instead of best alternatives.
  - Lexicographic, it requires a ranking of attributes, and selects the best alternative by choosing the one that has the best value for the first ranked attribute.
- *Multi-attribute utility theory* (MAUT) methods, these methods require that the stakeholders' articulate preference according to strict preference or in difference relations, this approach provides a clear axiomatic foundation for rational decision making.
- *Multi-attribute value theory* (MAVT) is considered a special case of MAUT, where there is no uncertainty in the consequences of an alternative; while in MAUT, it is explicitly incorporated. Both approaches can use a simple formulation (additive) to asses the value (or utility) of a given alternative (see Eqs. 10 and 11).

$$V(a_j) = \sum_{i}^{\text{all criteria}} w_i v_i(x_i(a_j)) \tag{10}$$

$$V(a_j) = \prod_{i}^{\text{all criteria}} \left(v_i(x_i(a_j))\right)^{w_i}, \tag{11}$$

where $V(a_j)$ represents the value of alternative a_j, w_i are attribute/criteria weights and $v_i\,(\cdot)$ are single attribute functions. If $u_i\,(\cdot)$ is used instead of the calculation of the utility of the option ($U(a_j)$) can be assessed. Typical examples of $v_i\,(\cdot)$ are found in Eqs. 12 and 13.

$$v_i(x_i(a_j)) = \frac{x_i(a_j)}{x_i^*} \tag{12}$$

$$v_i(x_i(a_j)) = \frac{x_i(a_j) - x_i^0}{x_i^* - x_i^0} \tag{13}$$

Equation 12, scales scores according to distances the origin and the best/highest option (x_i^*), while Eq. 13 scales scores relative to the distances between lowest (x_i^0) and highest scores (x_i^*).

- Outranking methods: to use this methodologies the decision maker is forced to express his preferences when comparing one alternative to other. If such binary relations hold, then by performing pairwise comparisons between each pair of alternatives under consideration for each criterion the decision of which alternative is best can be achieved. Several frameworks are available, such as ELECTRE, PROMETHEE, MELCHIOR and ORESTE. All these methods have calculation methods reflecting the idea that beyond a certain level, bad performance on one criterion cannot be compensated for by good performance on another criterion; a non-compensatory approach to decision-making is adopted, but they lack strong theoretical foundations [1].
- Other methods
  - The *analytical hierarchy process* (AHP) proposed by Saaty [47] calculates criteria scores (weights) through pairwise comparison using a pre-specified 1–9 point scale that quantifies verbal expressions of strength of importance between attributes or preference between alternatives. Having evaluated all comparisons, weights are calculated via a so-called principal eigenvalue method; consistency of preferences can also be assessed via an index also provided by the method.
  - In the technique for order by similarity to ideal solution (TOPSIS), proposed by Hwang and Yoon [48], the best solution is selected according to the alternative that has the shortest distance (euclidean) from the 'utopian' best possible alternative, formed by the best possible scores for each attribute. The same metric but measured fixing the worst (nadir) possible alternative also provides an alternatives ordering.

The selection of the MCDA technique used depends on each case, but the central question regarding sustainability is whether a compensatory or non-compensatory approach should be used. As Seppala et al. [1] exemplifies *can clean air compensate for dirty water*? in this sense the authors explicitly consider conditions under which non-compensatory methods have to be used.

- *Single, all-important indicator*: if there is one criterion whose importance is deemed to be overriding; then by that definition any compensation is forbidden, consequently lexicographic methods should be used. This is related to the question of strong-sustainability (non compensation) and weak-sustainability (compensation is allowed).
- *Criteria of ranked importance combined with performance uncertainty*: quantified uncertainty can aid decision makers in setting threshold values of difference and confidence that are required to distinguish between alternatives (eliminating possible ties). After ties are eliminated non-compensatory methods requiring rank order should be used.
- *Performance thresholds*, in this case the assessment of such thresholds can help in identifying situations where compensation does not hold.

In the case of selection of the appropriate multi-criteria decision framework, the most important question that requires to be answered is: are the criteria able to compensate each other? This question separates broadly the methods to be used, in the case that compensation is allowed then several techniques are available which depend on the amount of effort that the decision maker is willing to spend in devising his/her preferences.

4 Application of MCDA and MOO to Electricity Production

The case study selected aims at showing the capabilities of superstructure scenario based optimisation using different criteria considers the use of alternate fuels such as petroleum coke and residual biomass, as well as the evaluation of the use of natural gas as fuel for the combined cycle (as in a NGCC power plant). Several different metrics can be calculated from the results of an LCA as discussed in chapter " Modelling Superstructure for Conceptual Design of Syngas Generation and Treatment" and different economic and engineering-based metrics are complementary to give a global techno-economic and environmental point of view.

The flexibility of IGCC power plants has been already commented in chapter "Main Purification Operations", and evaluated from different points of view simultaneously with several configurations of carbon capture and storage (CCS) installations in the literature. In this sense we can mention its modelling including CO_2 removal and considering the improvement of the efficiency [49], or reduction of operating costs [50, 51], and different types of coal as feedstock have also been evaluated [52]. NGCC power plants have been evaluated as well for a large amount of scenarios [53, 54]. Life cycle assessment (LCA) has been already used as an environmental impact quantification tool for IGCC power plants [17, 55, 56].

The considered scenarios here to show the superstructure capabilities are the base case (SC1), the IGCC configuration with feedstock composition based on 45% coal, 45% petcoke and 10% orujillo (SC2), and the NGCC configuration with two layouts, differing in costs (SC3) (see Table 3). The analysis of different energy conversion performances is done from the point of view of operating cost, energy conversion efficiency and environmental impact.

- Process efficiency is a key parameter to evaluate the final product to the market and the energy integration of the plant, as already discussed. The energy conversion efficiency is calculated in this case study using Eq. 7 in chapter "Modelling Superstructure for Conceptual Design of Syngas Generation and Treatment".
- The economic metric considered is the COE of the plant, as described in Sect. 2.1.

Moreover, as a means of exemplifying the use of the ISO14040 guidelines, the study is organised following the ISO standard 4-steps as follows, for the environmental impact point of view:

- *Goal definition and scope*. The study is focused on the environmental contribution change attained by the different raw material composition feeds and associated

Table 2 Summary of different scenario LCI results, expressed in (kg/MJ)

	SC1	SC2	SC3
Inputs			
HardCoal\ES	0.0550	0.0499	0
PetCoke\RER	0.0550	0.0499	0
Limestone milled\RER	0.00382	0.00359	0
Natural gas\RER	0.00030	0.00030	0.05129
H_2O decarbonised\RER	0.551	0.556	0.739
Consumables (Inorganics\RER)	0.000601	0.000606	0.000072
Ouputs			
CO_2	0.2233	0.2119	0.1373
SO_2	3.12E-05	2.55E-05	7.08E-06
NO_x	9.99E-05	8.68E-05	1.77E-04
Particles	4.20E-07	4.23E-07	6.62E-07
Solid inert waste	0.02960	0.02960	0.00015
Water to waste treatment	0.55978	0.55978	0.66555
Sulphur produced	0.00323	0.00321	0

topological flowsheet changes. The simulated co-gasification plant and natural gas power production considers extraction and processing of raw material, up to the power production to the grid, which constitutes a '*cradle to gate*' approach. Regarding system boundary, it is worth mentioning that emissions from waste water treatment plants and inert solid disposal (ashes) are considered. Moreover, sulphur obtained from Claus plant is considered to be an environmental credit, by reducing its production from virgin materials. All scenarios are compared based on a functional unit of 1 MJ. Regarding total power plant production a similar time horizon was set and the same production hypothesis as in the economic aspects were done.

- The LCI is gathered for the different simulated flowsheets, considering the different superstructure topologies. Simulation results are used to estimate flue gas and plant wide emissions. The use of simulation software makes mass balances and energy balances to be met without requiring further data checks; making this approach a fairly robust one. This step constitutes a conservative approach (i.e., it overestimates emissions), given that the industry complies with all legal emission requirements, which in our analysis are disregarded. For all other echelons studied, which are related to production/extraction of raw materials and transport, their corresponding LCIs are retrieved from the Ecoinvent database. In the current case, production of: coal, petcoke, natural gas and other commodities (sulphuric acid, sodium hydroxide and others) are required given their consumption for plant operation, as shown in Table 2. Ecoinvent LCI information was gathered considering that the plant is located in Spain (ES), consequently geographical data for that region was considered when available, while the European average (RER) was selected otherwise. Table 2 results show that lower fuel consumption is achieved for the case of IGCC co-gasifying biomass. Solid wastes are considered to be inert and sent to

Table 3 Key performance indicator results for the considered scenarios

Metric	SC1	SC2	SC3
CED (MJ_{eq})	5.0	4.5	2.8
CExD (MJ_{eq})	9.2	8.3	3.1
EF (m^2a)	0.70	0.69	0.41
IPCC GWP ($kgCO_{2eq}$)	0.27	0.27	0.16
IMPACT 2002+ (μPts)	69.8	65.7	40.3
COE (€/kWh)	0.0619	0.0593	0.0537
TCR (€/kW)	3,119	2,987	1,026
TOC (€/kW_{year})	273	261	108
Eff_{global} (%)	42.1	41.6	53.12

a landfill, while waste water effluent is sent to a waste water treatment plant. Consumables are considered to be inorganic chemicals and a proxy LCI is used considering the average production for first 20 most important inorganic compounds. Regarding sulphur by-product, it is common practise to allocate emission and consequently impacts based on some criteria. However, in this case it has been decided to use a LCI of sulphur production and consider its production from virgin materials, consequently lowering the overall impact of IGCC operation.

- Afterwards, the LCIA, based on the previously gathered LCI (see Table 2), is performed. Several environmental impact indicators are available, regarding different environmental issues, for the case of electricity generation relevant environmental issues involve the estimation of impacts related to climate change using global warming potentials (GWPs), regional acidification using acidification potentials (APs), and virgin resources consumption using abiotic depletion potentials (ADPs). These metrics can be calculated using different ready to use LCIA methodologies such as the CML 2 [32] or the Impact 2002+ [31]. Other commonly used metrics are the ecological footprint (EF) [57] and the cumulative energy or exergy demand (CED or CExD) [58]. These metrics have an aggregating effect and serve in many cases of proxy of more complex environmental metrics. In the proposed methodology, the former metrics are calculated using Simapro [59], which allows gathering of LCI information from the Ecoinvent database [60] and its subsequent organisation for LCIA calculation. Table 3 shows the LCIA results for CED, CExD, EF and CO_2-eq emission as well as the Impact 2002+ overall environmental impact values.
- Finally, the LCA *results interpretation* is done. It is found that operation using NG instead of solid fuels reduces the CED and CExD. In the first case the operation using IGCC requires approximately 60% more than the total amount of exergy required in the NGCC. In the case of CExD, nearly three times more energy is required for IGCC than for NGCC; which clearly shows the low quality of energy produced from IGCC compared to NGCC. The differences found in the case of EF are not as great as in the CExD case and the same happens for the CO_2-eq. Analyzing the Impact 2002+ results, it is found that the lowest impacts refer to the case of operation using NGCC followed by IGCC

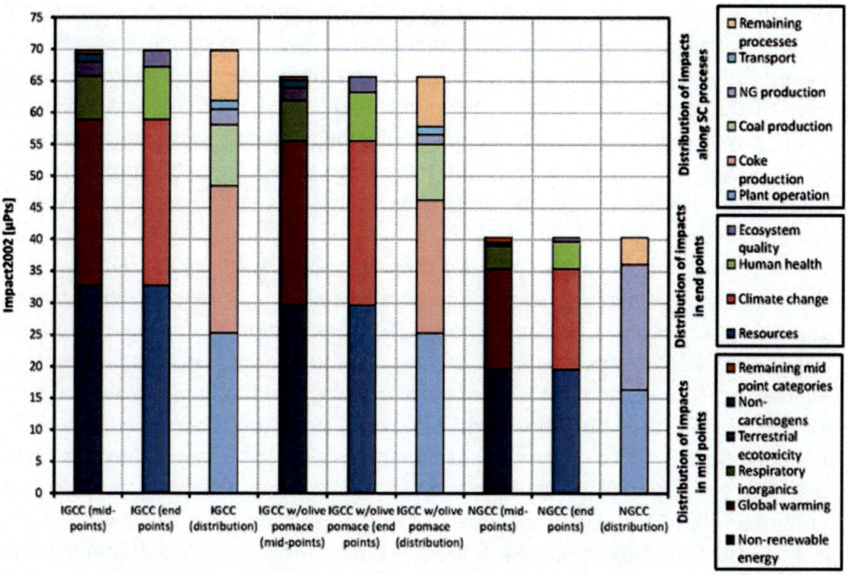

Fig. 3 Results for environmental impact estimation using Impact 2002+ assessment methodology. The impact is shown distributed along mid- and end-point indicators as well as along different SC echelons

operation using olive pomace together with coal and coke as feedstock. Figure 3 shows the results of Impact 2002+ assessment methodology distributed along the different mid and endpoint categories of the LCIA methodology and along the different supply chain echelons. In the case of mid-point categories the largest impacts are associated to: non-renewable energy and GWP, which are mimicked by the endpoint categories: resources and climate change. Small differences are found for the endpoint categories human health and ecosystem quality, which cannot be traced directly to a single mid-point impact as in the case of resources and climate change. Regarding endpoint impacts, in the three electricity production cases, human health impact and ecosystem quality accounted for less than 13% of the total impact, while the remaining was almost evenly partitioned between resources and climate change impacts. In the case of IGCC operation the biggest overall impact is related to the plant operation itself. In second and third place, raw material production is found. Despite the fact that both IGCC scenarios use the same percentage of coal and coke, their associated impact is different. Coal production is found more environmental friendly than coke production. In all the former environmental metrics, the life cycle stages associated to most impact are: raw materials production (coal, coke and NG, respectively for each scenario), for the case for resources, while climate change is because of the IGCC/NGCC echelon.

Table 3 presents the summary of all considered performance indicators. Economic values are referred to €$_{2007}$. The environmental metrics are related to environmental and raw material impacts; as described in the previous section. NGCC operation is found to have the lowest COE and the highest efficiency among the considered scenarios. This mode of operation has two values for the price of the energy; the larger one has been calculated with NG as a mode of an IGCC power plant. On the other hand, the lowest price only contemplates the usage of NG, so, only the investment associated to the CC is contemplated for the total plant investment. TCR and TOC values for NGCC mode consider this second configuration.

LCA results are in clear favour of the usage of NGCC instead of IGCC, and also show that the co-gasification of biomass also reduces the overall environmental impact. As expected, CED and EF are good proxy metrics for the case of electricity generation where raw material usage and climate change impacts are most important. Moreover for the case of these metrics which are mostly related to raw material origin, they heavily weight the use of non-renewable fuels such as coal and coke. In this sense coke which is commonly considered a residue of refineries has been assigned 3% of emissions associated to overall refinery crude oil consumption [60, 61], and consequently its energy and exergy demands are high. This fact is also found by using the Impact 2002 metric, where toxicological effects (human health or ecosystem quality) are small.

5 Conclusion

This chapter has presented the outline for tackling the MCDA problem; in this sense different metrics have been discussed in terms of economic, engineering and environmental aspects. The last section shows the application of commercial simulation and LCA commercial software for gathering the environmental interventions which coupled allow for techno-economic assessment and LCA of different energy conversion processes. Costs, efficiency and environmental impact have been considered as key parameters for selection criteria. In this way, calculation guidelines for these parameters have been developed for comparing explicit solutions for IGCC and NGCC power plants. This procedure and its implementation have been validated with data from the ELCOGAS power plant.

References

1. Seppala J, Basson L, Norris G (2002) Decision analysis frameworks for life-cycle impact assessment. J Ind Ecol 5(4):45–68
2. Azapagic A, Perdan S (2005) An integrated sustainability decision-support framework. Part I: problem structuring. Int J Sustainable Dev World Ecol 12:98–111
3. Azapagic A, Perdan S (2005) An integrated sustainability decision-support framework. Part II: problem analysis. Int J Sustainable Dev World Ecol 12:112–130
4. Sharratt P (1999) Environmental criteria in design. Comput Chem Eng 23(10):1469–1475

5. Tanzil D, Beloff BR (2006) Assessing impacts: overview on sustainability indicators and metrics. Environ Qual Manage Summer 2006:41–56
6. Constable D, Jimenez-Gonzalez C, Lapkin A (2008) Process metrics. In: Lapkin A, Constable D (eds) Green chemistry metrics: measuring and monitoring sustainable processes. Chap. 6. Wiley, Chichester, West Sussex, pp 228–247
7. Azapagic A, Perdan S (2000) Indicators of sustainable development for industry: a general framework. Process Saf Environ Prot 78(4):243–261
8. Pintaric ZN, Kravanja Z (2006) Selection of the economic objective function for the optimisation of process flow sheets. Ind Eng Chem Res 45(12):4222–4232
9. Biegler L, Grossmann I, Westerberg A (1997) Systematic methods of chemical process design. Prentice Hall, New Jersey
10. Peters M, Timmerhaus K (1991) Plant design and economics for chemical engineers, 4th edn. McGraw-Hill, Singapore
11. Koukouzas N, Katsiadakis A, Karlopoulos E, Kakaras E (2008) Co-gasification of solid waste and lignite–a case study for western macedonia. Waste Manage 28:1263–1275
12. Ordorica-Garcia G, Douglas P, Croiset E, Zheng LG (2006) Technoeconomic evaluation of IGCC power plants for CO_2 avoidance. Energy Convers Manage 47:2250–2259
13. Frey HC (1991) Probabilistic modeling of innovative clean coal technologies: implications for research planning and technology evaluation. PhD thesis, Department of Engineering and Public Policy, Carnegie Mellon University, Pittsburgh, Pennsylvania
14. Frey HC, Akunuri N (2001) Probabilistic modeling and evaluation of the performance, emissions, and cost of texaco gasifier-based integrated gasification combined cycle systems using Aspen. Technical report. Computational Laboratory for Energy, Air, and Risk Department of Civil Engineering North Carolina State University Raleigh, NC
15. Electric Power Research Institute [EPRI] TAG(tm) - Technical Assessment Guide, Volume 1: Electricity Supply - 1986 (1986) EPRI P-4463-SR. Electric Power Research Institute, Inc
16. Ulrich GD, Vasudevan PT (2004) Chemical engineering process design and economics: a practical guide, 2nd edn. Sheridan Books, Michigan
17. Pérez-Fortes M, Bojarski AD, Velo E, Puigjaner L (2010) IGCC power plants: conceptual design and techno-economic optimisation. In: Clean energy: resources, production and developments. Energy science, engineering and technology, Nova Science Publishers, Hauppauge NY
18. Hamelink CN, Faaij APC (2002) Future prospects for production of methanol and hydrogen from biomass. J Power Sources 111:1–22
19. van Vliet PPR, Faaij APC, Turkenburg WC (2009) Fischer–Tropsch diesel production in a well-to-wheel perspective: a carbon, energy flow and cost analysis. Energy Convers Manage 50:855–876
20. Metz B, Davidson O, Coninck H, Loos M, Meyer L (2005) Special report on carbon dioxide capture and storage. Chap 3. International Panel on Climate Change (IPCC), Switzerland
21. Bare J, Norris G, Pennington D, Mc Kone T (2003) TRACI, The tool for the reduction and assessment of chemical and other environmental impacts. J Ind Ecol 6(3–4):49–78
22. Finnveden G, Hauschild M, Ekvall T, Guinee J, Heijungs R, Hellweg S, Koehler A, Pennington D, Suh S (2009) Recent developments in life cycle assessment. J Environ Manage 91(1):1–21
23. Stefanis S, Pistikopoulos E (1997) Methodology for environmental risk assessment of industrial nonroutine releases. Ind Eng Chem Res 36(9):3694–3707
24. Allen D, Shonnard D, Prothero S (2002) Evaluating environmental performance during process synthesis. In: Allen D, Shonnard R (eds) Green engineering: environmentally conscious design of chemical processes. Prentice Hall, New Jersey, pp 199–249
25. Sinclair-Rosselot K, Allen D (2002) Flowsheet analysis for pollution prevention. In: Allen D, Shonnard R (eds) Green engineering: environmentally conscious design of chemical processes. Prentice Hall PTR, New Jersey, pp 309–359
26. Mackay D (2001) Multimedia environmental models: the fugacity approach, 2nd edn. CRC Press LLC, Boca Raton

27. Sonnemann G (2002) Environmental damage estimations in industrial process chains: methodology development with a case study on waste incineration and a special focus on human health. PhD thesis, Chemical Engineering Department, Universitat Rovira i Virgili, Tarragona, Spain
28. Allen D, Shonnard D (2002) Green engineering: environmentally conscious design of chemical processes, Prentice Hall PTR, New Jersey, Ch.2 pp 35-62
29. Cameron I, Raman R (2005) Process systems risk management, vol 6 process systems engineering. Elsevier, Amsterdam
30. de Haes H, Jolliet O, Finnveden G, Hauschild M, Krewitt W, Muller-Wenk R (1999) Best available practice regarding impact categories and category indicators in life cycle impact assessment, background document for the second working group on life cycle impact assessment of SETAC-Europe (WIA-2). Int J Life Cycle Assess 74:66–74
31. Humbert S, Margni M, Jolliet O (2005) IMPACT 2002+: user guide draft for version 2.1. Technical report, Industrial Ecology & Life Cycle Systems Group, GECOS, Swiss Federal Institute of Technology Lausanne (EPFL), Lausanne, Switzerland
32. Guinee J, Gorree M, Heijungs R, Huppes G, Kleijn R, de Koning A, van Oers L, Sleeswijk A, Suh S, de Haes H, de Brujin H, van Duin R, Huijbregts M, Lindeijer E, Roorda A, van-der Ven B, Weidema B (2001) Life cycle assessment. An operational guide to the ISO standards. Part 3: scientific background ministry of housing spatial planning and the environment (VROM) and centre of environmental science. Leiden University (CML), The Netherlands
33. Heijungs R, Guinee J, Huppes G, Lankreijer R, de Haes H, Wegener A, Sleeswijk A, Ansems M, Eggels PG, van Duin R, de Goede HP (1992) Environmental life cycle assessment of products. Center for environmental studies (CML), vol 1 and 2. Leiden University, The Netherlands
34. Hauschild M, Potting J (2004) Spatial differentiation in life cycle impact assessment: the EDIP-2003 methodology. Guidelines from the Danish EPA. The Danish Ministry of the Environment, Denmark
35. Wenzel H, Hauschild M, Alting L (1997) Environmental assessment of products, Vol. 1: methodology tools and case studies in product development. Chapman & Hall, London
36. Bare J (2002) Developing a consistent decision-making framework by using the US EPA's TRACI. Technical report. Systems Analysis Branch, Sustainable Technology Division, National Risk Management Research Laboratory, US Environmental Protection Agency Cincinnati, United States
37. Goedkoop M, Spriensma R (2001) The eco-indicator 99: a damage oriented methods for life cycle impact assessment, methodology report. Technical report. Pré Consultants BV, Amersfoort, The Netherlands
38. Steen B (1999) A systematic approach to environmental priority strategies in product development (EPS). Version 2000—General system characteristics, CPM report 1999:4. Technical report. Centre for Environmental Assessment of Products and Material Systems (CPM). Chalmers University of Technology, Technical Environmental Planning, Göteborg, Sweden
39. Heijungs R, Guinee J, Kleijn R, Rovers V (2007) Bias in normalization: causes, consequences, detection and remedies. Int J LCA 12(4):211–216
40. Huijbregts M, Breedveld L, Huppes G, de Koning A, van Oers L, Suh S (1995) Normalisation figures for environmental life-cycle assessment: the Netherlands (1997/1998), Western Europe (1995) and the world (1990 and 1995). J Clean Prod 11(7):737–748
41. Gandibleux X, Sevaux M, Sörensen K, T'Kindt V (2004) Metaheuristics for multiobjective optimisation. Series: lecture notes in economics and mathematical systems, vol 535. Springer, Berlin
42. Wiecek MM, Ehrgott M, Fadel G, Figueira JR (2008) Multiple criteria decision making for engineering. Omega-Int J Manage Sci 36:337–339
43. Cano-Ruiz J, McRae G (1998) Environmentally conscious chemical process design. Annu Rev Energy Environ 23:499–536

44. Messac A, Ismail-Yahaya A, Mattson C (2003) The normalized normal constraint method for generating the pareto frontier. Struct Multidisciplinary Optim 25:86–98
45. Steuer RE (1986) Multiple criteria optimisation: theory computation and application, Chap 14. Wiley series in probability and mathematical statistics—applied. John Wiley & sons, New York
46. Das I, Dennis JE (1998) Normal-boundary intersection: a new method for generating the pareto surface in nonlinear multicriteria optimisation problems. SIAM J Optim 8:631–657
47. Saaty T (1980) The analytic hierarchy process. McGraw-Hill, New York
48. Hwang C, Yoon K (1981) Multiple attribute decision making. Springer-Verlag, Berlin
49. Descamps C, Bouallou C, Kanniche M (2008) Efficiency of an integrated gasification combined cycle (IGCC) power plant including CO_2 removal. Energy 33:874–881
50. Desideri U, Paolucci A (1999) Performance modeling of a carbon dioxide removal system for power plants. Energy Convers Manage 40:1899–1915
51. Kanniche M, Bouallob C (2007) CO_2 capture study in advanced integrated gasification combined cycle. Appl Therm Eng 27:2693–2702
52. Chen C, Rubin ES (2009) CO_2 control technology effects on IGCC plant performance and cost. Energy Policy 37:915–924
53. Amann JM, Kanniche M, Bouallou C (2009) Natural gas combined cycle power plant modified into an O_2/CO_2 cycle for CO_2 capture. Energy Convers Manage 50:510–521
54. Delattin F, de Ruyck J, Bram S (2009) Detailed study of the impact of co-utilization of biomass in a natural gas combined cycle power plant through perturbation analysis. Appl Energy 86:622–629
55. Corti A, Lombardi L (2004) Biomass integrated gasification combined cycle with reduced CO_2 emissions: performance analysis and life cycle assessment (LCA). Energy 29:2109–2124
56. Fiaschi D, Lombardi L (2002) Integrated gasifier combined cycle plant with integrated CO_2–H_2S removal: performance analysis, life cycle assessment and exergetic life cycle assessment. Int J Appl Thermodynamics 5(1):13–24
57. Huijbregts M, Hellweg S, Frischknecht R, Hendriks J (2007) Ecological footprint accounting in the life cycle assessment of products. Ecol Econom 64:798–807
58. Huijbregts M, Rombouts L, Hellweg S, Frischknecht R, van-de Meent D, Ragas A, Reijnders L, Struijs J (2006) Is cumulative fossil energy demand a useful indicator for the environmental performance of products? Environ Sci Technol 40:641–648
59. de Schryver A, Goedkoop M, Oele M (2006) Introduction to LCA with simapro 7. Technical report. Pre-Product Ecology Consultants, Amersfoort, The Netherlands
60. Ecoinvent (2008) The ecoinvent database V2.0. Technical report. Swiss Centre for Life Cycle Inventories, Switzerland
61. Jungbluth N (2007) Erdöl. In: Dones R (ed) Sachbilanzen von energiesystemen: Grundlagen für den ökologischen vergleich von energiesystemen und den einbezug von energiesystemen in ökobilanzen für die schweiz, Ecoinvent Report No. 6-IV. Swiss Centre for Life Cycle Inventories, Duebendorf

Selection of Best Designs for Specific Applications

Aarón D. Bojarski and Mar Pérez-Fortes

Abstract The present chapter assembles all the previous chapter's concepts and methodologies with the aim of selecting a given process design. The developed superstructure is used to simulate scenarios with different feedstocks and process topologies for co-production of electricity and hydrogen. The most representative output data are shown, and the described points of view, discussed in chapter "Global Clean Gas Process Synthesis and Optimisation", are here evaluated under a techno-economic and environmental assessment. Sixteen scenarios are considered encompassing four different feedstocks combined with four different plant topologies; electricity generation with syngas, electricity generation with hydrogen, hydrogen production and hydrogen production with PSA flue gas profit in the CC.

Notation

AC	Annual costs
C_{ij}	Scenario using feedstock *i* and production option *j*
CC	Combined cycle
CCS	Carbon capture and storage
COE	Cost of electricity
CGE	Cold gas efficiency
EI	Environmental impact
EOR	Enhanced oil recovery
ER	Equivalence ratio
IGCC	Integrated gasification combined cycle

A. D. Bojarski (✉) · M. Pérez-Fortes
Universitat Politècnica de Catalunya, ETSEIB,
Diagonal 647, 08028, Barcelona, Spain
e-mail: aaron.david.bojarski@upc.edu

M. Pérez-Fortes
e-mail: mar.perez-fortes@upc.edu

L. Puigjaner (ed.), *Syngas from Waste*, Green Energy and Technology,
DOI: 10.1007/978-0-85729-540-8_11,

KPI	Key performance indicator
LCA	Life cycle assessment
LCI	Life cycle Inventory
LCIA	Life cycle impact assessment
MCDA	Multi-criteria decision analysis
MO	Multi-objective
NGCC	Natural gas combined cycle
PF	Pareto front
PRENFLO	Pressurised entrained flow
PSA	Pressure swing adsorption
r_d	Discount rate
T_{gasif}	Temperature of gasification
TCR	Total capital requirement
TOC	Total operating costs
TOPSIS	Technique for order by similarity to ideal solution
TOT	Turbine outlet temperature

1 Introduction

Summarising the previous chapters, the present context leads to the need of more efficient technologies, greenhouse gases mitigation and the requirement of profiting renewable sources as well as looking for solutions for waste disposal. The proposed superstructure includes these current needs by combining the use of gasification in an integrated plant, with coal and petcoke and biomass wastes as inputs, and the option of carbon capture and storage (CCS). Regarding the construction of a gasification plant as an investment project which has to be environmental friendly, it is necessary to know where the trade-offs of this type of plant are, and which are the optimal solution to be converted in a real project. Even if the gasification technology has been already extensively used, and all cleaning units have been widely applied too, as well as the carbon capture technology, the optimal combination of them is still designed challenge. Several works have already tried to measure the global performance of this type of plants.

For instance, the work of Hamelink and Faaij [1] evaluates technical and economic parameters of gasification plants to produce methanol and hydrogen, taking into account future prospects. Even if they have not developed a superstructure understood as in this book, they use an Aspen Plus simulation to obtain energy and mass balances of interest for the economical evaluation. When large-scale production is assessed, biomass necessity is an important item for operation costs when long distances should be covered as discussed and concluded in chapter "Raw Materials Supply". Hydrogen and methanol can result as conventional fuels alternatives; nevertheless the main bottleneck lies in the distribution infrastructure, mainly in the hydrogen case. Chiesa et al. [2] consider the production of hydrogen and electricity from coal; they have evaluated different scenarios, considering CO_2 venting or CO_2 capture; electricity production with conventional gas turbines, with turbines with for burning syngas and H_2 and with steam cycle (thus, pure H_2) as final syngas usages. Process intensification of acid species is also included by removing CO_2 and sulphur acid species in the same unit operation. They propose different analyses considering performance and emissions using simulation of real commercial units. In their economic analysis; performed by Kreutz et al. [3], it is interesting to appreciate that one of the barriers found for a wide H_2 economy is a lack of cost-effective storage and a large interested market on it. Also the CO_2 storage capacity and CO_2 transportation have to be addressed in an efficient way to promote such solution.

The specific problem of CCS in different plant types is tackled in Rubin et al. [4]: for Natural Gas Combined Cycle plant (NGCC), IGCC plant and pulverized coal plant taking into account the final transport and storage of CO_2 with different possibilities: geologic or saline storage are used for Enhanced Oil Recovery (EOR). They found, while comparing coal gasification and combustion with CCS, that costs are very sensitive to the coal quality. Moreover, depending on the coal quality combustion or gasification plants are the cheapest options while IGCC plants are the most penalised for the extra energy consumption of the CCS system. The most relevant contribution of Chen and Rubin [5] is the consideration of uncertainty in the cost of CCS in an IGCC power plant by considering again, coal quality and the CCS removal efficiency.

Finally, in the work of Brown et al. [6], a thermo-economic analysis of a biomass energy conversion system, at mid-scale, from wood gasification is performed. The final gas is used to produce electricity through an internal combustion engine-combined cycle, or a gas turbine-combined cycle. The model is based on a process simulation superstructure to select the optimal process unit and operating conditions. Optimisation results show that the energy conversion efficiency is maximised using the internal combustion engine option, at operating conditions that also favour low tar contents. Nevertheless, from the economic point of view, the option with the gas turbine shows better results, however, those operating conditions favour tar formation in the gasifier.

After this brief introduction, the aim of this chapter is to analyse the previously developed models in a combined fashion as described in chapter "Modelling Superstructure for Conceptual Design of Syngas Generation and Treatment", while using the metrics and methodologies in chapter "Global Clean Gas Process Synthesis and Optimization" to measure the different dimensions of different designs. In this sense this chapter is structured as follows; Sect. 2 analyses the overall model behaviour for different feedstocks which might include biomass waste. Next, Sect. 3 deals with different feedstocks coupled to flowsheet topological changes which tackle the co-production of electricity and hydrogen, evaluating the selected KPI's. To end, Sect. 4 shows the results in terms of Pareto efficient solutions, and Conclusions.

2 Overall Model Behaviour for Co-Gasification in Terms of Feedstock Changes

Recalling the most important assumptions that the model has, as seen in chapters "Modelling Syngas Generation" and "Main Purification Operations", biomass gasification in the entrained bed gasifier involves no tars formation and no technological limitation, because of the gas equilibrium approach adopted. Other assumption adopted is related to the IGCC power plant layout, which does not require changes for using biomass/waste specific pollutants, concretely, concerning the syngas cleaning units. Furthermore, the grand majority of operating variables remain invariable because T_{gasif} and TOT are adapted to each case study to accomplish with the plant premises concerning heat integration.

The use of coal, petcoke, orujillo (see their proximate and ultimate analysis in Table 3 in "Modelling Syngas Generation") and mixtures of them (addition of a percentage of orujillo to the base mixture formed by coal and petcoke in the same mass proportion) is considered for the co-production of hydrogen and electricity. The effect of fuel composition on syngas composition is studied by varying the H_2O ratio and the ER (defined in chapter "Modelling Syngas Generation"). This analysis is performed first for pure fuels (coal, petcoke and biomass), and later on fuel blends considering petcoke, coal and biomass based blends.

In Figure 1, the H_2 production in the gasifier is assessed when it is fed by pure raw materials. In all cases the maximum amount of H_2 is found for the highest ratios of H_2O (see *red stars* in Fig. 1), which is an analogous behaviour to the one found in the coal–petcoke blend base case composition (see Fig. 9 in chapter "Modelling Superstructure for Conceptual Design of Syngas Generation and Treatment"). Orujillo, which has the highest proportion of oxygen and hydrogen in its composition, achieves the highest proportion of H_2 with the lower ER values. Petcoke is the second one, being its value very close to the one found for orujillo, while coal is the one that implies higher ER. Please recall that higher ER values are also related to higher consumption of purified air in the ASU.

Figures 2 and 3 compare ternary blends in terms of cold gas efficiency (CGE) (see Eq. 31 in chapter "Modelling Syngas Generation") and Eff_{CC} (see Eq. 6 in chapter " Modelling Superstructure for Conceptual Design of Syngas Generation and Treatment"). First, it is appreciated that for the CGE results, the overall tendency is the same in all the blends proposed, even though the absolute value is higher for mixtures with less proportion of biomass. The former finding is because of the effect of petcoke, which the fuel with the highest HHV (see Table 3 in chapter " Modelling Syngas Generation"). It can also be seen that the effect of H_2O ratio is only appreciable at lower ER values, where low H_2O ratios produce higher CGEs, the trend is similar for all biomass content blends, note that Fig. 11 in chapter " Modelling Superstructure for Conceptual Design of Syngas Generation and Treatment" which contemplates the use of a binary blend without biomass produces the highest CGE also for low values of H_2O ratio. Second, Eff_{CC} has slightly

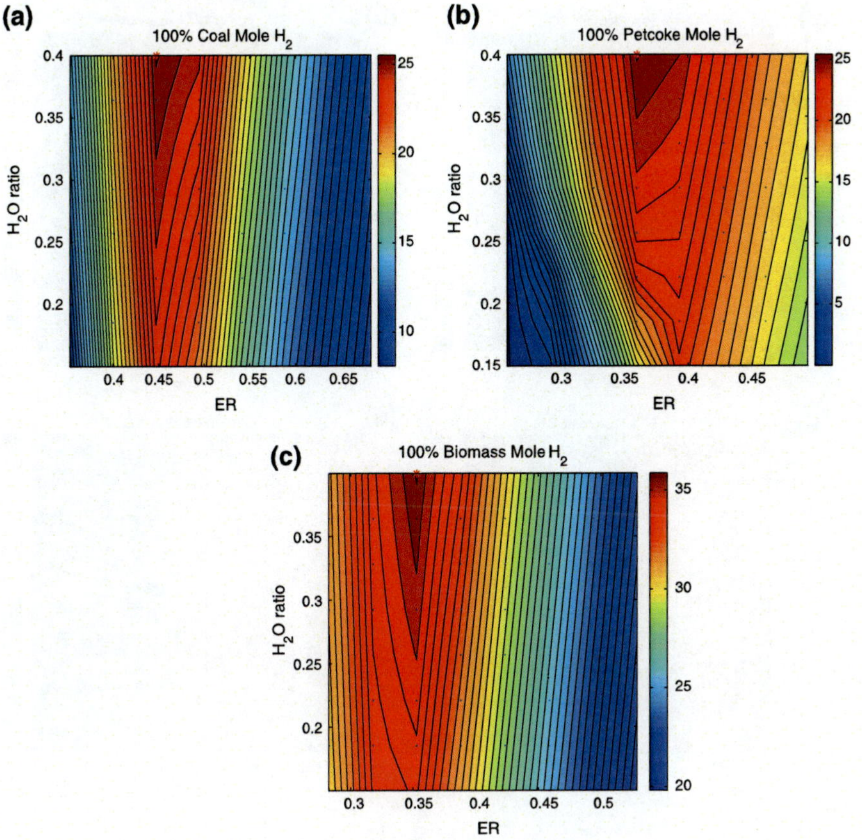

Fig. 1 Iso-lines for H_2 production in terms of molar fraction in dry basis considering feed of pure fuels (coal, petcoke and orujillo)

higher values when higher percentages of biomass are considered. For this parameter the H_2O ratio does not have any significant effect, and parallel isolines are found. Please note that CGE and Eff_{CC} follow opposite trends when the ER is increased, clearly showing that a trade-off is required for setting the ER value for maximising the global efficiency (see Eq. 10 in chapter "Modelling Superstructure for Conceptual Designof Syngas Generation and Treatment"). Also note that highest values of Eff_{CC} are associated to the highest values of ER (see *red stars* in Fig. 3).

As previously discussed in chapter "Modelling Syngas Generation", Sect. 4.1, and from the previous results it has been found that this model is a suitable representation of fuel blends where its overall behaviour can be set proportional to the fuel amount, disregarding possible synergetic effects. Therefore, in next sections only scenarios entailing pure fuels are considered, given that blends combining such fuels are expected to have an intermediate behaviour; these blends are exemplified by a blend of 50%/50% coal and petcoke.

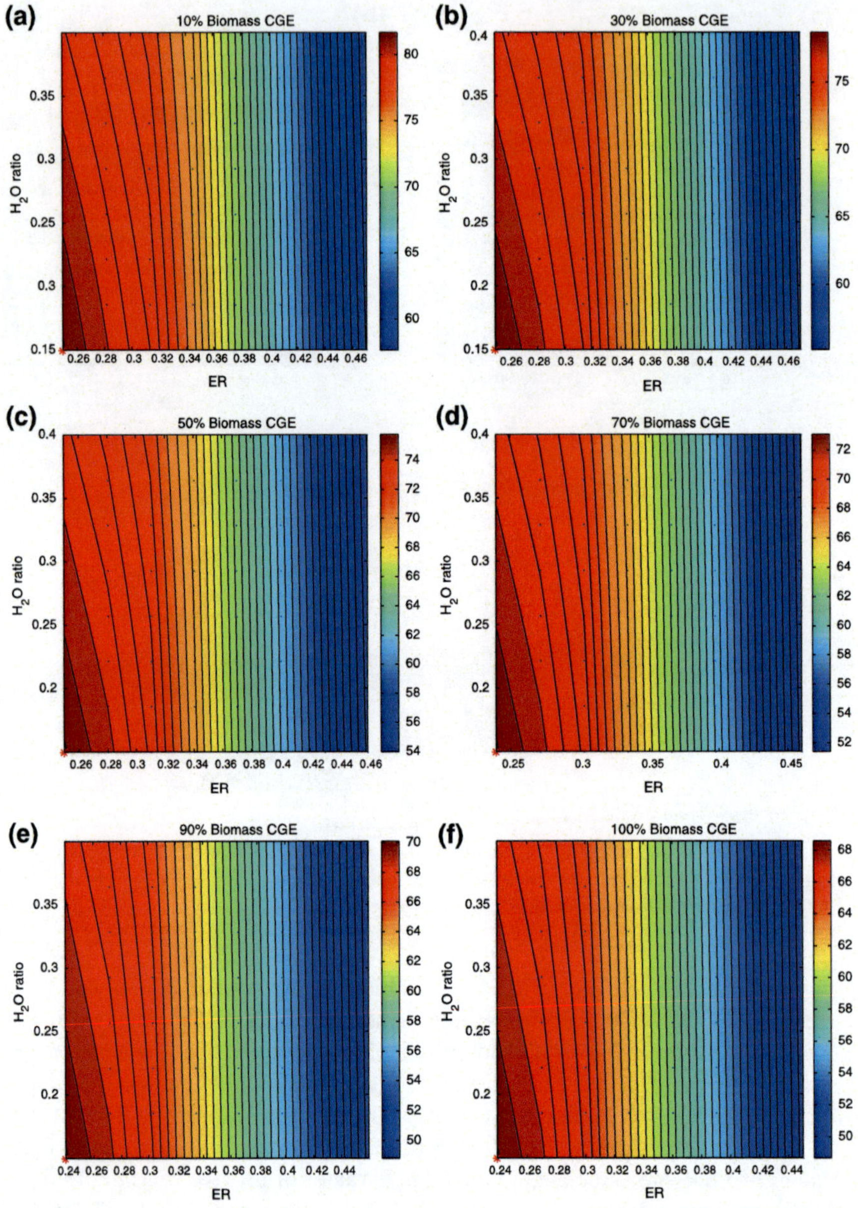

Fig. 2 Iso-lines for CGE values (%) for different ternary blends. *Small blue dots* indicate simulation results

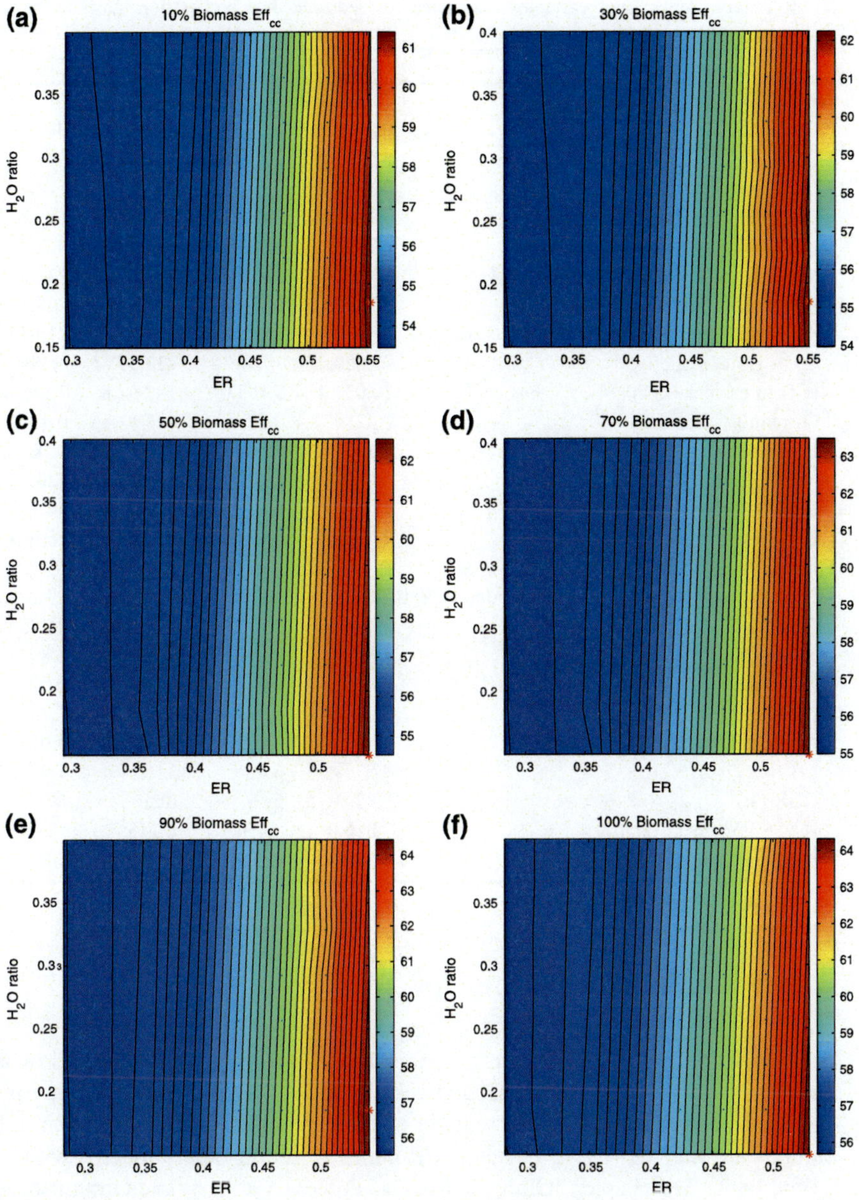

Fig. 3 Iso-lines for Eff_{CC} values for different ternary blends. *Small blue dots* indicate simulation results

Table 1 Input and output variables associated to feedstock dust preparation, gasification and syngas cleaning

	C1j	C2j	C3j	C4j
Inputs				
Coal (mf)	0.5	1	0	0
Petcoke (mf)	0.5	0	1	0
Olive pomace (mf)	0	0	0	1
Feed flowrate (kg/h)	106,856.2	106,856.2	106,856.2	106,856.2
Limestone (kg/h)	3,703.8	3,703.8	3,703.8	3,703.8
Natural gas (kg/h)	300.0	376.6	223.4	242.6
Air (kg/h) to gasifier	362,995.9	274,555.2	442,957.3	268,437.0
Air (kg/h) (feeding system and Claus plant)	28,940.0	23,582.0	34,743.0	21,645.7
O_2 to gasifier (kg/h)	93,090.4	70,409.8	113,596.5	68,840.7
H_2O to gasifier (kg/h)	10,031.4	6,195.0	14,062.8	7,886.5
O_2 ratio gasifier (ER)	0.42	0.52	0.37	0.41
H_2O ratio gasifier	0.16	0.16	0.16	0.16
NaOH (kg/h)	445.5	335.1	559.3	426.1
H_2SO_4 (kg/h)	57.3	43.1	71.9	54.8
H_2O to venturi scrubber (kg/h)	17,739.1	13,343.0	22,267.5	16,964.3
Outputs				
Combustion gas from feed preparation (kg/h)	30,458.2	30,360.6	29,177.7	29,289.8
CO_2 (kg/h)	2,775.0	2,775.0	2,775.0	2,775.0
NO (kg/h)	2.89	2.89	2.89	2.89
NO_2 (kg/h)	0.02	0.02	0.02	0.02
H_2O (kg/h)	8,900.9	8,803.3	7,620.4	7,732.5
Ashes/slag (kg/h)	26,405.5	47,636.2	3,982.2	9,914.4
LHV_{syngas} (MJ/kg)	9.5	6.6	11.4	7.9
CGE (%)	75.2	59.9	82.9	60.7
H_2O to treatment (kg/h)	7,310.0	8,454.3	6,715.7	13,591.8
Liquid sulphur (kg/h)	3,317.8	975.5	5,739.7	102.3

mf proportions given for mass flows

3 Overall Model Behaviour for Topological Changes

The topological changes considered refer to the feed of syngas to the CC or the CCS train, considering in this second case the possibility of pure hydrogen production to be used in the gas turbine, or of further purified hydrogen stream to be sold to the market. These changes are done by setting appropriately with the splitter units. See chapter "Global Clean Gas Process Synthesis and Optimization" for more details.

In all the topological changes, several input variables remain constant for case studies with the same feedstock. These variables encompass flows, temperatures and flow ratios. Table 1 summarises these variable values for the four different fuels. For the simulated scenarios the following naming convention has been used: Cij, where the ith value represents the fuel used: 1 coal/petcoke blend, 2 coal, 3 petcoke and 4 residual biomass; and j represents the different topological options (thus, the final product). The following topological changes were considered

Option 1: All syngas produced is sent to combined cycle for electricity production

Option 2: All syngas is sent to produce hydrogen which is burnt at a combined cycle with a hypothetical hydrogen turbine (which behaves similarly to a gas turbine)

Option 3: All syngas is sent to produce hydrogen, further purified and delivered to market. PSA purge gas is not integrated

Option 4: All syngas is sent to produce hydrogen, further purified and delivered to market. PSA purge gas is sent to the combined cycle for electricity generation

Consequently C34 is the scenario where petcoke is used as fuel and where all syngas is sent to produce purified H_2.

As it can be seen in Table 1, the feedstock mass flowrate, fed to the gasifier, is maintained constant in all the different input feeds. Nevertheless, because of the humidity conditions of each feedstock, and because of their LHV, oxygen requirements to obtain the stipulated T_{gasif} (according to PRENFLO gasifiers, 1,450°C is the objective value, see chapter "Modelling Syngas Generation") are different in each case (consequently, ER is different in each case). As a design condition, the steam ratio has been maintained constant in all the case studies. Given that the limestone flowrate is considered dependant on the global feed mass flowrate, it remains invariable for all scenarios. Natural gas is variable because it depends on the feed humidity, as well as the other consumables, while water fed to venturi scrubber depends on the syngas flowrate and on the gasifier H_2S formation. Concerning the invariable outputs for all topological scenarios for specific feeds, the gas composition from the dryer is reported, consequence of the abovementioned natural gas combustion, and is used for drying and heating the fuel feedstock. Ashes and slag flowrates are directly related with the ashes and the limestone flows. The LHV for the syngas just after the gasifier is different for each feed, being higher for the petcoke, followed by the base case, the orujillo and the coal scenario, respectively. It is directly related with the LHV of the materials; the same occurs with the CGE. Water to treatment refers to the water that goes out from the system after the sour water stripper. It directly depends on the inlet water stream to venturi scrubber. Liquid sulphur depends on the feedstock composition, thus on the sulphur percentage (according to the ultimate analysis).

For the former cases, Ci1 and Ci2, co-production is studied by setting and feeding a given proportion of the syngas produced to the CC or to hydrogen recovery (see Sect. 3.1). Sixteen scenarios are simulated performing all possible combination options between the different feeds and topological changes considered. Each one of the scenarios is analysed using the three points of view described in chapter "Global Clean Gas Process Synthesis and Optimization" (economic, engineering and environmental), and later compared for optimisation by means of their KPI values: COE, Eff_{total} and environmental impact (EI), using the Impact 2002+ metric.

3.1 Economic Point of View

As mentioned in the previous chapter, the metric used for comparison between scenarios and as KPI is the COE. In this specific case it has been defined as the energy price that should be used to pay back the power plant investment. As hydrogen production, and co-production are of concern, the denominator of the equation (Eq. 1 in chapter "Introduction") that describes the COE calculation, should be taken into account for both productions, electricity and H_2 measured in units of energy. Consequently, we have defined the kWh_{eq} that includes both parameters.

COE calculation requires of investment estimation as well as cash flows related to plant operation costs. Typical methodologies for investment estimation (i.e., total capital requirement—TCR) are based on scaling parameters. In the special case of gasification and power-plant-related unit operations, the methodologies of Hamelink and Faaij [1] and van Vliet et al. [7] are widely used. For the TCR estimation in our case, all the units considered in our flowsheet are included in the work of Hamelink and Faaij [1].

The life time of the plant has been considered to be 25 years, and the interest rate (r_d) is 5% (which is a typical figure for this type of processes). Annual costs (AC) are the 4% of the TCR, as proposed in the followed methodology. Prices for the coal and petcoke are the same than in chapter "Raw Materials Supply" (45 €/tons and 75 €/tons, respectively), and for the olive pomace, it is assumed to be 65 €/tons. The annual load considered is 7,200 h. Costs are reported in $€_{2007}$. For scenarios comparison, *the reference plant* has been compared under two situations:

- The plant feed flowrate in mass basis is constant for all scenarios (as reported in Table 1). Therefore, the plant output (kWh_{eq}) is different in all case studies.
- The output of the plant is the same for all the cases. Thus, the feed flowrate is variable (and all the subsequent flowrates are related to it). The reference value is the one obtained for the base case: C11.

In both cases, the TCR is influenced by the reference plant chosen. See in Figs. 4 and 5 the breakdown of TCR for the two references mentioned. Case studies are grouped by its topological configuration, the so-called options.

Regarding Fig. 4, for Ci1 and Ci2 case studies, where syngas and H_2 are sent respectively to the GT, the most important investment comes from the electricity generation section, which encompasses gas and steam turbines instalment, HRSG and WHB. Nevertheless, it decreases its importance as the LHV of the raw material decreases (see C21, with coal, and C31, with petcoke), gaining prominence for the gasification and ASU costs. In H_2-based scenarios, the most important investment comes from the hydrogen production section, or from the gasification and ASU island, depending on the feed type. On the one hand, for Ci3

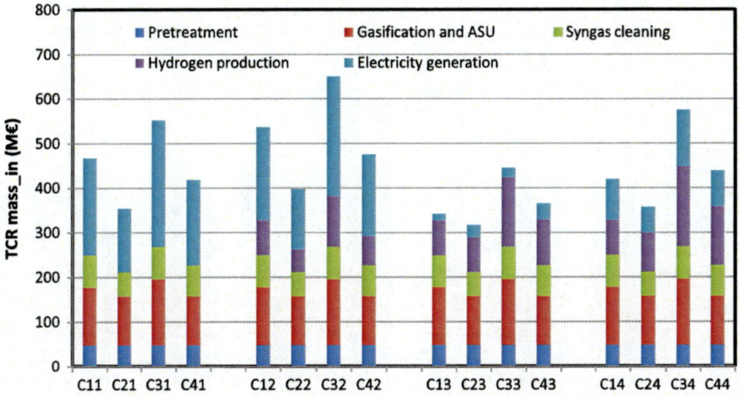

Fig. 4 Breakdown of investment costs for 106.8 tons/h of feedstock in

cases the electricity production investment comes from the WHB and the ST cycle that is operated alone not in CC mode (because GT and HRSG are not used). On the other hand, for Ci4 cases, the electricity generation section gains relevance, but is still less significant than the hydrogen generation one. In these Ci4 scenarios, TOT has been modified to be adapted to the PSA LHV, which is around 350°C. Thus, the HRSG is working with less heat recovery than when operating with syngas and hydrogen. It is worth noticing that the water consumption of the WGS reactors is important if compared with other water consumptions such as the gasifier water feed. The former water consumption penalises the net power produced in scenarios Ci3 and Ci4. Note that because of the reference plant criteria of fixed amount of fuel, pre-treatment costs are the same for all the cases, independently of the feedstock type. Moreover, gasification, ASU and syngas cleaning costs remain the same for all the same feedstock cases (C1j, C2j, C3j and C4j). The difference between cases remains in the final syngas application which is highly influenced by the LHV of the feedstock.

According to Fig. 5, to maintain the same plant output value, the TCR of scenarios Ci2 are noticeable larger than the rest of case studies. They are 35–36% higher than their equivalent scenarios, compared to Ci1 (CC fuelled by syngas). In these reference plant criteria, the case studies show a different behaviour than the previous ones, given that they have mass flowrates of feedstocks increased or decreased to be adapted to the final output, showing as a consequence different tendencies when comparing one figure with the other. Comparing Fig. 5 with Fig. 4, for scenarios Ci1, Ci2 and Ci4 the electricity generation section has approximately the same cost (as the amount of electricity to produce is the same). All in all, the final product generation plays the main role in TCR contribution, followed by syngas generation.

In Fig. 6, the different feedstock costs (FC) are plotted when varying the input feed to be adapted to the reference plant output. They follow the same tendency as the TCR. Higher feedstock needs are appreciated for Ci2 cases. Hydrogen production scenarios continue with their tendency of having lower costs, taking advantage of the high calorific value of the pure H_2 produced, which is not

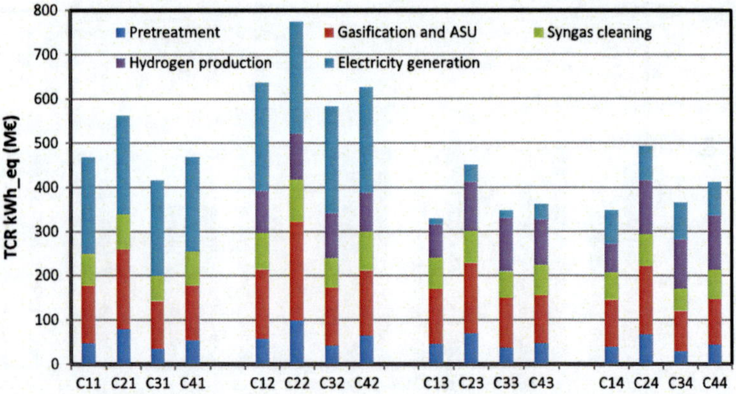

Fig. 5 Breakdown of investment costs for 2.05 GWh_{eq} as product

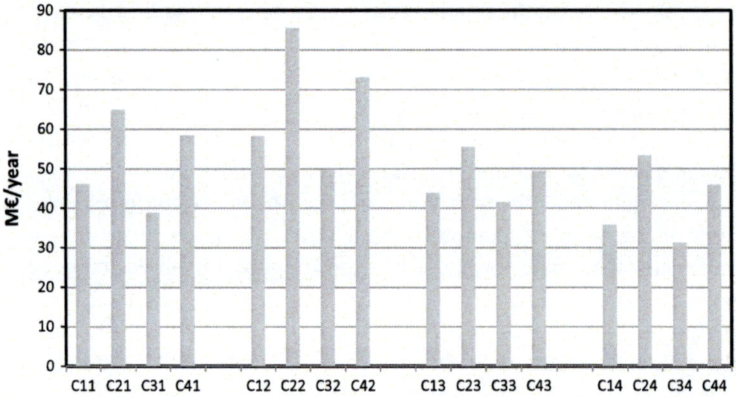

Fig. 6 Feedstock costs per year for different scenarios, considering 2.05 GWh_{eq} as product

'penalized' by the combustion process. Please note that when compared to C11 scenario, all others behave by increasing or decreasing their feedstock costs according to their LHV.

Figure 7 compares the different COE values obtained for the 16 case studies, and for the two reference scenarios. Lower COE values are found for hydrogen production cases (Ci3 and Ci4) when compared to their respective Ci1 and Ci2. The highest COE values are found for coal- based cases, while petcoke and biomass shows nearly similar values.

3.1.1 CO_2 Metrics

The costs of CO_2 captured and avoided are calculated for all the scenarios according to Eqs. 3 or 4 in chapter "Global Clean Gas Process Synthesis and Optimisation".

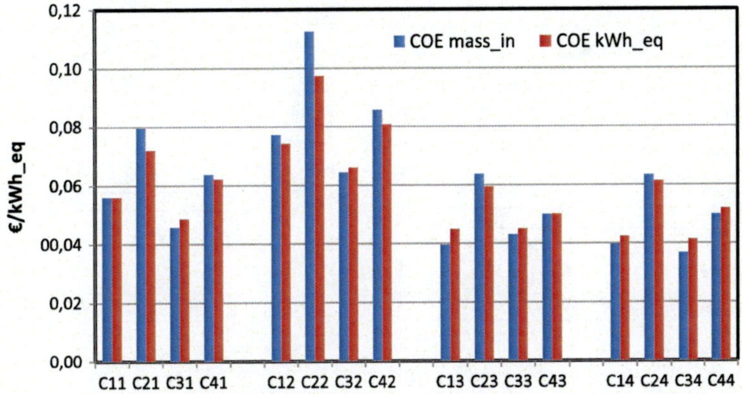

Fig. 7 COE values for different scenarios considering both situations: fixed inlet flow (106.8 tons/h) and fixed amount of product (2.05 GWh_{eq})

It is important to remark the differences between both points of view: in the first case, the CO_2 value is the value of emissions. In the second case, the value of CO_2 is the value of the captured flow. Both of them referred to the net plant capacity. Scenarios Ci1 and Ci2 are of concern in this analysis given that both produce electricity while the second captures CO_2. In all cases, the energy requirements for pumping CO_2 and its storage are not considered in cases Ci2, and the increase of COE is because of investment changes (see Figs. 4 and 5) and lower power production is due to burning H_2 (see Fig. 9). In the model proposed the cost of CO_2 captured and avoided are the same given that the extra consumption because of the energy required operating the CCS technology is not assessed. The values for each type or feedstock are 0.025 €/ton CO_2, 0.0585 €/ton CO_2, 0.0175 €/ton CO_2 and 0.0345 €/ton CO_2 for C1j, C2j, C3j and C4j. These figures show that the use of coal will require higher CO_2 market prices or higher government subsidies.

3.1.2 Co-Production Analysis

In the model proposed co-production is possible, hence the analysis of the prices of H_2 and electricity influence on the plant revenue are mandatory.

In the analysis shown in Fig. 8, the production of electricity and H_2 are plotted, multiplied by their assumed market price (0.15 €/kWh, and 3 €/kg; as reported in chapter "Raw Materials Supply") which could be assumed as a 'fictitious' revenue in the co-production scenarios, where raw material costs are considered proportionally distributed according to the syngas to CC split. Thus, a syngas split fraction of 0 corresponds to Ci3 scenarios, while a split fraction of 1 represents Ci1 cases. Raw material costs are reported in former sections. In all cases the fraction of syngas going to the CC is varied considering the same inlet

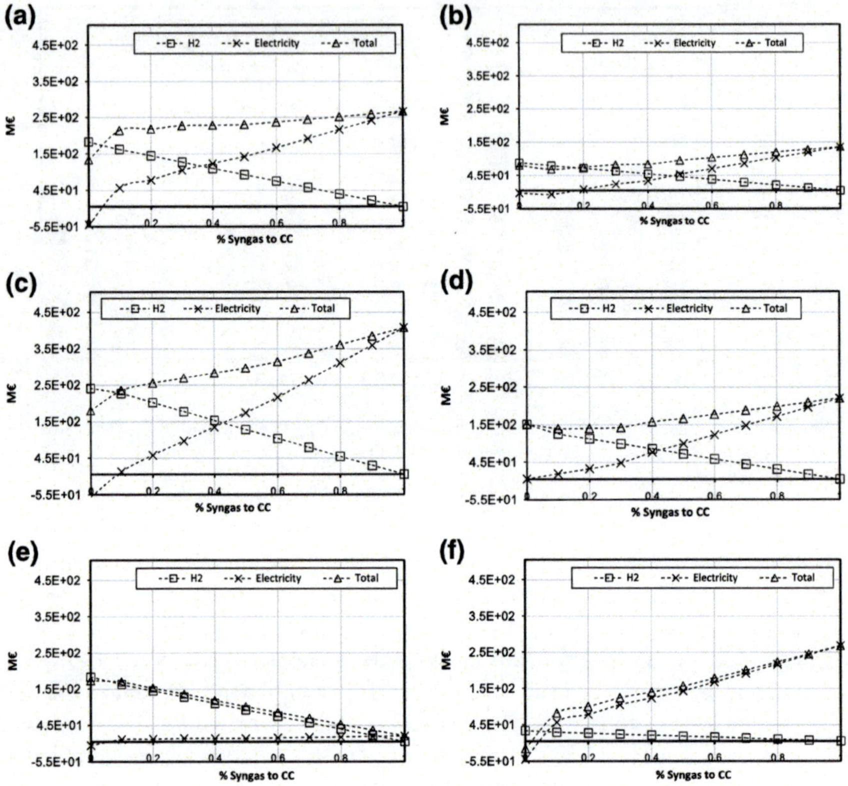

Fig. 8 Analysis for H_2 and electricity prices on co-production scenarios. **a** Base case, **b** Coal, **c** Petcoke, **d** Orujillo, **e** Base case, assumption 1, **f** Base case, assumption 2

mass flowrate. Thus, the same heat recovery is not achieved in all the different cases; consequently the different graphs do not have a linear behaviour for small fractions of syngas to CC. It is directly noticeable that higher LHV in feeds implies higher revenues.

Because of the assumed prices, higher revenues are found when producing the maximum amount of electricity for all feedstocks. Clearly different feedstock use involves different revenues, which are also influenced by the raw material price. Note that the revenues of hydrogen and electricity cross, for different split fractions for different feeds, in all cases near the 0.4–0.5 values.

The last two graphs in Fig. 8 show different price assumptions. For instance, assuming a price of 0.03 €/kWh and maintaining the price of H_2 at 3 €/kg (see Fig. 8, assumption 1), the clear preferable option is to produce always H_2, even at low-split fractions. On the other hand, by decreasing the price of the H_2 to 1 €/ton while maintaining the other at 0.15 €/kWh, the inverse situation is observed (Fig. 8, assumption 2). These graphs can aid in deciding the actual split and the

Table 2 Power produced for all considered scenarios

Option	Scenario	Net power produced (MW)	H_2 produced (MW)	Net equivalent power (MW_{eq})
Electricity production from syngas	C11	285.3	0.0	285.3
	C21	151.9	0.0	151.9
	C31	423.2	0.0	423.2
	C41	243.9	0.0	243.9
Electricity production from H_2	C12	225.6	0.0	225.6
	C22	115.3	0.0	115.3
	C32	331.5	0.0	331.5
	C42	195.1	0.0	195.1
H_2 production without PSA purge recovery	C13	−45.9	346.1	300.2
	C23	−7.1	184.7	177.6
	C33	−55.3	452.0	396.7
	C43	0.7	287.8	288.5
H_2 production with PSA purge recovery	C14	26.4	340.7	367.2
	C24	3.4	181.3	184.7
	C34	56.7	468.1	524.8
	C44	26.5	283.8	310.3

raw material preferred that are required to meet certain electricity and H_2 demands, if co-production is allowed.

3.2 Plant Engineering Point of View

As discussed in chapters "Global Clean Gas Process Synthesis and Optimization" and "Modelling Superstructure for Conceptual Design of Syngas Generation and Treatment", other point of view that can be used is the application of efficiency metrics. The net power as electricity and the H_2 equivalent energy content are reported in Table 2 for all scenarios.

It is interesting to note that in the case of scenarios where PSA purge is sent to the CC a higher amount of power is produced as electricity, while when this gas is discharged into the atmosphere net consumption of electricity is required for the case of abiotic resource use (C13, C23, C33). To check which of the scenarios uses the available energy in the fuel in a more efficient way, the total and global efficiencies are calculated as in Eqs. 4 in chapter "Global Clean Gas Process Synthesis and Optimization, 7 and 8 in chapter "Modelling Superstructure for Conceptual Design of Syngas Generation and Treatment", and the results are shown in Fig. 9. Please note that the fuel's flow is fixed for all topological options, see Table 1.

Interestingly the production of electricity based on H_2 is the least efficient use of the available energy in the fuels, as shown by the total efficiency values for scenarios Ci2. It is also interesting to note the effect of purge gas use in the CC by comparing the values of total efficiency for the case of scenarios Ci3 and Ci4, where higher values are found for the latter scenarios.

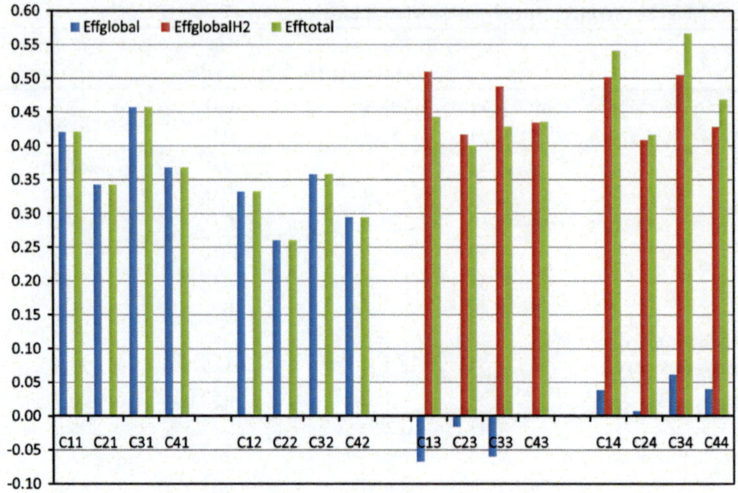

Fig. 9 Total and global efficiencies for the different considered scenarios

3.3 Environmental Point of View

Each one of the former scenarios was compared in terms of environmental impact following the ISO standards as in chapter "Global Clean Gas Process Synthesis and Optimization" for compiling a LCA. Regarding the functional unit, given that co-production is feasible in some of those options, the functional unit used is 1 MJ of equivalent energy which could be from electricity or from H_2. In this sense, environmental impacts which are proportional to mass/energy flows are normalised by the total amount of energy produced in each scenario. We recall that the composition of the base case coincides with the base composition of ELCOGAS; it corresponds to a mixture of coal and petcoke, 50%/50% in mass basis, where coal is supposed to come from the local ENCASUR mines. The petroleum coke is from the Puertollano REPSOL refinery, obtained as a by-product, with high sulphur content, as described in chapter "Modelling Syngas Generation". The residual biomass used is olive pomace or *orujillo*, which is the solid residue that remains after pressing olives, a very abundant industrial residue in the south of Spain. The system boundaries have been drawn from cradle-to-gate with the following considerations:

- Transport of fuel materials has been disregarded, mimicking the geographical considerations of ELCOGAS.
- CO_2 disposal does not consider its transportation and injection to formation, but it considers its liquefaction.
- Hydrogen transport is not considered, and it is available at high pressure at plant.

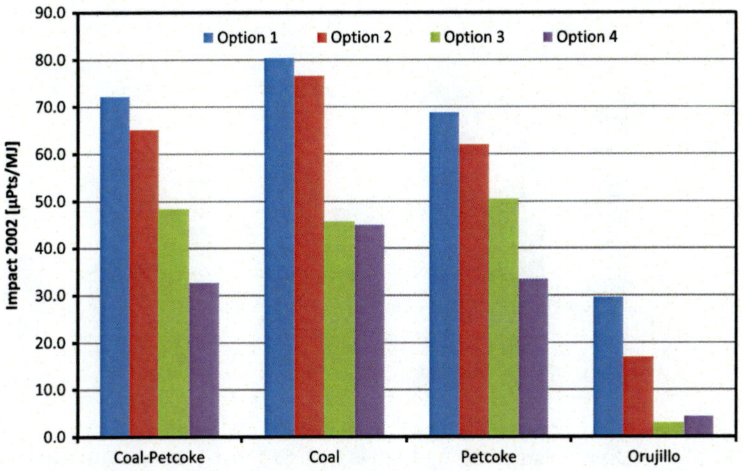

Fig. 10 Environmental impact for each scenario

- Electricity consumption for the cases where net power production was found negative is considered to be from the Spanish grid. Natural gas use is represented using Spanish data.
- No allocation of the emissions is done for other possible co-products such as CO_2 and sulphur. Moreover electricity and Hydrogen production are added up considering their energy content together and considered indistinctively.
- Waste water treatment and slag disposal are considered, and emissions associated to their treatment are included within boundaries.

In all cases the life cycle inventories (LCIs) for the use of raw materials, electricity and others are gathered from the Ecoinvent database [8].

Using the results gathered in Table 1 together with the results from each scenario the LCIs are gathered and on those values the LCIA can be calculated. In this case the environmental impact assessment technique used is the Impact 2002+ [9].

Figure 10 shows the results of the total environmental impact for each scenario. The highest values are obtained for the case of electricity production using coal (C21), while the lowest is found for the cases C43 and C44 where biomass is used for the production of hydrogen. By comparing cases Ci2 and Ci1 it can be seen that the production of electricity directly from syngas is less environmental friendly than its production with H_2, which is mainly because of the use of a hydrogen turbine more environmental friendly than the direct combustion of syngas, and the reduction of climate change impacts because of the CO_2 capture.

The environmental impact can be subdivided in terms of endpoint categories; consequently impacts to different areas of protection can be grasped: human health, ecosystem quality, climate change and natural resource use (as discussed in chapter "Environmental Impact Estimation" in Sect. 2.3.3). Furthermore, the same figure can be distributed along the different activities in the supply chain: raw

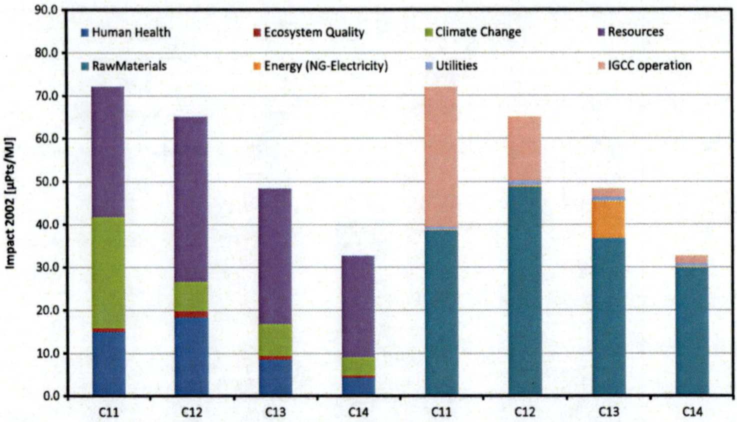

Fig. 11 Environmental impact for coal–petcoke blend-based options (C_{1j}) distributed along endpoint categories and supply chain activities

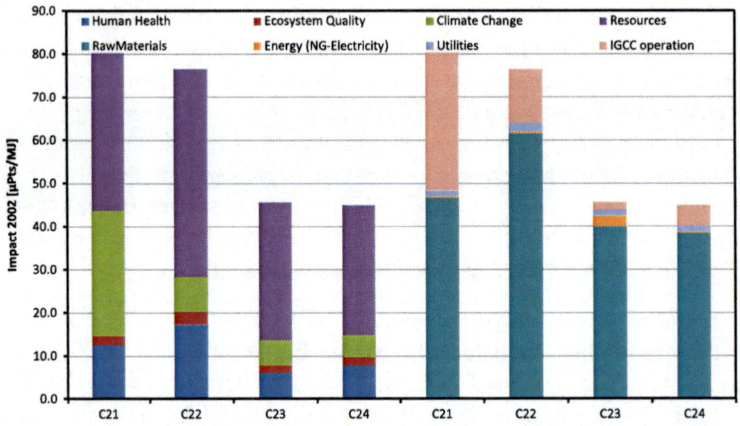

Fig. 12 Environmental impact for coal blend-based options (C_{2j}) distributed along endpoint categories and supply chain activities

materials (fuel consumption), energy consumption (electricity from grid and natural gas), utilities and IGCC operation. The utilities category encompass the consumption of: limestone, NaOH, H_2SO_4, water pre- and post-treatment and slag disposal, while IGCC operation gathers all the impacts associated directly to the IGCC, which in this case are those related to air emissions (CO_2, CO and other species).

In terms of environmental impact to different areas of protection Figs. 11–14 show the clear reduction in the climate change category obtained by shifting from electricity directly from syngas to its production via H_2, where CO_2 is captured. In the case of abiotic fuels use the areas of protection more impacted are the global

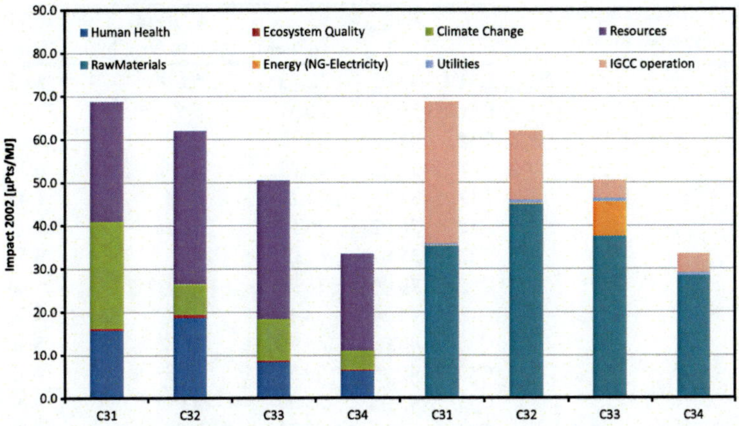

Fig. 13 Environmental impact for petcoke-based options (C_{3j}) distributed along endpoint categories and supply chain activities

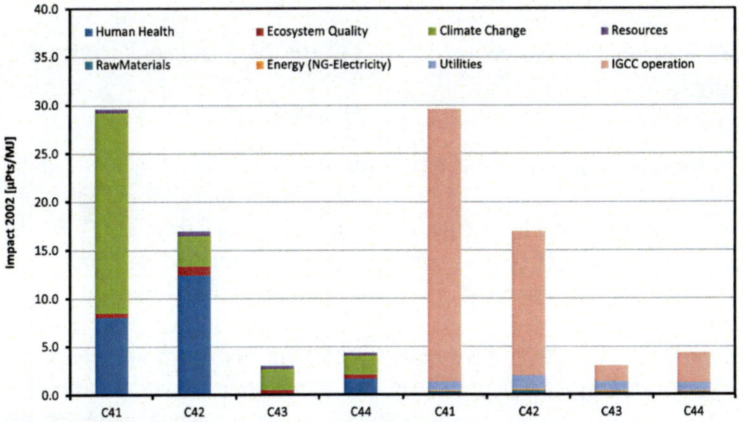

Fig. 14 Environmental impact for orujillo-based options (C_{4j}) distributed along endpoint categories and supply chain activities. Please note that vertical scale in this case has been reduced for better representation

climate change and the resource use, while in the case of the biomass use, the resource use impact is nearly negligible (see Fig. 14). Regarding the items of the SC responsible for the environmental impact, it can be seen that in the case of abiotic resource use (coal and petcoke), the production of fuels under the raw materials category encompasses the largest portion of the impact, while in the case of biomass use IGCC operation is the most important aspect (see Fig. 14). In all cases utilities use impacts are very little, while energy consumption is noticeable in the case of options C13, C23 and C33, where electricity from the grid is required.

Table 3 Scenarios results for the different metrics considered

	Eff_{total} (%)	EI (μPts/MJ)	COE (a) (cts €/kWh$_{eq}$)	COE (b) (cts €/kWh$_{eq}$)
C11	42.05	72.13	5.62	5.62
C21	34.28	80.35	7.97	7.21
C31	45.72	68.72	4.58	4.89
C41	36.83	29.62	6.38	6.22
C12	33.25	65.12	7.73	7.43
C22	26.02	76.59	11.26	9.75
C32	35.82	62.01	6.45	6.61
C42	29.46	16.99	8.57	8.08
C13	44.24	48.42	3.97	4.51
C23	40.08	45.76	6.38	5.96
C33	42.85	50.53	4.33	4.53
C43	43.56	**3.05**	5.01	5.02
C14	54.12	32.74	3.98	4.26
C24	41.67	45.00	6.35	6.16
C34	**56.70**	33.46	**3.67**	**4.16**
C44	46.86	4.38	5.01	5.21

4 Performance Assessment in Terms of Pareto Efficient Solutions

As previously discussed chapter "Global Clean Gas Process Synthesis and Optimization", Sect. 3, one of the possible ways of tackling with multicriteria decision-making problems is the generation of Pareto frontiers (PFs), by classifying alternatives in two sets: dominating and dominated solutions. This action allows for focusing attention on a lower number of scenarios from where a decision can be made.

In this case three KPIs have been selected for consideration: energy efficiency use by considering the Eff_{total} values, economic concerns are measured using the COE, as calculated in previous sections, while environmental concerns are tackled by using the environmental impact (EI) per unit of energy produced. In the case of COE two considerations were done (a) one where the feedstock input is fixed disregarding the amount of power produced, while in the second case (b) the feedstock amount is modified to attain a certain amount of total power produced. The following Table 3 summarises the results for the scenarios considered.

Bold values in Table 3 show the best scenario for each metric, showing that scenario C43 is the best when considering the environmental impact and the COE-based on feed, while C34 is the best when maximising efficiency and minimising COE calculated considering a fixed amount of energy. Given that there is no coincidence between the selected best scenarios for each metric, a given decision rule has to be applied. These rules were discussed in chapter "Global Clean Gas Process Synthesis and Optimization", Sect. 3.2, where different MCDA techniques were presented. One possibility is to apply the technique for order by similarity to

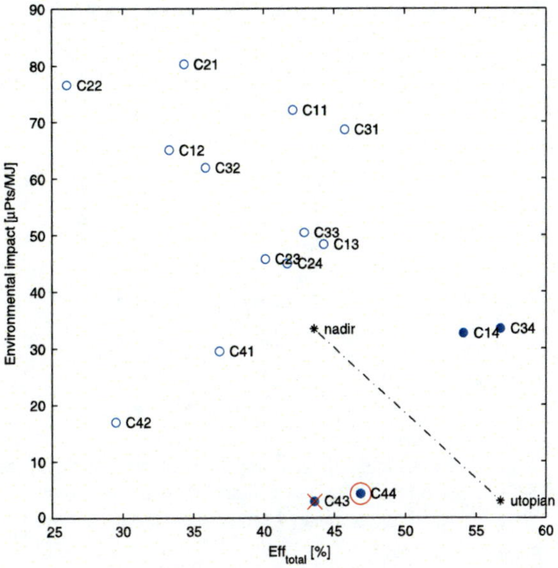

Fig. 15 Pareto front considering the efficiency and environmental impact of the scenarios. *Blue circles* show dominated scenarios, *blue dots* show dominating solutions and *stars* denote utopian and nadir points. *Red cross* shows scenarios farthest from nadir and *red circle* shows scenarios closest to utopian point

ideal solution (TOPSIS). This technique, proposed by Hwang and Yoon [10], uses the euclidean distances from each scenario to the 'utopian' and 'nadir' solutions are to select better solutions; two possibilities can be selected: the solution from the Pareto fsront (PF) which is closest to the utopian point or the one farthest from the nadir point. Please note that the utopian and nadir points are defined considering the values of the solutions only pertaining to the PF. Regarding the utopian solution it would be formed by and hypothetical scenario that would achieve the best possible value for all criteria considered. The analysis of efficient solutions in terms of efficiency and environmental impact can be seen in Fig. 15.

The Pareto front for the case of efficiency and environmental impact considers scenarios: C43, C44, C14 and C34, clearly showing that from these points of view option 4 for blends, petcoke and biomass dominates all other scenarios. The closest scenario to utopian point is C44, while the farthest from nadir is C43, showing that biomass is the fuel to use when considering these two objectives. In the case of efficiency and COE, two Pareto analyses can be performed; one considering each COE, results are shown in Fig. 16.

In the case of comparing COE and efficiency, scenario C34 dominates all others showing that petcoke use for producing hydrogen is the most efficient and more profitable in terms of COE. In the case of comparing COE and environmental impact, the trade-offs are shown in Fig. 17.

The Pareto front for the case of COE and environmental impact considers scenarios: C43, C44, C14 and C34, which coincides with the one found for the case of efficiency and environmental impact. Note that in the case of the calculation of COE based on the same amount of energy scenario C44 is not included in the PF. In both cases the solution is that TOPSIS will select C43, which is closest to utopian point and also is farthest from nadir.

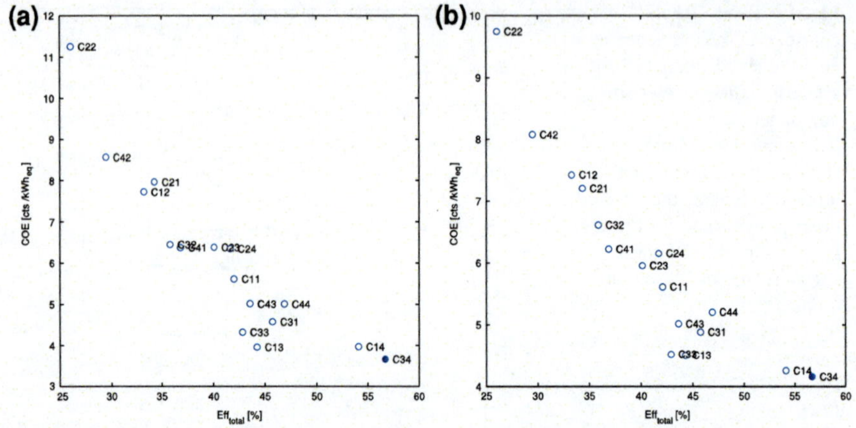

Fig. 16 Pareto front considering the scenarios efficiency and cost of energy. *Left figure* shows COE based on fixed feedstock input, whereas *right* shows COE based on fixed amount of energy. *Blue circles* show dominated scenarios and *blue dots* show dominating solutions

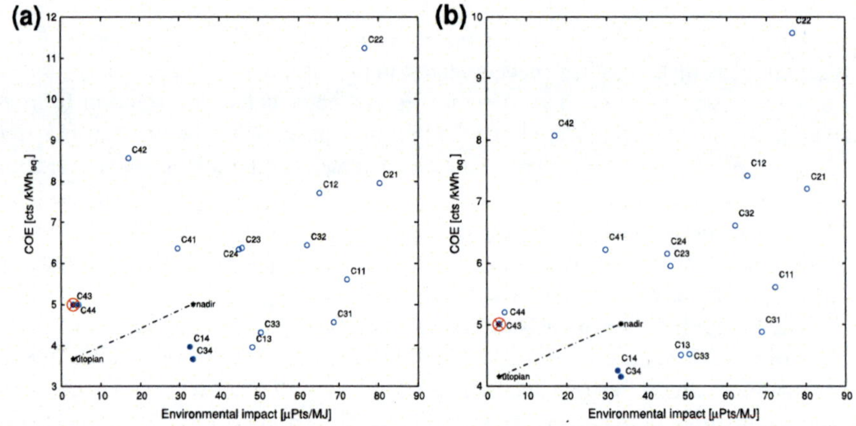

Fig. 17 Pareto front considering the environmental impact and cost of energy of the scenarios. *Left figure* shows COE based on fixed feedstock input, whereas *right* shows COE based on fixed amount of energy. *Blue circles* show dominated scenarios, *blue dots* show dominating solutions and *stars* denote utopian and nadir points. The *red cross* shows scenarios farthest from nadir and *red circle* shows scenarios closest to utopian point

5 Conclusion

The most representative output data is shown, and the described points of view, discussed in chapter "Global Clean Gas Process Synthesis and Optimization", are here evaluated under a techno-economic and environmental assessment. Sixteen scenarios are considered encompassing four different feedstocks' combined with four different plant topologies; electricity generation with syngas, electricity generation with hydrogen, hydrogen production and hydrogen production with PSA flue gas profit in the CC. This chapter has discussed the selection of processing options in terms of different metrics. The approach used considers all objectives to be equally important, but this can be easily modified if different trade-offs are required. The former approach is readily used as a decision-making aid tool.

References

1. Hamelink CN, Faaij APC (2002) Future prospects for production of methanol and hydrogen from biomass. J Power Sources 111:1–22
2. Chiesa P, Consonni S, Kreutz T, Williams R (2005) Co-production of hydrogen, electricity and CO_2 from coal with commercially ready technology. Part A: performance and emissions. Int J Hydrogen Energy 30:747–767
3. Kreutz T, Williams R, Consonni S, Chiesa P (2005) Co-production of hydrogen, electricity and CO_2 from coal with commercially ready technology. Part B: economic analysis. Int J Hydrogen Energy 30:769–784
4. Rubin ES, Chen C, Rao AB (2007) Cost and performance of fossil fuel power plants with CO_2 capture and storage. Energy Policy 35:4444–4454
5. Chen C, Rubin E (2009) CO_2 control technology effects on IGCC plant performance and cost. Energy Policy 37:915–924
6. Brown D, Gassner M, Fuchino T, Maréchal F (2009) Thermo-economic analysis for the optimal conceptual design of biomass gasification energy conversion systems. Appl Therm Eng 29:2137–2152
7. Van Vliet OPR, Faaij APC, Turkenburg WC (2009) Fischer–Tropsch diesel production in a well-to-wheel perspective: a carbon, energy flow and cost analysis. Energy Convers Manag 50:855–876
8. Ecoinvent (2008) The Ecoinvent database V2.0. Technical Report, Swiss Centre for Life Cycle Inventories, Switzerland
9. Humbert S, Margni M, Jolliet O (2005) IMPACT 2002+: user guide draft for version 2.1. Technical report from Industrial Ecology & Life Cycle Systems Group, GECOS, Swiss Federal Institute of Technology Lausanne (EPFL). Lausanne, Switzerland
10. Hwang C, Yoon K (1981) Multiple attribute decision making. Springer-Verlag, Berlin

Examples of Industrial Applications

Pilar Coca, Mar Pérez-Fortes and Aarón D. Bojarski

Abstract The description of the 335 MWe_{ISO} coal-based Puertollano IGCC power plant as example of industrial application of IGCC technology together with its main lessons learnt are summarised in this chapter. This chapter includes process description, syngas analysis, main operational real data (power production and emissions), and main causes of unavailability as well as R&D lines. These are mainly focused on improvement of IGCC technology taking into account efficiency increase and emissions reduction. So, first results of the CO_2 capture and H_2 co-production $14MW_{th}$ pilot plant installed in the Puertollano IGCC are included.

Notation

ASU	Air separation unit
CC	Combined cycle
CCS	Carbon capture and storage
CENIT	National Strategic Consortia for Technical Research (Consorcios Estratégicos Nacionales en Investigación Técnica)
CIEMAT	Spanish Research Centre (Centro de Investigaciones Energéticas, Medioambientales y Tecnológicas)

P. Coca (✉)
ELCOGAS, S. A.Ctra. Calzada de Calatrava a Puertollano Km. 27,
Apdo. Correos 200, 13500 Puertollano – (Ciudad Real), Spain
e-mail: pcoca@elcogas.es

M. Pérez-Fortes · A. D. Bojarski
Universitat Politècnica de Catalunya, ETSEIB, Diagonal 647,
08028 Barcelona, Spain
e-mail: mar.perez-fortes@upc.edu

A. D. Bojarski
e-mail: aaron.david.bojarski@upc.edu

L. Puigjaner (ed.), *Syngas from Waste*, Green Energy and Technology,
DOI: 10.1007/978-0-85729-540-8_12,

CIUDEN	Energy City Foundation (Fundación Ciudad de la Energía)
HP	High pressure
HRSG	Heat recovery steam generator
IGCC	Integrated gasification combined cycle
IGME	Geological and Mining Institute (Instituto Geológico y Minero de España)
IPCC	Intergovernmental Panel on Climate Change
INCAR	Coal Spanish Research Centre (Instituto Nacional del Carbón)
ISO	International Standards Organisation
LLB	Lurgi Lentjes Babcok
MBM	Meat and bone meal
aMDEA	Active methyl diethanol amine
MWe	Electrical mega Watt
MWh	Mega Watt-hours
MW_{th}	Thermal mega Watt
NGCC	Natural gas combined cycle
O&M	Operation and maintenance
Petcoke	Petroleum coke
PSA	Pressure swing adsorption
R&D	Ressearch and development
Syngas	Synthetic gas
UCLM	University of Castilla-La Mancha
VI FP	Sixth Framework Programme

1 ELCOGAS Description

1.1 The Company

ELCOGAS, S.A., the owner company of the Puertollano IGCC power plant was founded on 8th April 1992, as a mercantile company subject to Spanish legislation, with the objective of the construction and exploitation of the Puertollano integrated gasification combined cycle (IGCC) plant (Fig. 1).

This power plant is the largest IGCC plant in the world using solid fuel in a single pressurised entrained flow gasifier and is in commercial operation since 1998 with synthetic gas. Its design fuel is a mixture 50:50 of poor quality coal (high content of ash) and petcoke (high content of sulphur).

The founding members were European electrical companies along with the main combined cycle and gasification plant suppliers, Krupp Koppers and Siemens from Germany, in association with Babcock Wilcox Española, from Spain as manufacturer. The current members and their percentage of shares are shown graphically in the following Fig. 2.

Fig. 1 The Puertollano IGCC power plant view

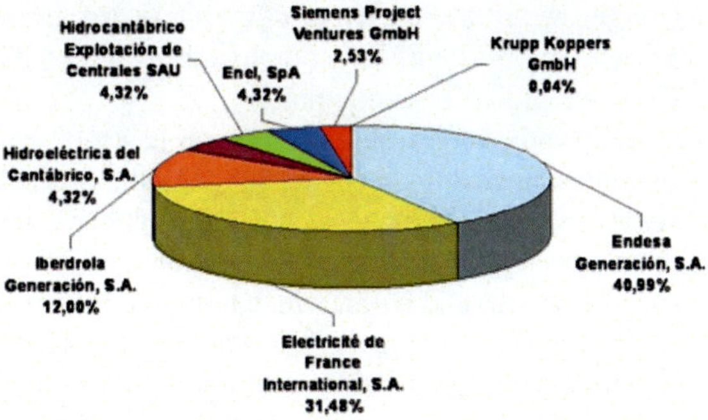

Fig. 2 ELCOGAS capital share (at 31st December 2010)

1.2 Process and Integration Description

Table 1 summarises the principal data for the ELCOGAS IGCC power plant. The Puertollano IGCC power plant consists of three main units: (i) the gasification unit (generating the synthetic gas) supplied by Krupp Koppers, (ii) the air separation unit (ASU) that produces nitrogen and oxygen supplied by Air Liquide and (iii) the combined cycle (CC) (producing electricity) supplied by Siemens. These main units division is analogous to the considered approach in chapter Modelling Superstructure for Conceptual Design of Syngas Generation and Treatment for superstructure conception. The solid fuel is dried, mixed and milled in the coal preparation system and then sent with pure nitrogen to the gasifier to produce synthetic gas. Then, the syngas obtained in the pressurised entrained flow gasifier is cooled down, cleaned and subsequently burnt as fuel in the gas turbine

Table 1 Summary of the ELCOGAS IGCC power plant main data

Design fuel: coal and petcoke (50 wt%.)				
	Coal	Petcoke		Mix
LHV (MJ/kg)	13.10	31.99		22.55
Electrical output				
	Gas turbine (MW)	Steam turbine (MW)	Total gross (MW)	Total net (MW)
ISO conditions	200	135	335	300
Site conditions	182.3	135.4	317.7	282.7
Efficiency (LHV)				
	Gross		Net	
Thermal efficiency	47.12%		42.2%	
Heat rate	7,647 kJ/kWh		8,538 kJ/kWh	

of the combined cycle plant. The synthetic gas is the result of several reactions between fuel (a mix of coal and petroleum coke) with oxygen/steam at high temperatures of up to 1,600°C. The required oxygen for the gasification process is produced in an integrated ASU (based on a cryogenic process), which also produces pure nitrogen for drying the pulverised fuel, for fuel transportation and for the safety inertisation of the different circuits with a purity of 99.99% and waste nitrogen with 98% purity to dilute clean gas from gasification unit before being burnt in the gas turbine combustion chamber. Figure 3 shows the ELCOGAS IGCC simplified flow diagram.

The synthetic gas obtained, which basically consists of CO and H_2, is subsequently subjected to an exhaustive cleaning process to eliminate the small parts of pollutants, fly ash, halogens, cyanides, sulphur compounds, etc. Then, the so-called clean gas, free of pollutants, is saturated, mixed with waste nitrogen (to reduce NO_x formation) and burnt, with a high-efficiency level, in the gas turbine of the CC electricity-generating unit. The gas turbine (model V94.3, 200 MW_e under ISO conditions) is capable of operating with both synthetic and natural gases. The gas turbine exhaust gases with residual heat are fed into a heat recovery steam generator (HRSG), producing steam that is used together with the steam produced in the gasification process to generate additional electricity in a conventional steam turbine (135 MW_e under ISO conditions) with condensation cycle. The demonstrated plant net efficiency is 42.2% under ISO conditions.

The design of the heat exchangers battery is particularly relevant in terms of efficiency, basically as regard steam production and consumption, incorporating two heat recovery boilers, one for the raw gas produced in the gasifier and the other for the turbine exhaust gases. Furthermore, the steam acts as a heat conductor for several uses in gasification, desulphurisation and air separation processes.

The Puertollano power plant was designed with a high-integration level that involves the integration of the three previously mentioned units:

- *Integration of the gasification island and combined cycle water–steam systems*: The water fed to the steam generators is pre-heated in a section of the combined

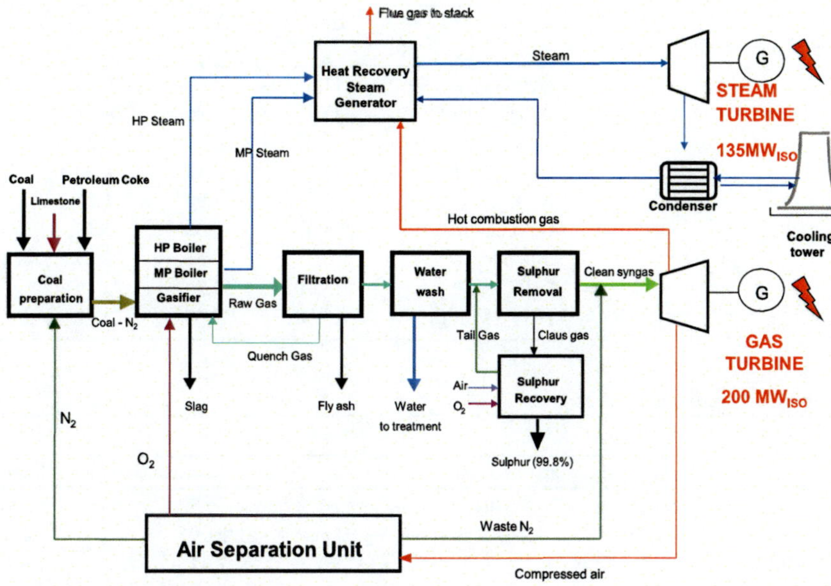

Fig. 3 Simplified flow diagram of the Puertollano power plant

cycle's HRSG and is sent to gasification where a saturated steam is produced as a result of the exchange of heat with the raw gas. This saturated steam is exported to the HRSG for superheating and expansion inside the steam turbine, generating additional electricity.

- *Nitrogen-side integration between ASU and combined cycle*: The waste N_2, a by-product of the ASU, is compressed and mixed with the syngas to reduce NO_x emissions and to increase the capacity of the gas turbine.
- *Air-side integration between ASU and combined cycle*: The compressed air required by the ASU is totally extracted from the gas turbine compressor.

The integration of water–steam systems is normal in all IGCC power plants in operation. On the other hand, integration between the ASU and CC is an option, which is much more frequently discussed. The highly integrated designs mean greater power plant efficiency, because the consumption of auxiliary systems for air compressors and ASU products is reduced. Nevertheless, these involve longer start-up times during which time, the back-up fuel (natural gas in most cases) is used. With regard to the IGCC power plants using coal that are in operation in Europe, highly integrated design has predominated because of its increased efficiency, whereas in the United States, with lower fuel prices, increased availability and flexibility, which a non-integrated design offers, has been preferred. Currently, the tendency is towards designs where the air required by the ASU comes in part from the gas turbine compressor and in part from a separate compressor. This provides the necessary flexibility for faster start-ups and an intermediate auxiliary consumption between the two options.

Table 2 Fuel characteristics (design data)

	Coal	Coke	Mix (50/50 wt/wt%)
Humidity (wt%)	11.8	7.00	9.40
Ashes (wt%)	41.10	0.26	20.68
Carbon (wt%)	36.27	82.21	59.21
Hydrogen (wt%)	2.48	3.11	2.80
Nitrogen (wt%)	0.81	1.90	1.36
Oxygen (wt%)	6.62	0.02	3.32
Sulphur (wt%)	0.93	5.50	3.21
LHV (MJ/kg)	13.10	31.99	22.55
HHV (MJ/kg)	13.58	32.65	23.12

Table 3 Syngas composition

	Raw gas		Clean gas		
	Average design	Design		Average design	Design
CO (%)	59.26	61.25	CO (%)	59.30	60.51
H_2 (%)	21.44	22.33	H_2 (%)	21.95	22.08
CO (%)	2.84	3.70	CO (%)	2.41	3.87
N_2 (%)	14.32	10.50	N_2 (%)	14.76	12.5
Ar (%)	0.90	1.02	Ar (%)	1.18	1.03
H_2S (%)	0.83	1.01	H_2S (ppm)	3	6
COS (%)	0.31	0.17	COS (ppm)	9	6
HCN (ppmv)	23	38	HCN (ppmv)	–	3

1.3 Fuel and Clean Gas Data

Main parameters of ELCOGAS fuel components, coal, petcoke and mixture are shown in Table 2.

Coal comes from the ENCASUR mine and the petcoke from REPSOL-YPF refinery being both of them located very close to the power plant and transported by trucks. It must be noted that the main features of mixture are its high content in ash and sulphur approximately 21 and 3.5%, respectively.

The composition of obtained syngas before and after the cleaning-up processes—dry dedusting, washing and desulphurisation systems—(called raw and clean gas, respectively) is shown in Table 3.

1.4 Environmental Advantages

ELCOGAS IGCC power plant can meet all projected environmental legislation, solving the compliance problems of electric power generation. Because it operates at higher efficiency levels than conventional fossil-fuelled power plants, ELCOGAS emits less CO_2 per unit of energy.

Table 4 ELCOGAS power plant milestones

1992	Main contracts
1993	Start of civil works at site
June 1996	First synchronisation of gas turbine
October 1996	Commercial operation of combined cycle with natural gas
June 1997	Performance test of air separation unit
March 1998	First switch over from natural gas to syngas in the gas turbine
November 2000	First 1,000,000 MWh produced with coal gas as IGCC
December 2010	Total: 21,052 GWh, IGCC: 14,437 GWh

ELCOGAS gaseous emissions (SO_2, NO_x) are small fraction of allowable limits, being NO_x emissions lower in IGCC operation than in NGCC (Natural Gas Combined Cycle) mode.

The water required to operate it is less than half of that required for pulverised coal plant with a flue gas scrubbing system. In addition, a complex wastewater treatment plant permits and meets European and Spanish legislation (European Union Directive [1] and AAI-CR21 [2], respectively).

The solid residues are, in its majority, vitrified (no leachable), resulting in useable by-products for construction industry. Sulphur recovery is approximately 99.9% because of tail gas recycling system.

1.5 Main Milestones

Since order of main contracts, up to more than 14 million of electrical MWh has been produced as IGCC, the main milestones can be summarised as follows in Table 4.

1.6 Operating Data: Power Production and Emissions

1.6.1 Power Production

A graphical summary of historical power production data is presented in the following figures and tables. Figure 4 presents the annual power production using syngas (mode IGCC) and natural gas (mode NGCC), as well as, the total gross power production, showing a great improvement since 1998 up to 2002. In the year 2003 and 2006, two gas turbine major overhauls (50,000 and 75,000 equivalent operating hours) were carried out, reducing the annual production. Other main causes of reduction of energy production were in 2004 and 2005 because of a gas turbine main generation transformer isolation fault, and in 2007 and 2008 because of the ASU waste nitrogen compressor coupling fault and poor

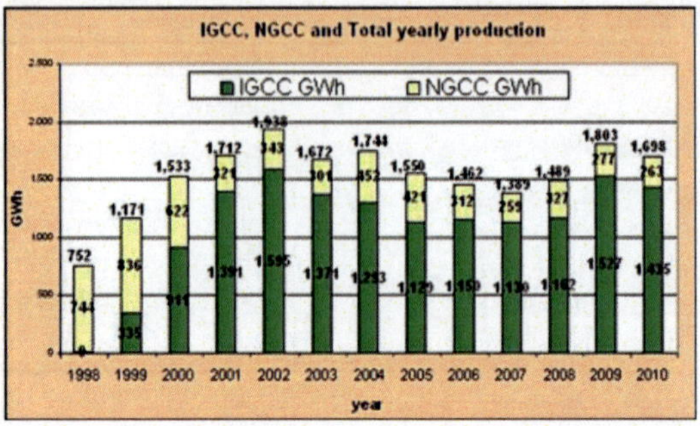

Fig. 4 IGCC, NGCC and total gross power production

Table 5 ELCOGAS main operational achievements

	Total maximum power	IGCC mode
Gas turbine (MW)	225.2	201.6
Steam turbine (MW)	147.1	147.1
Total gross (MW)	359.5	359.5
Maximum continuous operation of the ELCOGAS power plant	1,300.0 (h)	954.0 (h)

repair of MAN TURBO. A new increase in power production was produced in 2009. Table 5 shows the main operational achievements for the ELCOGAS power plant.

1.6.2 Emissions

The Puertollano power plant atmospheric emissions of SO_2, NO_x and particles comply with the European Union Directive 2001/80 EEC [1], which for the IGCC operations is much more restrictive than that applied to all other coal-fired power plant, as well as the ELCOGAS specific power plant regulation, as Fig. 5 shows including real average data from 2010.

2 Lessons Learnt in the Early Operating Years

The following can be highlighted as main lessons learnt in relation to the ELCOGAS operating experience achieved in the early years.

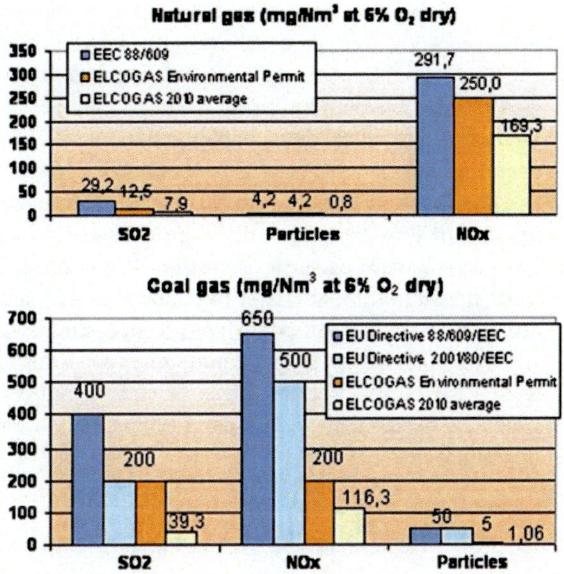

Fig. 5 ELCOGAS power plant emissions in IGCC and NGCC modes including 2009 average emissions data

2.1 *Inflexibility of the Operation Because of the Design Including Total Integration*

Although its advantages in relation to plant efficiency have been demonstrated, the total integration between the ASU and combined cycle involves, besides a greater level of complexity, a long and costly start-up sequence. In practice, these result in it operating as a base load power plant, maintaining a high-minimum technical load (60%). The regulation of the load being around 60–100% is indeed viable, with it being possible to offer a competitive response in relation to reaction times (3% load variation per minute).

Bearing in mind the additional cost involved in the high level of N_2 required during the commissioning process, one reaches the conclusion that an important saving can be made if in the new designs the concept of total integration is reconsidered in favour of an ASU capable of producing pure N_2 independently of the combined cycle operation.

2.2 *Main Causes of Limitations in Availability During First Operating Years*

Availability has not been substantially affected by problems that are intrinsic to the gasification process, but by the low level of reliability of more conventional units in any coal-fired thermal power plant.

At a global level, the most significant problem was related to the gas turbine, because of the burner overheating and the refractory tiles on the combustion chamber what implied an overhaul every 500 operating hours. Both problems were solved by burners modifications—carried jointly between ELCOGAS and Siemens—and after their implementation in 2003, overhauls are every 4,000 operating hours.

Other main problems were related to water leakages of gasifier membrane wall because of flow blockages, local erosion and distributors design, solids handling (slag and fly ash) because of erosion of components by local high velocities that were substituted by abrasion resistant materials and design and operating procedures were revised, pressure control and fluidisation stability of fuel dust conveying and feeding systems, candle filter performance because of poor engineering from LLB (Lurgi Lentjes Babcok) and COS hydrolysis alumina-based catalyst water carryover that were solved changing the catalyst by other one based on titanium oxide.

2.3 Alternative Fuels

Different tests were undertaken during early operating years to demonstrate IGCC fuel flexibility (see Sect. 3.2 for additional fuel tests). So in 2000, different ratios of coal and petcoke were tested modifying their design percentage (50/50 wt/wt %) to 54–46%, 58–42%, 45–55% and 39–61% whose main results were:

- Carbon conversion ranges varied between 98.4 and 99.7%.
- Clean gas composition was kept very stable during the tests.
- The higher mixture ash content, the higher slag/ash separation.
- Emissions fulfilled European and Spanish legislative limits and ELCOGAS emission permit in the whole range of tested mixture.

In addition, in 2000, some tests with meat and bone meal (MBM) were undertaken, using a total of 93 tons of MBM. The main conclusions obtained from these tests were:

- Co-gasification IGCC technology is the best to eliminate MBM, without environmental impact in emissions and with high energy efficiency
- A method for controlled dosage of MBM to the gasifier was defined, checking that the method was valid for inserting a material different from usual fuel in the gasifier.
- Protection of MBM from humidity during storage and handling is the most important factor for preventing problems of transport in grinding train.
- There were not great differences with usual operation in fuel preparation, sluicing systems, fly ash dedusting systems and slag discharge system. The expected behaviour of MBM as fusion agent (because of its high Ca content) could not be confirmed.

Table 6 Summary of possible improvements in new IGCC designs

System/ equipment	Reduction in production variable cost	Reduction in investment cost
Coal preparation	Hot gas generator fed with synthesis gas	Mixing equipment removal
Pressurised coal dust feeding	Fluidification vessel optimisation for N_2 consumption reduction	Concrete building for coal storage and lock hopper system removal
Gasifier	Fine slag recycle	Auxiliary burner removal HP surfaces decrease, increasing raw gas speed
Slag extraction	Filtration system replacement by settling system	Slag water circuit simplification. One lock hopper and extractor removal
Dry filtration	Design, material and candle filter cleaning system improvement	Fly ash recycle removal (lock hoppers, distribution and discharge systems)
Fly ash extraction	Watery system removal	Conveying vessel removal
Gas washing and gas stripping	–	Controlling filter removal
Sulphur removal	SuperClaus plant assessment	Equipment dimension decrease by using enriched air
Air separation unit	Storage capacity increase for liquid N_2	Liquid O_2 storage removal. O_2 purity control flexibility
Gas turbine	New gas turbine with higher efficiency	Higher output gas turbine, scale benefits

- Two effects were clear: The chloride concentration increased with MBM percentage increase and the fouling of HP boiler gasifier tended to decrease.

2.4 Improvements for Future Designs

The large amount of knowledge acquired during the design, construction and operation of the Puertollano IGCC power plant enables ELCOGAS to define a series of improvements to be incorporated in the design of a new IGCC plant. Those with a high financial value, which enable substantial cost reductions to be made, are summarised in Table 6.

If a new and optimised design for the ELCOGAS IGCC plant was to be created, using the same combined cycle technology, but incorporating the above-mentioned improvements and other less significant improvements resulting from the operating experience of the first operating years, a saving in terms of the investment cost of between 20 and 25% would be made in relation to the cost of the current power plant.

However, the improvement that would have the greatest impact on the installed cost would be the use of more advanced gas turbines, which would enable IGCC units with larger capacities and higher efficiency levels to be developed, with a significant reduction in investment costs because of the benefits of size. The use of these gas turbines, together with the improvements noted, can lead to a completely competitive IGCC power plant with regard to costs. Therefore, it will be fundamental to ensure improved performance with regard to the gas turbines, which must be based on the experience demonstrated by the supplier in syngas applications.

3 Towards Zero-Emissions IGCC Power Plants: ELCOGAS R and D Lines

During the early operating years, ELCOGAS obtained important achievements demonstrating the potential of IGCC technology, including its advantages and disadvantages and identifying its main improvement and optimisation lines.

The future of the ELCOGAS IGCC is based on the opportunity that to have an operative IGCC plant, the R&D activities should be related to fuel flexibility (real tests with different coals, petcoke, biomass, wastes, etc.), multi-production (electricity, hydrogen, synthetic gasoline, biodiesel, etc.) and zero emissions (reduction of emissions, CO_2 capture, etc.).

So, since 2007, ELCOGAS has defined an R and D investment plan to develop IGCC technology to decrease the environmental impact of power production (towards zero emissions) as main target. ELCOGAS presents a yearly results report of that R and D plan to Spanish Government for evaluation. Main lines of this R and D plan are (apart from dissemination):

- CO_2 emission reduction in utilisation of fossil fuels
- H_2 production by gasification of fossil fuels
- *Diversification* of raw fuels and products
- Other *environmental* improvements
- IGCC processes *optimisation*

In the following sections, a brief summary of main activities is described.

3.1 Optimisation of IGCC Processes

This activity is oriented to improve availability and costs, being its main ongoing tasks:

- *Gasification island materials*: life extension.
- *Study of syngas corrosion processes*: *optimisation and tests of materials.*
- *Elimination of water leakages* in the membrane at reaction chamber.
- *Ceramic filter system*: Many tests with alternative filters have been done with poor results, so final assessment is to install a new filter system provided by Pall-Schumacher.
- Gas turbine reliability improvement and life extension.

Table 7 Olive waste (orujillo) and ELCOGAS common fuel average composition analysed by ELCOGAS laboratory

Parameter (*dry base)	Orujillo average composition (Laboratory of ELCOGAS)	ELCOGAS fuel design
Moisture (%)	13.13	9.40
Ash (%)	8.51*	20.68
Volatiles (%)	68.89*	15.92
C_{fixed} (%)	22.52*	54.00
LHV (kcal/kg)	3,695	5,386
C* (%)	49.40	59.21
H* (%)	5.96	2.80
N* (%)	1.44	1.36
S* (%)	0.14	3.21
Cl^- (mg/kg)	2,735	200

- Improvement of integration with ASU: installation of a start-up compressor.
- Analysis of O&M specific availability incidents.

3.2 Diversification of Raw Fuels and Products

The aim of this activity is to demonstrate IGCC fuel flexibility by undertaking tests with alternative fuels. Main tests recently undertaken are described in the following paragraphs.

3.2.1 Co-Gasification Tests of Olive Wastes

Within the Spanish project PIIBE (CENIT Programme) [3], whose aim was to impulse biofuels technologies in Spain, ELCOGAS coordinated the sub-project about biodiesel from gasification by real co-gasification up to 10% of biomass and syngas characterisation (F-T process in laboratory). The selected biomass to be tested in the IGCC was olive waste (orujillo).

Table 7 shows the average composition of the received orujillo and the common fuel (coal, petcoke and limestone) used in the power plant.

All tests carried out using orujillo as fuel, including the duration as well as the operating hours of them, are shown in Table 8. So, more than 3,600 tons of orujillo was co-gasified in more than 1,100 operating hours (Table 8).

Main conclusions from the co-gasification tests can be summarised as follows:

- The technical viability of co-gasification up to 10% has been demonstrated.
- Operation has been within design ranges.
- *Biomass handling*:
  - Orujillo should not be stored for a long time, because the biomass absorbs humidity.

Table 8 Battery of co-gasification tests in ELCOGAS (2007-2009)

Co-gasification test month/year	Orujillo dosage ratio in weight (%)	Tons of orujillo (t)	Test duration (h)
August 2007	1	7.4	9.5
September 2007	2	20.00	7
November 2007	2	81.86	28.5
August 2008	4	100.42	21
October 2008	4	299.36	79
November 2008	4	252.36	54
February 2009	2	518.86	291.3
March 2009	6	395.86	64.4
March 2009	2	512.38	289
June 2009	8	383.90	46
July 2009	2	136.86	40
September 2009	2	295.48	135
September 2009	10	656.68	62
	Total	3,661.42	1,126.7

– Orujillo goes easily stodgy if a large quantity is stored in the feed hopper before its consumption.

- *Grinding system:* During the 8 and 10% tests, the increase of the mills consumption and the pressure difference was detected.
- *Gasifier load*: No influence on the gasifier load arises from the orujillo co-gasification when 1, 2, 4 and 6% tests were carried out. More difficult to maintain it in 8–10% tests because of the mills load.
- *Clean gas*: Orujillo co-gasification has no impact on the clean gas quality; its characterisation is similar to those relating to ELCOGAS common operation.
- *Emissions*: The 8 and 10% addition of orujillo seems to have an influence in the SO_2 emissions (although orujillo has no content in sulphur), but always within limits. AAI-CR21 [2] establishes these limits in 200 mg/Nm^3 for the SO_2 at IGCC mode, with 6% of oxygen.

Other alternative fuels, such as shredder fibres and wastes from paper industry are currently under study to be tested as alternative fuels to the design fuel, coal and petcoke.

3.3 CO_2 Emission Reduction Using Fossil Fuels

Main tasks to develop this R and D line are related to:

- IGCC efficiency optimisation.
- Analysis of viability to improve efficiency based on critical assessment of Puertollano IGCC design. CARNOT project (EU programme) [4]: Pre-engineering studies for a new IGCC plant; a critical assessment of Puertollano plant design was done together with Siemens and Krupp and a detailed pre-engineering energy and

Table 9 CARNOT project results (pre-engineering studies for a new IGCC plant)

Parameter	ELCOGAS design		CARNOT design		CARNOT design with CO_2 capture	
Fuel type	Coal and petcoke (50 wt%)		Coal and petcoke (50 wt%)		Coal and petcoke (50 wt%)	
LHV (MJ/kg)	22.55		24.41		24.41	
	ISO	Site	ISO	Site	Site	
Gas turbine (MW)	200	182.3	291.0	281.1	279.4	
Steam turbine (MW)	135	135.4	193.6	186.8	171.6	
Total gross (MW)	335	317.7	484.6	467.9	451.0	
Total net (MW)	300	282.7	414.2	399.8	342.3	
	Gross	Net	Gross	Net	Gross	Net
$Efficiency_{LHV}$ (%)	47.12	42.2	52.1	44.5	45.2	34.3
$Heat\ rate_{LHV}$ (kJ/kWh)	7,647	8,538	6,908	8,086	7,959	10,486

mass balance of the future plant with and without CO_2 capture and H_2 production, based on ELCOGAS plant experience was done too.

- Auxiliary consumption optimisation. New revision.
- Development of tools to improve efficiency. Supervision online of main equipment efficiency, they are installed and in tests.
- Integration optimisation. Improvement of controls to adjust heat and mass balances in real operation.
- Net efficiency in the case called 'with CO_2 capture' includes the process of 100% of produced syngas to capture CO_2 as well as CO_2 compression for geological storage.

Table 9 summarises main results of this critical assessment.

- CO_2 capture for CCS with IGCC. ELCOGAS participates in the following funding projects:
- $ALCO_2$: Study to determine viability of CO_2 geological storage in the proximity of Puertollano IGCC plant [5]. A study based on documentation was done, determining the existence of two areas with high probability and 12 with some probability. *Status*: closed, developed during 2003–2004.
- PSE-CO_2: To explore H_2 production and CO_2 capture, from coal and petcoke, integrated with electricity production in an existing commercial IGCC, by installing a pilot plant of 14 MW_{th} that takes syngas from main plant [6]. *Status*: Ongoing, period 2005–2011.

Currently, ELCOGAS largest investment is focused on the PSE-CO_2 project. It is the first IGCC plant in the world to have an integrated pilot plant of industrial scale (14 MW_{th}) to obtain H_2 and CO_2 ready for geological storage, integrated with electricity production. Project was presented to the VI FP in 2004 for the call of studies for HYPOGEN plant (HYdrogen and POwer GENeration from fossil

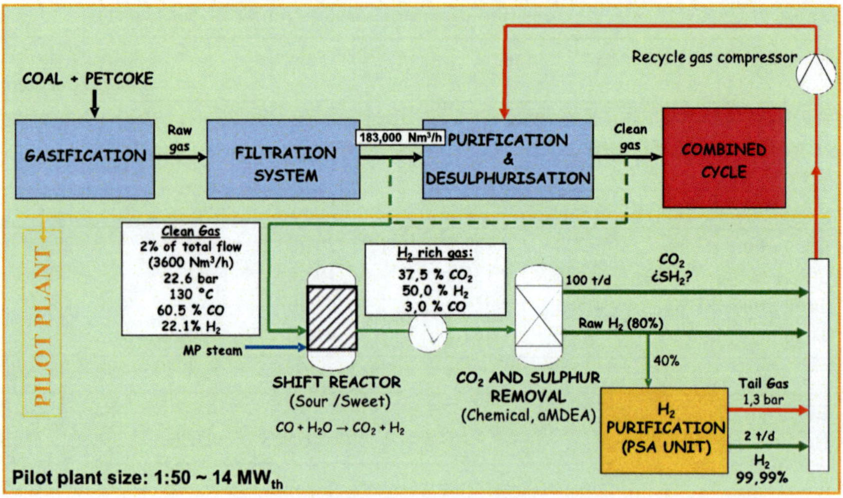

Fig. 6 Flow diagram of the CO_2 capture and H_2 production pilot plant

fuels with CCS) [7] and it was rejected with the main argument of 'premature', but it is being supported by Spanish and Castilla-La Mancha governments.

The PSE-CO_2 project is part of a Spanish national initiative called 'Advanced Technologies of CO_2 Conversion, Capture and Storage' that includes other related projects; being project #1 the building up of ELCOGAS itself:

- Project #2 explores CO_2 capture with oxyfuel technology, led by CIUDEN (*Fundación ciudad de la Energía*).
- Project #3 deals with study and regulation of geological storage in Spain, led by IGME (*Instituto geológico y minero de España*)
- Project #4 focuses on public awareness of CCS technologies, led by CIEMAT (*Centro de Investigaciones Energéticas, Medioambientales y Tecnológicas*).

Main targets of the PSE-CO_2 project are: (1) to demonstrate the feasibility of CO_2 capture and H_2 production in an IGCC that uses solid fossil fuels and wastes as main feedstock and (2) to obtain economic data enough to scale it to the full Puertollano IGCC capacity in syngas production. The participants are ELCOGAS (coordinator), University of Castilla-La Mancha (UCLM), CIEMAT (Spanish research centre) and INCAR (coal Spanish research centre), being the original budget 18.5 M€.

The process of the 14 MW_{th} pilot plant consists of a shifting unit to convert CO into CO_2, a CO_2 separation unit—based on absorption processes with amines—and a H_2 purification unit (PSA) being all of them are commercial processes. Auxiliary systems and full control are integrated in the existing IGCC, supplied by Zeus Control. The syngas—approximately 3,600 Nm^3/h, dry base—can be fed into the pilot plant desulphurised, i.e., downstream of IGCC desulphurisation unit (called sweet gas) or upstream of this unit (called sour gas). Main differences between both gases are H_2S and COS contents. So, sour and sweet catalysts will be tested to obtain technical and economic yields at full scale, obtaining CO_2 capture costs at

Fig. 7 CO_2 capture and H_2 co-production pilot plant: location

different purity grades. Figures 6 and 7 show the pilot plant flow diagram and its location in the Puertollano IGCC power plant.

The detailed description of the *pilot plant three steps* are shown in the following paragraphs:

3.3.1 First Step: Conversion With Water Steam

The aim of this phase is to modify the clean gas composition to increase its CO_2 and H_2 content. The syngas from the existing IGCC is desulphurised in a sulphur removal reactor (Zn oxide-based adsorber) and mixed with saturated medium pressure water steam. A static mixer is used to obtain a proper homogenisation of this mixture, and then being heated up to 310°C. Subsequently, this mixture is fed to a shift catalytic reactor supplied by Johnson–Matthey where conversion from CO and steam to CO_2 and H_2 is produced up to achieve 480°C. Downstream, there is an intermediate cooling down phase, reducing the mixture temperature up to 350°C, subsequently the mixture is fed to the second shift reactor, where it reacts up to 390°C achieving the appropriate conversion degree. Afterwards, the gas is first cooled down to 160°C in two pre-heaters, then up to 80°C in an aero-refrigerator and finally up to 45°C.

Sour and sweet catalysts will be tested to obtain technical and economic yields at full scale, obtaining CO_2 capture costs at different purity grades.

3.3.2 Second Step: CO_2 and H_2 Separation

The target of this step is to separate CO_2 and hydrogen, obtaining a hydrogen-enriched gas. So, an aMDEA (active Methyl Diethanol Amine) solution is used to

Table 10 CO_2 pilot plant main characteristics

Characteristics of clean gas (pilot plant inlet, sweet capture)			
Flow	3,600 Nm^3/h (dry base)	Pressure:	22.6 bars
CO (% volume)	60.5	Temperature:	137.6°C
H_2 (% volume)	22.1		
Characteristics of outlet flows (pilot plant outlet)			
CO_2 flow	100 tons/day	CO_2 capture percentage	>90%
Pure H_2 flow	2 tons/day	Purity of pure H_2	99.99%
Raw H_2 flow	5 tons/day	Purity of raw H_2	77.4%

capture CO_2. Downstream of this capture, the resulting gas is a hydrogen-enriched flow called raw hydrogen (77.4% of purity). This flow can be sent to the gas turbine or to be purified in the next pilot plant step. The aMDEA is regenerated—with CO_2 desorption—by means of temperature increase and pressure reduction. The expected CO_2 capture is higher than 90%. The regenerated aMDEA is conditioned (pressure increase and temperature decrease) to be re-used.

3.3.3 Third Step: Hydrogen Purification

Pure hydrogen (99.99% purity) can be obtained in this step from the raw hydrogen coming from the previous step. For this propose, 40% of raw hydrogen is purified by means of a PSA unit (pressure swing absorption) supplied by LINDE. Impurities such as CO_2, CO, N_2 and Ar are trapped in an adsorption multi-bed system while the hydrogen passes through it. This purification unit consists of four stages: (i) adsorption, (ii) decompression, (iii) regeneration and (iv) compression. It requires at least two adsorption beds, so while one bed is in the adsorption stage, the other one is carrying out the other three stages. However, four beds, consisting each of them in activated carbon, alumina and molecular sieve, are used to obtain the expected hydrogen purity.

The estimated capacity of this unit is 2 tons of hydrogen per day with 99.99% of purity, being the expected hydrogen recovery 95%. The tail gas generated in this step can be sent to the gas turbine or can be used as heat source in other processes.

Main features of the pilot plant are summarised in Table 10.

3.3.4 Current Status of the Pilot Plant:

- CO_2 fist captured: 13 September 2010
- Up to March 2011: sweet capture tests
- Up to June 2011: sour capture tests

Fig. 8 Laboratory-scale installation (UCLM, CIEMAT and INCAR)

3.3.5 Other Partners Activities in the PSE-CO_2 Project:

UCLM activities are related to water gas shift, mainly development of new catalyst (based on Co). Its lab installation (see Fig. 8) can operate up to 80 bar, which includes two reactors in series with cooling and can produce any kind of gas composition. INCAR tasks are related to research of CO_2/H_2 non-commercial separation processes based on solid adsorbents. Finally, CIEMAT participation is focused on using solid adsorbents, catalysts or membranes for gas treatment and commercial catalyst evaluation. Its installation (Fig. 8) can operate up to 750°C and 30 bar, it can manage flows between 5 and 20 Nm^3/h and it can work continuously for several days.

3.3.6 First Results of the Pilot Plant

First learning from the pilot plant is related to costs. The 14 MW_{th} pilot plant costs have been €13 million including its design, supply, construction, commissioning and start-up. Depending on the considered operational scenarios and taking into account the previous investment cost, the capture costs of the avoided CO_2 would be between 18 and 23 €/ton CO_2.

ELCOGAS's aim after the battery of planned tests (until March 2011) is to demonstrate that capture costs of the avoided CO_2 can be reduced to approximately 10 €/t CO_2.

3.3.7 Activities After PSE-CO_2 Project Completion

In addition to the objectives of the PSE project, once it is finished, ELCOGAS will have installed in its IGCC a *large pilot plant* that can be used as *industrial platform to other projects about syngas uses, CO_2capture and treatment processes and H_2purification and use*, which permits, among others, the following activities:

- Optimisation of catalyst for shifting reaction. Test with different catalysts.
- Development and demonstration of new processes for CO_2–H_2 separation.
- Demonstration of several processes for CO_2 treatment.
- Improvement of integration efficiency between the capture CO_2 and the IGCC power plant.

4 Conclusions

The world energy demand is expected to be doubled by 2050, being those based on fossil fuels what will experiment the biggest increase (especially in the Asiatic countries) because of the their lower costs. In view of this fact, diversification of fossil fuels use according to reserves—located all over the world and total available amount—and total life cycle are absolutely necessary to assure sustainability and supply guarantee. Because of this, coal is going to be one of the main energy sources and, therefore, an availability of clean coal technology is mandatory.

Based on real data and lessons learnt (e.g., lower emissions compared with other coal-based technologies and natural gas power plants and fuel flexibility) in more than 10 operating years of IGCC coal-based power plants, the IGCC technology is the best candidate to obtain clean energy from coal.

Fuel flexibility of IGCC technology has been demonstrated through the several undertaken co-gasification tests. Co-gasification with coal improves economics and efficiency of biomass fuels, encouraging renewable energy production.

IGCC power plants can be adapted as a multi-product plant to be adjusted to the market demand. So, from syngas, hydrogen can be produced—with lower production costs than from alternative sources—to be used in refinery as well as in fuel cell for power generation or transportation and several chemical can be generated such as ammonium, urea and methanol. IGCC refinery-based plants also show favourable commercial perspectives, avoiding the disposal of residues and supplying power, hydrogen and steam to the refinery.

In addition, because of the higher efficiency of IGCC plants, significant reductions in CO_2 emission can be achieved by replacing conventional coal-fired units by IGCC plants. Besides, CO_2 capture technology used in IGCC power plants (pre-combustion) is considered the best one because of estimated costs and also implies H_2 production. Both increase of efficiency and CO_2 reduction follow the Intergovernmental Panel on Climate Change (IPCC) recommendations to cut greenhouse gas emissions and to reduce the impact on global warming.

The existing IGCC power plants have the opportunity to contribute to the optimisation of IGCC technology. So improvements and processes that are being set out for designing new plants can be tested and developed even at commercial scale, leading to ultra-efficient and zero-emission energy plants based on gasification of low-cost fuels. It must be noted that ELCOGAS contribution with its CO_2 capture and hydrogen production pilot plant to obtain proven results at industrial scale about the real costs and feasibility of CCS.

References

1. European Union Directive 2001/80/EC: Limitation of emissions of certain pollutants into the air from large combustion plants (the LCP Directive)
2. AAI-CR21: ELCOGAS Environmental Permit. Autorización Ambiental Integrada
3. PIIBE Project: Proyecto de Investigación para el Impulso del Biodiésel en España (Spanish Research Project for Biodiesel Promotion)
4. CARNOT project: Pre-engineering studies for a new IGCC plant based on Puertollano ELCOGAS plant experience
5. ALCO$_2$ Project: Potential for CO_2 geological storage in the vicinity of ELCOGAS IGCC plant
6. PSE-CO$_2$ project: Proyecto Singular y Estratégico (Singular and Strategic Project for Advanced technologies of CO_2 conversion, capture and storage) - Subproyecto Tecnología de separación de CO_2 en pre-combustión
7. HYPOGEN: HYdrogen and POwer GENeration from fossil fuels with CCS Project

Industrial Data Collection

Aarón D. Bojarski, Carlos Rodrigo Alvarez Medina, Mar Pérez–Fortes and Pilar Coca

Abstract In this chapter, a general description of data-mining techniques is done in the context of IGCC operation. The different control philosophies applicable to IGCC operation are discussed together with different examples of data reconciliation based on process simulation. The problem of process monitorisation, as an example of data-mining application, is extensively discussed and an approach based on PCA is presented.

Notation

ASU	Air separation unit
CC	Combined cycle
CPV	Cumulative percent variance
DCS	Distributed control system
DR	Data reconciliation
ICA	Independent component analysis
MSPC	Multivariate statistical process control
NOC	Normal operating condition
OTC	Outlet temperature corrected
PC	Principal component

A. D. Bojarski (✉) · C. R. A. Medina · M. Pérez–Fortes
Universitat Politècnica de Catalunya, ETSEIB, Diagonal 647, 08028 Barcelona, Spain
e-mail: aaron.david.bojarski@upc.edu

C. R. A. Medina
e-mail: carlos.rodrigo.alvarez@upc.edu

M. Pérez–Fortes
e-mail: mar.perez-fortes@upc.edu

P. Coca
ELCOGAS, S.A.Ctra, Calzada de Calatrava a Puertollano, Km. 27 Apdo. Correos 200, 13500 Puertollano–(Ciudad Real), Spain
e-mail: pcoca@elcogas.es

L. Puigjaner (ed.), *Syngas from Waste*, Green Energy and Technology,
DOI: 10.1007/978-0-85729-540-8_13,

PCA	Principal components analysis
PIMS	Plant information system
PLS	Partial least squares
SPE	Squared prediction error

1 Introduction

A common characteristic of modern industries is the ability to generate and store a very large amount of data that are frequently used to describe the process behaviour. Generating and collecting data have become the most important topics in different and diverse areas of knowledge. Nevertheless, the ability of collecting such data has increased faster than the capabilities to analyze it. The biggest available data collection in any area is utterly useless by itself. The ability of users of those databases to extract useful information from them is a key part in modern plant practices to push plant-operating conditions to improve profit. The field of extracting useful information from databases is usually referred to as Knowledge Discovery in Databases (KDD). KDD can be understood as the non-trivial process of identifying novel data patterns that could prove to be useful and understandable [1] (p. 2). In the specialised literature, KDD is understood as a sequential process that involves the following stages:

- Analysis objective setting.
- *Data selection*: This selection is done according to the objectives followed by the analysis, and is mainly related to the aspects of how to access and store the data.
- *Data pre-processing*: This step is done to ensure the quality of data aiming at eliminating: noise, outliers and dealing with lost and unreliable data.
- *Data transformation* is done using different techniques; in most cases, it aims at reducing the dimensions considered (by projection), which might improve the identification of patterns.
- *Data mining*: This is the core step, where significant and well-defined patterns are looked for.
- Interpretation and validation.

This chapter discusses the issues related to the gathering data and its subsequent analysis in the context of KDD for the case of syngas production for energy use. Section 2 discusses the implementation of a control system that gathers and monitors the process variables, making special emphasis on an industrial case, the ELCOGAS power plant. Section 3 discusses the different methodologies associated to Data Reconciliation, where variables information are used to improve the knowledge of model parameters, while the ending Sect. 4 provides with a description of a data-mining procedure that could be implemented to data recorded on the previously discussed control systems.

2 Process Variables Recording and Control

In general, the control system of a plant is used to keep the plant working properly within the established limits (efficiency, net output, etc.), and as a consequence, this control system is usually the source of data for any analysis. The different measurement systems provide with signals that are stored by centralised software that acts as the data reservoir. Data selection and pre-processing can be done using different software, such as MS Excel, Matlab and other software suites.

The Puertollano IGCC plant is equipped with a distributed control system (DCS) using *Siemens Teleperm* XP for the whole plant except for both turbines that have a specific automation called *Simadyn*. This system has a modular structure and consists of the following subsystems:

- An automatic system for automatic function implementation at the lowest control level
- A communications network
- An operation control and monitoring system for operation processes and information interchange
- An engineering system for planning, configuration and start up

In a short time, both *Teleperm* and *Simadyn* are going to be replaced by an updated automation system T-3000 that is going to integrate both distributed control systems.

2.1 Variables Recording

Main features of the process—apart from pressure, temperature and flows—that are monitorised during operation are:

- In the gasifier, quality of gasification process, which is evaluated taking into account syngas composition, such as CO_2, H_2, CO, energy exchange in the immersion shaft and energy gasification reaction.
- In the combined cycle (CC), clean gas LHV, OTC (outlet temperature corrected) of the gas turbine exhaust gases and combustion stability. The OTC is a measure of exhaust gas temperature that reflects the temperature in the combustion chamber and is represented as the difference between the average temperature of the flue gas (measured at line) and outside air temperature.
- In the air separation unit (ASU), purity of oxygen, pure and waste nitrogen.

The former variables are fed to a plant thermo-economic model. Advanced thermo-economic diagnosis of the plant equipment performance is intended to improve the plant operation and maintenance procedures. Additionally, steady-state simulation tools assess the potential for design improvements and the impact of plant modifications.

The diagnosis system focuses on the effective management of both fixed inputs (e.g., plant layout, instrumentation, etc.) and operation and management (O&M) data (e.g., process data, operating modes, maintenance schedule, etc.) to screen the equipment performances and operating costs at steady-state conditions. The Plant Information Management System (PIMS), with its own interface to the DCS (Distributed Control System), laboratory and other data sources provide raw data that is sorted, validated and stored. The cost evaluation module calculates the potential savings that can be achieved by comparing the performance tests (based on actual plant measurements) and the state of reference model (best possible operating case), under given operating constraints.

The system conceptual design enables online applications and easy module update and/or modification. The main applications are:

- Plant raw process data validation and reconciliation ahead of energy/exergy calculations. Dynamic monitoring of plant instrumentation.
- Comprehensive map of resources thorough the plant at the process stream level.
- *Costs evaluation*: actual versus best possible operation. Allocation and causes of performance degradation.
- Assess the extra cost because of performance degradation versus corrective O&M actions impact.
- *Online monitoring*: plant heat rate, cost targets compliance and instrumentation status.

2.2 Control Levels

The current system's automation and control levels (Fig. 1) of the *Teleperm* XP are as follows:

- *Field level*: The lowest level where sensors are located and data are withdrawn. Its function is to receive signals from sensors and to transmit them to the higher levels or the actuators.
- *Automation level*: The automation level has two sub-levels:
  - *Individual control level*: Basic control of operations with analogue and binary signals. Actuators control in open loop and individual control in closed loop.
  - *Group control level*: Automatic functions, such as closed loop regulations, open loop control and signal protection management.
- *Process level*: Storage of process data and transfer of the dynamic information to the man–machine interface.

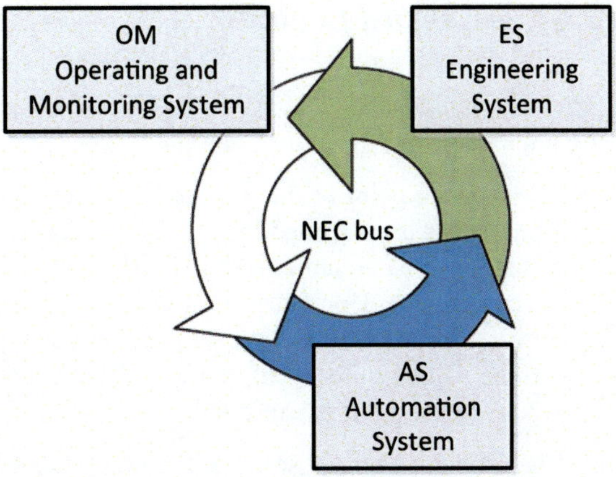

Fig. 1 Subsystems of the Teleperm XP

- *Operation and control level*: Association of the man–machine interface with the interaction supervision and configuration systems.

2.3 Plant Control Philosophy

The ELCOGAS power control operates on gas turbine leading mode, so both gasifier and ASU follow the gas turbine load. Therefore, the clean gas production by the gasification island depends on the gas turbine demand, being the objective to consume as much gas as produced to match the power demand of the grid. The ASU produces the necessary oxygen to obtain the required clean gas. The steam turbine works in slide control, taking the vapour produced by both the CC and gasifier boilers. However, a significant time delay (a couple of minutes) is needed to accordingly adjust the feed streams to the gasifier and particularly the coal flow rate. It is a major task of the proposed control concept to harmonize these diverging facts, taking into account the storage capacity of the gas path.

One of the main objectives of the control philosophy is to ensure stable operating conditions for the gasifier. For this target, the gasifier pressure is used as a principal controlled variable and is kept at a fixed value. The differential pressure along the gas path is dependent on gas production, and the gas inventory constitutes a reservoir that can be loaded and unloaded. The gasifier and the sulphur removal units contain a certain gas inventory determined by the geometrical volume, temperature and pressure. In case of a short-term gap between clean gas production and demand of the gas turbine, part of this inventory can be used.

2.4 Coordinated Control

The *coordinated control* between the CC and the gasifier is based on the following principles:

- The load set point of the gas turbine is sent to the load controller.
- The gasifier is controlled by a pressure control and uses the set point of the gas turbine load as an anticipating signal.
- To prevent any depressurization in case of any problems with gasifier load, the gasifier defines a maximum set point of clean gas flow to be taken by the gas turbine, which differs with the actual flow controls of the CC clean gas valve when the set point is exceeded.

The set point of the power plant load is given by the control room operator and is sent by the controller to each island. This set point is elaborated with a maximum gradient of 3% load/min, according to the ASU limitation and is used:

- By the CC to elaborate the load set point of the gas turbine and to determine the gas flow consumption actuating on the position of control elements (clean gas valve and Inlet Guide Vanes). This set point is adjusted, taking into account the steam turbine contribution.
- By the gasifier to calculate, with a proportional value, the clean gas flow and the oxygen flow. The set point of oxygen and nitrogen flows is elaborated by the gasifier for the ASU.
- On the other hand, the steam turbine is considered in case of uncontrolled steam turbine load variation (i.e., steam turbine trip or during load ramping), the control is designed to by-pass the steam turbine load influence in the elaboration of the gas turbine set point.

So, the power plant control operates on gas turbine leading mode, the gasifier—working in pressure control—and the ASU follows the gas turbine load, even if because of operational problems the actual load does not match the external set point.

In general, the former primary control loops try to use most of the information available in the process historical, however, they do not perform an efficient use of them and other techniques such as the ones described in next sections that could allow for extracting other information and knowledge from the raw control data.

3 Data Reconciliation

The problem of *data reconciliation* (DR) was first stated by Kuehn and Davidson [2] for a total material balance where all variables were measured. Several works in both data reconciliation and gross error detection areas have been published since then. Important improvements to the basic DR technique were presented in those years including treatment of unmeasured variables using a graph-theoretic approach and the introduction of the concepts of observability and redundancy.

Nonlinear data reconciliation was first addressed by Knepper and Gorman [3] using an iterative technique proposed originally with parameter estimation that employs a nonlinear regression. In general, because the nonlinear data reconciliation problems are basically nonlinear optimization problems, some well-known constrained nonlinear optimization methods have been used to solve them, for instance, Tjoa and Biegler [4] used sequential quadratic programming (SQP) technique to solve a combined data reconciliation and gross error detection problem. Several other approaches can be seen in Romagnoli and Sanchez [5] and Alvarez-Medina [6].

3.1 Data Reconciliation Basics

Generally speaking, data reconciliation can be defined as the process of adjusting (reconciling) process measurements, such as flow rates, temperatures, compositions, etc., to obtain a new set of estimates that are consistent with mass and energy balances, and sometimes, with some thermodynamic equilibrium equations as well. In the context of KDD, DR is used for data transformation, given that DR techniques receive pre-processed data and produce 'reconciled data', which could improve the detection of trends in data.

From the data reconciliation's point of view, in the absence of gross errors, a process measurement can be 'modelled' as follows:

$$\mathbf{y} = \mathbf{x} + \boldsymbol{\varepsilon}; \qquad \mathbf{y}, \mathbf{x}, \boldsymbol{\varepsilon} \in R^n \tag{1}$$

where $\mathbf{y}$ is a $(n \times 1)$ measurement vector, $\mathbf{x}$ is a $(n \times 1)$ vector of true variables values and $\boldsymbol{\varepsilon}$ is a vector of random measurement errors.

The statistical basis in DR relies basically in the assumptions made regarding vector $\boldsymbol{\varepsilon}$. It is usually assumed that:

1. Vector $\boldsymbol{\varepsilon}$ has a null expected value, $E(\varepsilon) = 0$, and it is normally distributed.
2. Random errors for successive measurements are independent, i.e., $E\left(\boldsymbol{\varepsilon}_i \boldsymbol{\varepsilon}_j^{\mathbf{T}}\right) = 0$, for $i \neq j$.
3. $\mathrm{Cov}(\boldsymbol{\varepsilon}) = \Psi = E\left(\boldsymbol{\varepsilon}_i \boldsymbol{\varepsilon}_i^{\mathbf{T}}\right)$ is known (or estimable) and positive definite.

The data reconciliation problem can be generally stated as the following constrained weighted least-squares estimation problem defined in Eq. 2.

$$\begin{aligned} \min_{x,u} \quad & \varepsilon^{\mathbf{T}} \boldsymbol{\Psi}^{-1} \varepsilon \\ \text{s.t.} \quad & \varphi(\mathbf{x}, \mathbf{u}) = 0 \\ & \mathbf{x}^{\mathbf{L}} \leq \mathbf{x} \leq \mathbf{x}^{\mathbf{U}} \\ & \mathbf{u}^{\mathbf{L}} \leq \mathbf{u} \leq \mathbf{u}^{\mathbf{U}} \end{aligned} \tag{2}$$

In general, the constraints sets are presented as a set of nonlinear algebraic equations, such as:

$$\varphi(\mathbf{x},\mathbf{u}) = 0 \qquad \mathbf{x} \in R^{\mathrm{n}} \quad \mathbf{u} \in R^{\mathrm{m}} \quad \varphi \in R^{\mathrm{i}}, \tag{3}$$

where $\mathbf{u}$ indicates the $(m \times 1)$ vector of unmeasured variables and $\varphi(\mathbf{x},\mathbf{u})$ is the constraints set. Because random errors have been assumed normally distributed, by solving the above problem, the maximum likelihood estimates of process variables are obtained. Different methodologies have been proposed for solving these problems depending on whether the constraints constitute a linear, bilinear or a nonlinear set of equations.

One of the most common calculations made in chemical engineering is total mass balances, because they enable the engineer to get a general idea of the process state. If the purpose is to reconcile total mass flows, or molar flows if reactions are not considered, a linear data reconciliation problem results. Two different situations can be found depending on if dealing with only measured or both measured and unmeasured variables.

In the case of having all the flowrates measured, then Eq. 2 can be rewritten as in Eq. 4, given that Eq. 3 is simply a set of linear relationships among the different measured flows.

$$\begin{array}{ll} \min\limits_{x} & (\mathbf{y}-\mathbf{x})^{\mathbf{T}}\mathbf{\Psi}^{-\mathbf{1}}(\mathbf{y}-\mathbf{x}) \\ \text{s.t.} & \mathbf{A}\mathbf{x} = 0 \end{array}, \tag{4}$$

where $\mathbf{A}$ is a $(i \times n)$ incidence matrix, which shows the relationship among flowrates and plant equipment. Matrix $\mathbf{A}$ elements (a_{in}) are 1 or -1 if the nth flow enters or leaves equipment i, whereas zero if the flow is not related. Because of the simplicity of the restrictions, the former problem can be solved applying the Lagrange multipliers method, the improved estimate of the process variable value $(\hat{\mathbf{x}})$ can be obtained as:

$$\hat{\mathbf{x}} = \mathbf{y} - \Psi\mathbf{A}^{\mathbf{T}}(\mathbf{A}\Psi\mathbf{A}^{\mathbf{T}})^{-\mathbf{1}}\mathbf{A}\mathbf{y} \tag{5}$$

For the case of unmeasured flow rates, the reader is referred to Crowe et al. [7] and Sánchez and Romagnoli [8], where the reader will find different implementations of linear and bi-linear DR problems.

In the general case, process operation is commonly modelled by nonlinear systems of algebraic equations. This nonlinearity comes from components or energy balances and it may also include thermodynamic relationships to explain some particular system's behaviour. Sometimes, process control systems require an accurate process model, and the most common linear or bilinear data reconciliation strategies are not applicable anymore. To these cases, the most general mathematical formulation of an optimization problem is as in Eq. 2. The constraints set $(\varphi(\mathbf{x},\mathbf{u}))$ is in general nonlinear and customarily difficult to solve.

The necessary conditions for an optimal solution come from the defined Lagrange function and are known as Karush–Kuhn–Tucker (KKT) conditions. The sufficient condition for obtaining a global minimum of the nonlinear problem is that both the objective function and the constraint set must be convex. Otherwise, there is no guarantee that the reached local optimum will be the global optimum.

As discussed in chapter "Modelling Superstructure for Conceptual Design of Syngas Generation and Treatment", most commonly used optimization software, such as GAMS or MatLab packages, includes several well-known optimization algorithms as SQP. Other techniques include solvers such as NPSOL which is especially effective for nonlinear problems whose functions and gradients are expensive to evaluate.

Because of the complexity included in the set of constraints of Eq. 3, process simulation environments are used to solve them with ease, as discussed in chapters "Modelling Syngas Generation, Main Purification Operations and Modelling Superstructure for Conceptual Design of Syngas Generation and Treatment". Consequently, the application of DR techniques requires frameworks that use the process simulation data together with other optimisation algorithms. In such frameworks, the simulation environment usually acts as a server of the optimisation algorithm by providing the value of Eq. 3 for the different values of the variables being optimised. The following examples try to clarify on this sense.

3.2 Examples of Data Reconciliation in obtaining Synthesis Gas

3.2.1 Case A: DR Applied to a Claus Desulphurisation Process

Recovery of sulphur from industrial waste gases is an important problem from the environmental point of view, and as it was discussed in chapter "Main Purification Operations", it is of paramount importance in the case of IGCC power plant operation. The rise of sulphur volumes in waste gases together with tightening emission regulations leads to the increase of sulphur-recovering needs [9].

In gasification plants, the main source of sulphur recovered is hydrogen sulphide produced in the gasification step as a sub-product. The most widely used method to treat H_2S is based on oxidation of hydrogen sulphide into sulphur by adding oxygen, further details are discussed in chapter "Main Purification Operations".

The global sulphur recovering in Claus plants containing recirculation is around 99.8% of the sulphur content in the input gas. All the equipments and process units involved in the model are put together as a whole operational unit. Six streams are considered to link the Claus plant to the rest of the whole process equipments and facilities, see Fig. 2.

The proposed strategy described in Eq. 2 has been applied to the above described Claus plant to reduce the noise level in the inputs and output streams. The set of constraints are modelled using the process simulator (in this case, AspenHysys). These constraints include: mass and energy balances together with the corresponding thermodynamic models for the calculation of phase distribution. Clearly, the simulation environment eases the implementation of such framework

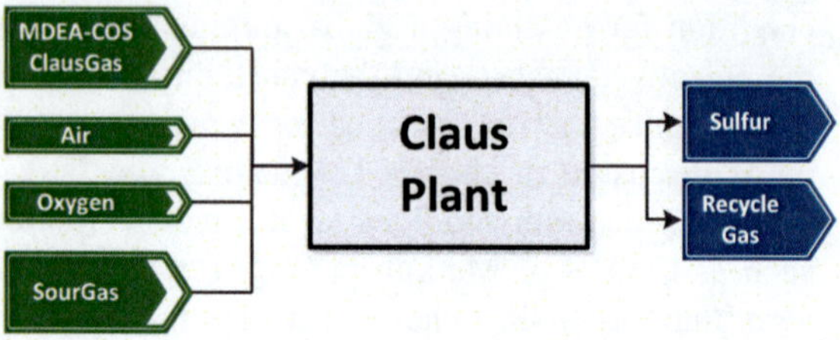

Fig. 2 Considered Claus plant scheme, for mass streams reconciliation

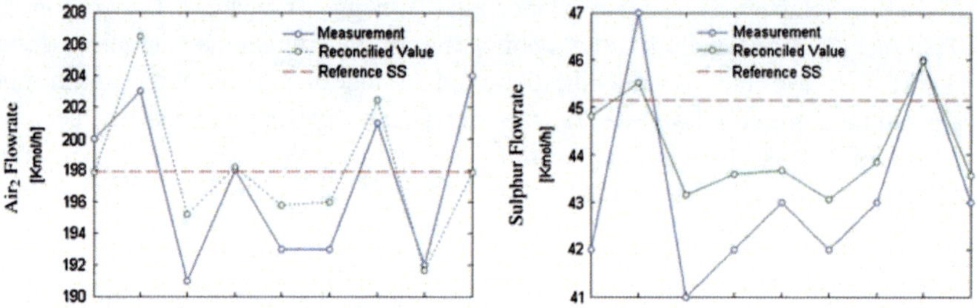

Fig. 3 Reconciled total molar flows for the Claus Plant

by providing the simulation models that account for the former concerns. For this case, different simulated noisy measurements around a stationary operation point are used. The optimiser is a SQP-based algorithm implemented in the Matlab optimisation toolbox.

The results obtained are shown in Fig. 3. Molar flows for two streams are shown along the considered stationary point and the noisy measurements. Reconciled flowrates show lower discrepancies compared with the stationary value.

3.2.2 Case B: DR Used for Parameter Estimation in a Coal Gasification Reactor

The considered gasification process is a pressurised entrained flow gasification reactor, as described in chapter "Modelling Syngas Generation", Sect. 2, but considering it to be a kinetic reactor as described in Perez-Fortes et al. [10]. The feed to the gasifier is composed of dry raw carbon material that is fed means of pneumatic transportation with nitrogen at high pressure. Because nonconventional compounds are used to model this reactor, a set of reaction extensions has been created to cope with such system and to incorporate the stoichiometry related to

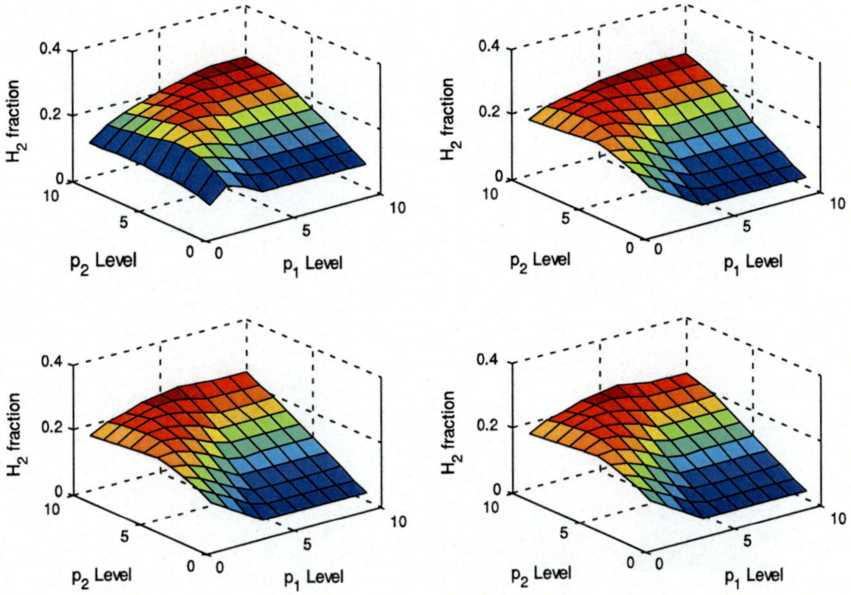

Fig. 4 H_2 fraction parametric dependence at different levels of parameters p3-p9

the involved reactions. These extensions enable to define sets of user-defined parameters that pile up several different physical parameters to overcome the lack of some of them. A set of nine parameters p_1–p_9 are defined in such a way that each one includes parameters such as pre-exponential factors for each considered reaction, average char particle size, reactor porosity (in terms of fraction of particles and gas), defined ratios in the feed composition and others. Four different main reactions are considered:

- CHAR + O_2 → Reaction products.
- CHAR + H_2O → Reaction products.
- CHAR + H_2 → Reaction products.
- CHAR + CO_2 → Reaction products.

As a first step, a sensibility analysis of the output molar fractions to the different parameters is carried out to estimate which are the most influential set of parameters. Figure 4 shows the effect of the parameters p_1 and p_2 at different levels and for different combinations of the remaining ones in the fraction H_2 in the outlet stream. Only two different behaviours can be noticed from the curves in Fig. 4. The first one is in the upper-left corner that corresponds to low levels of parameter p_3 that is related to the pre-exponential factor of the CHAR + H_2 reaction. The remaining three surfaces show a very similar behaviour among them, and only slight differences can be noticed at higher p_2 values.

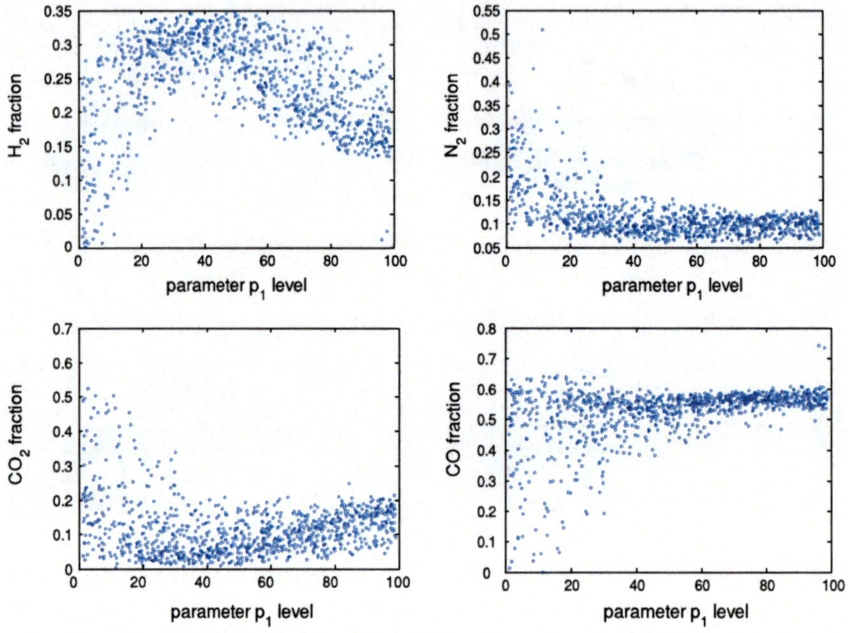

Fig. 5 H2, CO2, CO and N2 fractions against p1 for different values of the remaining

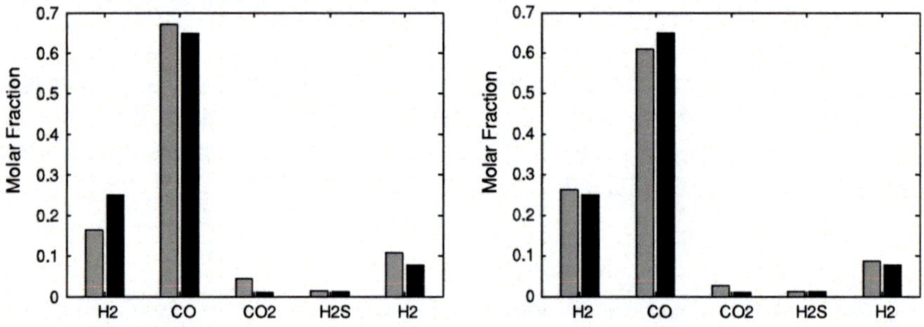

Fig. 6 H2, CO, CO2 and N2 reactor effluent molar fractions. Left side before parameters adjust, right side: after parameters adjust

Other sensitivity analysis studied the effect on H_2, CO_2, CO and N_2 fractions against p_1 at different levels in the remaining parameters. It can be seen that the system behaviour regarding parameter p_1 is highly nonlinear for some of the variables but no so much for others, see Fig. 5.

Two different raw material compositions and their corresponding output composition are used as variables to be reconciled by solving the DR and parameter estimation problem. Figure 6 shows the results obtained showing lower discrepancies in the model results when the parameter estimation is performed, see specially the H_2 fraction.

The first example showed how different input variables relationship should be changed to produce a lower noise in the model predictions, whereas the second example showed how different model parameters can be tuned to improve the model-predicting capabilities. In both examples, DR has been used to improve the quality of data, so patterns and trends can be identified with ease.

4 Data Mining

The former section used historical data to improve the reliability of model predictions by estimating model parameters or by modifying the possible input variables values. Many other different points of view, which do not rely on a model to extract information, are available; and they are broadly known as data-mining techniques.

Data mining has recently emerged as a new field whose main objectives are developing and refining concepts and tools that may sometimes have been developed in other disciplines to provide the user of data bases with powerful tools to transform data into useful knowledge [1, 11].

Knowledge is one of the most valuable assets in almost any human activity. In business, for example, having a good command of all the related issues (purchasing, marketing, design, production, maintenance and distribution) enables a company to differentiate itself from competitors and to compete efficiently and effectively to the best of its ability.

There are many possible definitions of data mining but a very simple and accurate one is given by Han and Kamber [12]: 'Simply stated, data mining refers to extracting or "mining" knowledge from large amounts of data'. In this definition, the existing difference between data and knowledge is clearly highlighted.

4.1 Statistical Process Control

During the recent years, many research works have been devoted to the development and improvement of a set of statistical tools that enable the process operator to determine whether the process is working as expected. Online process performance monitoring and product quality prediction in real time ensure safe and profitable operation because they provide the opportunity to take corrective actions before the effects of excursions from normal operation ruin compromise plant operation.

Common operation policies include monitoring just a few process variables that are thought as critical variables and whose operation window is usually well defined between an upper and a lower bound. Although this is a common industrial practice,

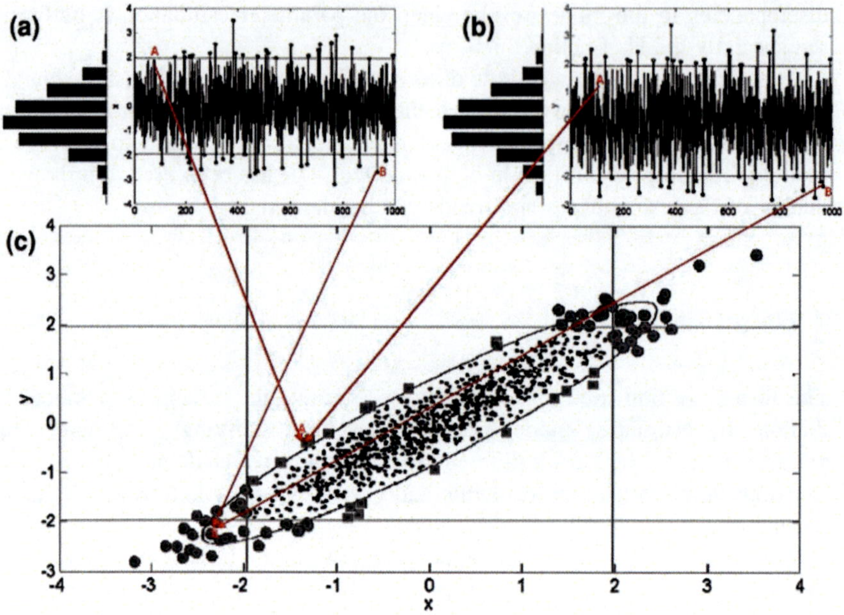

Fig. 7 Differences between univariate and multivariate control charts

it was extensively demonstrated that it is not an efficient and consistent approach when it comes to determining whether the process is or not in control.

Process state is commonly determined by multivariate observation vectors, and therefore, a consistent approach, which considers the multivariate nature of the system, has to be applied to classify the process state. The importance of considering a vector of variables as a whole instead of a set of independent measurements is easy to understand considering the simple case showed in Fig. 7.

Let us consider the case of two normal distributed variables that are simultaneously measured in a process (see Fig. 7a, b). Two particular observation vectors are highlighted in Fig. 7a and b. First, let us analyze point A. This observation would have been considered as '*in control*' observations if the two charts (a) and (b) were considered to monitor the process because it falls in the confidence region for a given significance level (α). On the other hand, when the bivariate nature of the process is taken into account, see Fig. 7c, it fires an '*out of control alarm*'. The opposite situation happens to point B. In this case, this point fires an '*out of control alarm*' for the separate charts but behaves as '*in control point*' in the bivariate case. The case shown in Fig. 7 is a classic example used to reveal the importance of considering multivariate processes as such, instead of simply neglecting the relationships among variables [1, 13].

The knowledge of the former kind of behaviour has attracted the interest of many research groups that have conducted extensive works intended to find effective methodologies to deal with this problem. However, considering the multivariate nature of an industrial process is not a trivial task. Working with many variables could make the monitoring-related tasks much more complicated.

During recent years, many successful applications of multivariate statistical process control (MSPC) have been presented for monitoring and fault diagnosis purposes in the case of continuous and batch processes. A common industrial practice is to monitor the batch progress by exploiting the information contained in a historical database of successful batches using projection techniques, such as principal components analysis (PCA), partial least squares (PLS) and independent component analysis (ICA).

Most of the former techniques allow for estimating the normal operating condition (NOC). The knowledge of such condition enables both to develop an empirical model of the process in terms of latent variables and to calculate statistical confidence limits that will be used to test the progress of new batches; both tasks are preformed offline. Once the model is set for online monitoring, new process observations are analyzed to decide whether the process is under control. Each new observation is projected into the reduced space defined by the latent variables, and the corresponding statistics is calculated and compared against their critical values. Among the former-mentioned techniques, we focus on PCA, given its broad use and its ease of implementation.

4.2 PCA-Based Techniques Applied for 'Data Mining'

The PCA-based approach for monitoring and fault diagnosis purposes is one of the most popular methods when predictions of not measured variables are not needed. The basic concepts and schemes to implement a PCA-based monitoring strategy are explained along this section. It is the reader's duty to find and deal with the particularities related to the adaptation of this strategy to some alternative schemes, such as moving windows, exponential weighting or dynamic-PCA.

Originally, the main application of PCA was related to exploratory data analysis because of its ability to compress the meaningful information in a few dimensions usually enabling a graphic analysis of the data behaviour. It was this compression and reduction capability that reinforced the interest of many researchers in using PCA-based approaches to overcome some of the main problems observed when implementing monitoring schemes in the original (measurement) space. In this sense, PCA-based approaches are specially suited for recognising if a given pattern is present in a set of data.

4.2.1 Principal Components Analysis: Basics

PCA is just one of several latent variables projection techniques. From the mathematical point of view, obtaining the principal directions is merely a space projection problem in which the projection space fulfils some special requirements.

The principal component space is an 'ordered space' in which each direction is hierarchically ordered in terms of its variance. In addition, principal directions are requested to be orthogonal. The principal components directions come as a solution of optimization problem in which it is desired to find a set of orthogonal vectors that maximize the data variance along them. The solution of this problem can be analytically obtained [13] by performing a singular value decomposition of the system covariance matrix ($\mathbf{\Sigma}$), as in Eq. 6.

$$\mathbf{\Sigma} = \bar{\mathbf{P}}\bar{\Lambda}\bar{\mathbf{P}}^T, \tag{6}$$

where $\bar{\Lambda}$ is the eigenvalues diagonal matrix and $\bar{\mathbf{P}}$ is the corresponding eigenvectors matrix. The columns of $\bar{\mathbf{P}}$ define the principal component directions. They are arranged in descending order in terms of variance that is given by the corresponding eigenvalue [13]. There are many efficient algorithms that can be applied to calculate $\bar{\Lambda}$, also known as loading vectors matrix. One of the most popular is the NIPALS algorithm presented by Wold [14]. Once the loading vector matrix has been calculated, the data matrix can be rewritten as in Eq. 7.

$$\mathbf{X} = \bar{\mathbf{T}}\bar{\mathbf{P}}^T, \tag{7}$$

where $\bar{\mathbf{T}}$ is the scores matrix that contains the coordinates of matrix of $\mathbf{X}$ in the latent space.

Usually, the actual covariance matrix ($\mathbf{\Sigma}$) is unknown and the sample covariance matrix ($\mathbf{S}$) is used instead. An important result that the reader should always consider is that, as it was mentioned before, $\mathbf{S}$ is to $\mathbf{X}$ as $\bar{\Lambda}$ is to $\bar{\mathbf{T}}$.

$$\mathbf{S} = \frac{1}{(I-1)}\mathbf{X}^T\mathbf{X} = \bar{\mathbf{P}}\bar{\Lambda}\bar{\mathbf{P}}^T \tag{8}$$

$$\bar{\Lambda} = \frac{1}{(I-1)}\bar{\mathbf{T}}^T\bar{\mathbf{T}} = \mathrm{diag}\left([\lambda_1,\ \lambda_2,\ \cdots \lambda_J]^T\right), \quad \forall j = 1, 2, \ldots, J \tag{9}$$

It is clear from Eq. 9 that principal components are essentially uncorrelated because $\bar{\Lambda}$ is a diagonal matrix.

As it was mentioned before, the most useful property of PCA is its capacity to condense most of data variability in just a few first directions. As a consequence, it is possible to represent the data using just a few directions without losing much of the meaningful information in data. PCA-based monitoring strategies exploit this property to reduce the original space from R^J to R^R (with $R < J$) splitting the original space in two well-differenced domains, the first is referred as the retained space (R^{R}) and the second as the excluded space (R^{J-R}). According to this, the loading and score matrices can be reordered as in Eqs. 10 and 11.

$$\bar{\mathbf{P}} = \begin{bmatrix}\mathbf{P} & \tilde{\mathbf{P}}\end{bmatrix} \tag{10}$$

$$\bar{\mathbf{T}} = \begin{bmatrix}\mathbf{T} & \tilde{\mathbf{T}}\end{bmatrix}, \tag{11}$$

where $\bar{\mathbf{P}} \in R^{J \times J}$ is the loading matrix used when considering the whole latent space, $\mathbf{P} \in R^{J \times R}$ is the loading matrix for the retained space and $\tilde{\mathbf{P}} \in R^{J \times (J-R)}$ contains the excluded principal directions. Similarly, $\bar{\mathbf{T}} \in R^{I \times J}$ defines the coordinates of $\mathbf{X}$ in the full principal components space. Finally, $\mathbf{T} \in R^{I \times R}$ and $\tilde{\mathbf{T}} \in R^{I \times (J-R)}$ are the coordinates of $\mathbf{X}$ in the retained and excluded space, respectively. Reordering Eq. 7, we obtain:

$$\mathbf{X} = \bar{\mathbf{T}}\bar{\mathbf{P}}^T = \mathbf{T}\mathbf{P}^T + \tilde{\mathbf{T}}\tilde{\mathbf{P}}^T. \tag{12}$$

In literature, it is usually found that $\mathbf{T} = \mathbf{XP}$, and therefore, $\hat{\mathbf{X}}$ and $\mathbf{E}$ result:

$$\hat{\mathbf{X}} = \mathbf{XPP}^T \tag{13}$$

$$\mathbf{E} = \mathbf{X}\left(\mathbf{I} - \mathbf{PP}^T\right), \tag{14}$$

where $\mathbf{E}$, is the residuals matrix that is generated by the exclusion of the principal directions contained in $\tilde{\mathbf{P}}$.

One central point when using projection-based techniques is the goodness of the representation in the reduced space. This point is closely related to the minimum number of directions that have to be used to achieve a reasonable data description in the PC's space. One characteristic parameter in multivariate statistics is the total variance of the data ($\sigma_T = tr(\boldsymbol{\Sigma})$). Equation 15 shows how to estimate σ_T in both the original and the PC's spaces:

$$\sigma_T = tr(\boldsymbol{\Sigma}) = \sum_{j=1}^{J} \sigma_{jj}. \tag{15}$$

It is easy to note from Eq. 15 that each principal direction explains just a fraction of σ_T, the percentage explained by the rth component is:

$$\sigma^r_{\%,T} = \frac{\lambda_r}{J} \times 100, \quad \forall\, r = 1, 2, .., J. \tag{16}$$

The percentage of the total variance explained in the R-dimensional space (PC's 1 to R) is:

$$\sigma^{r=1:R}_{\%,T} = \sum_{r=1}^{R} \lambda_r \times \frac{100}{J}. \tag{17}$$

It is also possible to estimate the fraction of the variance of a particular variable in the R-dimensional by using Eq. 18.

$$\sigma^{r=1:R}_{j\%} = \sum_{r=1}^{R} \frac{\lambda_r p^2_{j,r}}{s_{j,j}}. \tag{18}$$

There exist many ways to determine the number of principal components to be retained (R). Many of these are closely related to the afore-mentioned goodness-related indexes. The most simple may result the cumulative percent variance (CPV) criteria in which one should retain as many PC's as needed to match a previously defined percentage of reconstruction for σ_T. Further information regarding this and other criteria can be found elsewhere [13, 15].

4.3 PCA-Based Monitoring Scheme

The set up of a statistical process monitoring scheme is usually presented as a two-stage procedure. Stage I, also referred as '*off line stage*', includes the main calculations to build the PCA model and to estimate the threshold values for the hypothesis testing. Although it is not frequently mentioned, a central step is the correct pretreatment of the data set that will be used to calculate the model.

Pretreatment activities may include—but are not limited to—measurement scaling, outlier detection and data clustering and classification. Data pretreatment is not usually covered in research papers dealing with process monitoring; nevertheless, the success of the whole monitoring strategy strongly depends on it. Pretreatment-related activities by itself do constitute a whole subject, and are out of the scope of this chapter. Descriptions of the most commonly used tools for data pretreatment, its purposes and limitations can be found elsewhere [13]. The main objective of data pretreatment is to obtain a data set that can be considered as a good sample of the process operation in normal operation condition (NOC) because they will be used to build the PCA model and to determine the normal region.

Let us consider $\mathbf{X}$ to be the process data matrix after an appropriate pretreatment. As it is usually the case, process observations are made of various measured variables with different measurement units and variability ranges. As a consequence, a very common step during stage I is variable scaling or standardization. This procedure results useful for both putting measurements a common unit-less scale and avoiding undesired side effects that can arise from working with variables in very different scales and ranges. Once $\mathbf{X}$ has been properly scaled, the PCA model is calculated as it is explained in previous chapter "Modelling Syngas Generation" in Sect. 2.1, and the normal operation regions have to be determined.

The former PCA model serves as basis for comparison of new metrics, consequently, the adequacy of these approaches heavily depends on how well this PCA model represents the plant behaviour.

It is a very common practice in PCA-based monitoring approaches to use two complementary statistics to follow the process evolution. Typical statistic metrics are the Hotelling's T^2 [16] and the squared prediction error (SPE).

The T^2 statistic metric is widely employed in multivariate systems analysis, and it was proposed as a generalisation of Student's t distribution to the multivariate case. It is based on the Mahalanobis distance, which confers it the capacity for evaluating changes in the system's correlation structure and to decide if one observation can be considered part of a given data population. Consequently, it is useful for devising if observed process deviations can or cannot be considered within the normal process variability. The T_k^2 for a given observation vector $\mathbf{x}_k^0 \in R^J$ can be calculated as in Eq. 19.

$$T_k^2 = \left(\mathbf{x}_k^{0^{\mathrm{T}}} - \bar{\mathbf{x}}^{\mathrm{T}}\right)\mathbf{S}^{-1}\left(\mathbf{x}_k^0 - \bar{\mathbf{x}}\right), \tag{19}$$

where $\bar{\mathbf{x}}$ is the population mean and matrix $\mathbf{S}$ is its covariance.

The SPE-statistics, (referred as Q-statistic in some cases), is defined as the summation of the squared errors between the original signal and the reconstructed signal using the retained PCs, see Eq. 20.

$$\mathrm{SPE}_k = \|\mathbf{e}_k\|^2 = \left\|\left(\mathbf{I} - \mathbf{P}\mathbf{P}^{\mathrm{T}}\right)\mathbf{x}_k\right\|^2. \tag{20}$$

SPE-statistics accounts for the distance between the measured observation point and its projection into the reduced PCA space. If the SPE value exceeds the threshold values for a given process realization, the last is said to show a completely new behaviour not shown by the data in the NOC. Threshold values for both statistics strongly depend on the assumptions and simplifications that the analyst makes. Detailed procedures for the calculation of the control limit values can be found in literature see, for example, Nomikos and MacGregor [17].

Stage II of the PCA-based monitoring is the actual use of the PCA model on the plant current operating conditions and analyses its possible discrepancies compared with the NOC calculated on stage I. In this sense, new data points (x_k) allow for calculating the new values for T^2 and SPE. These values are compared against the respective control limit values $\left(T_{\mathrm{lim}}^2 \text{ and } \mathrm{SPE}_{\mathrm{lim}}\right)$ such that the following inequalities hold.

$$T_k^2 < T_{\mathrm{lim}}^2 \tag{21}$$

$$\mathrm{SPE}_k < \mathrm{SPE}_{\mathrm{lim}}. \tag{22}$$

If some or both of the former inequalities do not hold (for a certain number of consecutive observations), then, the system is said to be out of normal operation conditions. These comparisons are typically done using monitoring-control charts where the T^2 and SPE values are plotted together with the control limit values. Latent projections, projection residuals, T^2 and SPE values for each new observation are calculated as follows:

$$\mathbf{t}_k = \mathbf{P}^T\mathbf{x}_k \tag{23}$$

$$T_k^2 = \mathbf{t}_k^{\mathrm{T}}\mathbf{\Lambda}^{-1}\mathbf{t}_k \tag{24}$$

$$\mathbf{e}_k = \left(\mathbf{I} - \mathbf{P}\mathbf{P}^T\right)\mathbf{x}_k \tag{25}$$

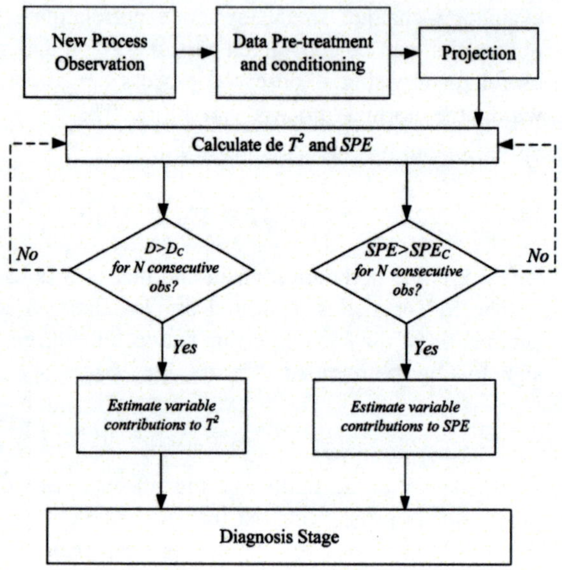

Fig. 8 General procedure for application of PCA to monitoring (Stage II)

$$\mathrm{SPE}_k = \|\mathbf{e}_k\|^2. \tag{26}$$

When an alarm signal is fired, identification stage begins. The purpose of this stage is to determine the variable, or set of variables, that appears to be the ones with major deviations. Variable identification is usually carried out by analysing the so-called variable contributions to the out of control statistic. Variable contributions to T^2 and SPE statistics are calculated as in Eq. 27.

$$c_{j,k}^{T^2} = \mathbf{t}_k^T \mathbf{S}^{-1} \left[x_{j,k} \mathbf{P}_{(j,:)} \left(\mathbf{P}^{\mathrm{T}} \mathbf{P} \right)^{-1} \right]^T, \tag{27}$$

where $c_{j,k}^{T^2}$ is the contribution of the jth variable in the kth observation, $\mathbf{P}_{(j,:)}$ is the jth row in $\mathbf{P}$ and $\mathbf{S}$ is the covariance of $\mathbf{T}$. Variable contribution to SPE-statistics is calculated as in Eq. 28.

$$c_{j,k}^{\mathrm{SPE}} = e_{j,k}^2. \tag{28}$$

A thorough analysis about variable contributions, its limits and interpretation can be found on Westerhuis et al. [18]. Figure 8 presents the general structure for stage II when a monitoring scheme as described above is implemented.

4.3.1 Application of PCA to IGCC Plant Data

As an application example, the previous PCA-monitoring strategy is used to set an on-linemonitoring strategy for an IGCC plant. A set of 12 process variables is chosen as the one having the most meaningful and important information related to

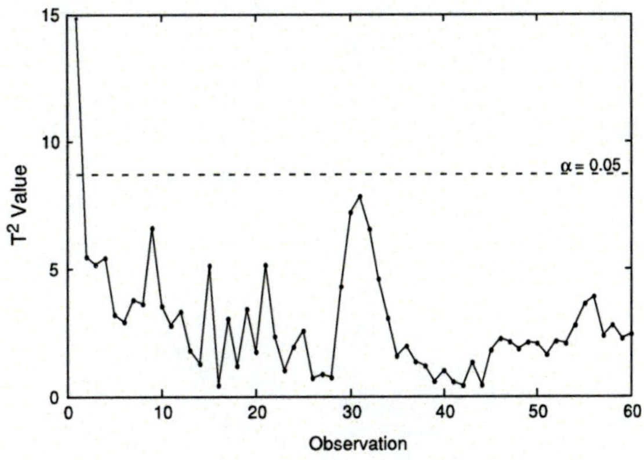

Fig. 9 T2 Chart for observations contained in NOC

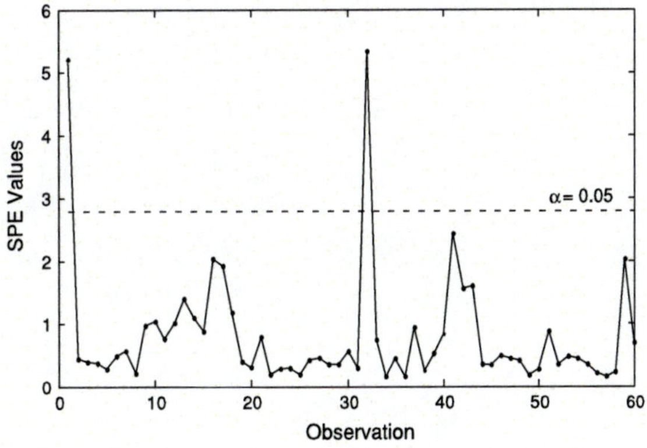

Fig. 10 SPE Chart for observations contained in NOC

the plant operation. After a proper data pre-treatment and off-line analysis, a set of 60 process realizations is taken as the reference population defining the NOC. The profiles for the T^2 and SPE statistics for these normal data are shown in Figs. 9 and 10.

In addition, an extra data trend containing slight excursions from the normal operational data is considered to picture the detection and identification capabilities of the whole strategy. The principal deviations included in this data are because of an abnormal value in the coal feed. Figures 11 and 12 show the profiles for T^2 and SPE, respectively. It is easy to notice that SPE shows more points out of the control region than T^2.

In this case, the process excursion is noticed just by one of the statistics and therefore, only its contribution plot should be analyzed to find the variable (or set of variables) that shows the deviation. It is a common practice to set that a given number of consecutive observations should fall outside of the control region before an alarm signal is fired. In this case, this number was set to consider five consecutive measurements. When contribution plots are analyzed, it is easy to note

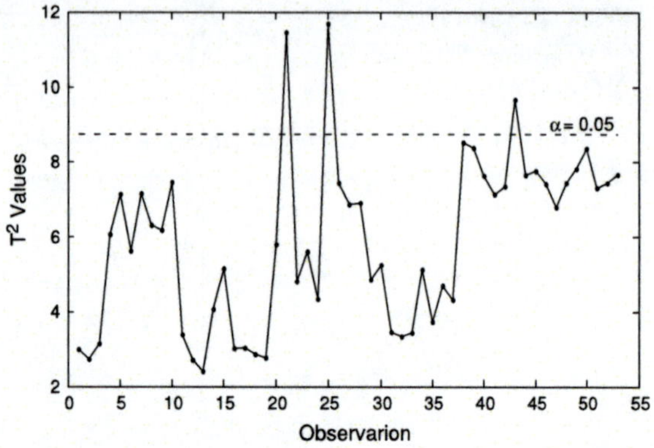

Fig. 11 T2 Chart for the considered faulty data trend

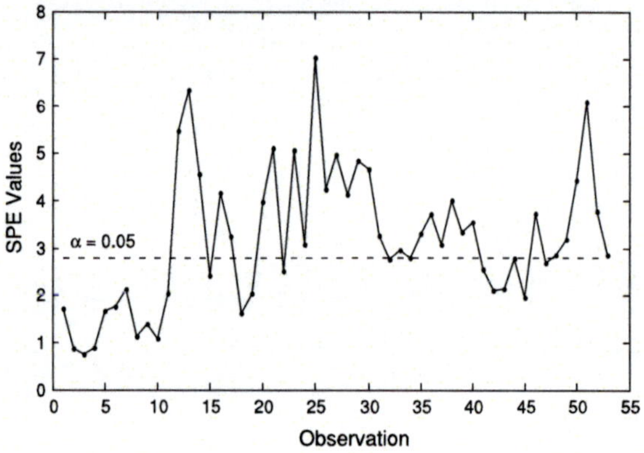

Fig. 12 SPE Chart for the faulty data trend

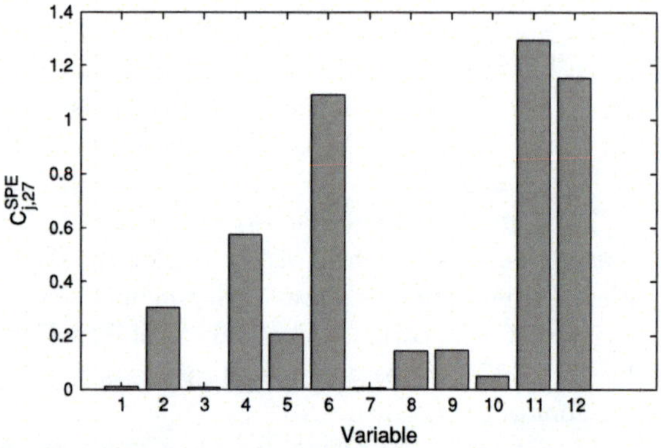

Fig. 13 SPE variable contribution for observation # 27

that variable 11 (relation O_2/Coal) has the most important effect, closely followed by variable 12 (Steam/Coal), see Fig. 13.

Clearly PCA helps in identifying the variables that might be responsible for plant excursions and allows improving the knowledge of the process. However, it has to be emphasised that PCA-based monitorisation schemes depend on the adequacy of the process representation by the selected principal components.

5 Conclusion

The chapter has discussed the possible control schemes present in IGCC plants, together with the generalities related to the variables which are of usual concern. The problem of data reconciliation using process simulation has been exemplified using different models and approaches for improving the reliability of measurements and for the actual estimation of parameters. An overview of possible data-mining techniques, mainly those related to PCA, has been provided. The problem of applying PCA to process monitoring has been addressed and it was discussed in detail.

References

1. Tona-Vásquez RV (2006) Estrategias de análisis y exploración de datos como soporte a la operación y supervisión de procesos químicos (in Spanish). PhD thesis, Universitat Politecnica de Catalunya, Barcelona, Spain
2. Kuehn D, Davidson H (1961) Computer control at American oil I: mathematics of control. Chem Eng Prog 57(6):44–51
3. Knepper JC, Gorman JW (1980) Statistical analysis of constrained data sets. AIChE J 26:260–264
4. Tjoa I, Biegler L (1992) Reduced successive quadratic programming strategy for errors-invariables estimation. Comput Chem Eng 16(6):523–533
5. Romagnoli J, Sanchez M (1999) Data processing and reconciliation for chemical processes operation. Academic Press, New York
6. Alvarez-Medina CR (2010) Integración de reconciliación de datos y monitoreo en procesos de polimerización (in Spanish). PhD thesis, Universidad Nacional del Sur, UNS, Bahia Blanca, Argentina
7. Crowe C, García-Campos Y, Hrymak A (1983) Reconciliation of process flow rates by matrix projection part I: linear case. AIChE J 29:881–888
8. Sanchez M, Romagnoli J (1996) Use of orthogonal transformations in data classification-reconciliation. Comput Chem Eng 20(5):483–493
9. Zagoruiko AN, Matros YS (2002) Mathematical modelling of Claus reactors undergoing sulfur condensation and evaporation. Chem Eng J 87(1):73–88
10. Perez-Fortes MM, Bojarski AD, Velo E, Nougués JM, Puigjaner L (2009) Conceptual model and evaluation of generated power and emissions in an IGCC plant. Energy 34:1721–1732
11. Sequeira SE (2003) Real time evolution (RTE) for on-line optimisation of continuous and semi-continuous chemical processes. PhD thesis, Universitat Politecnica de Catalunya, Barcelona, Spain
12. Han J, Kamber M (2006) Data mining: concepts and techniques, 2nd edn. Morgan Kaufmann, Massachusets, USA
13. Jackson JE (1991) A user's guide to principal components. John Wiley & Sons, New York, USA
14. Wold H (1966) Estimation of principal components and related models by iterative least squares. In: Krishnaiah PR (ed) Multivariate analysis. Academic Press, New York, pp 391–420

15. Valle S, Li W, Qin SJ (1999) Selection of the number of principal components: the variance of the reconstruction error criterion with a comparison to other methods. Ind Eng Chem Res 38(11):4389–4401
16. Hotelling H (1931) The generalization of Student's ratio. Ann Math Stat 2:360–378
17. Nomikos P, MacGregor JF (1995) Multivariate SPC charts for monitoring batch processes. Technometrics 37(1):41–59
18. Westerhuis JA, Gurden SP, Smilde AK (2000) Generalized contribution plots in multivariate statistical process monitoring. Chemom Intell Lab Syst 51:95–114

Index

L. Puigjaner (ed.), *Syngas from Waste*, Green Energy and Technology,
DOI: 10.1007/978-0-85729-540-8, © Springer-Verlag London Limited 2011